Quantum

NANOSTRUCTURE PHYSICS AND FABRICATION

Proceedings of the International Symposium

College Station, Texas   March 13–15, 1989

NANOSTRUCTURE PHYSICS AND FABRICATION

Proceedings of the International Symposium

College Station, Texas   *March 13–15, 1989*

Edited by

Mark A. Reed

Central Research Laboratories
Texas Instruments Incorporated
Dallas, Texas

Wiley P. Kirk

Department of Physics
Texas A&M University
College Station, Texas

ACADEMIC PRESS, INC.
Harcourt Brace Jovanovich, Publishers

Boston   San Diego   New York
Berkeley   London   Sydney
Tokyo   Toronto

ACADEMIC PRESS, INC.
1250 Sixth Avenue, San Diego, CA 92101

United Kingdom Edition published by
ACADEMIC PRESS, INC. (LONDON) LTD.
24-28 Oval Road, London NW1 7DX

Library of Congress Cataloging-in-Publication Data

Nanostructure physics and fabrication : proceedings of the
   international symposium, College Station, Texas, March 13–15, 1989 /
   edited by Mark A. Reed, Wiley P. Kirk.
      p.   cm.
   "Contains contributions presented at the First International
Symposium on Nanostructure Physics and Fabrication . . . [held] on the
campus of Texas A&M University from March 13 to 15, 1989" — Pref.
   Includes bibliographies and index.
   ISBN 0-12-585000-X (alk. paper)
   1. Electronic structure--Congresses.   2. Quantum electronics-
-Congresses.   3. Superconducting quantum interference devices-
-Congresses   I. Reed, Mark A.   II. Kirk, Wiley P.
III. International Symposium on Nanostructure Physics and
Fabrication (1st : 1989 : Texas A&M University)
QC176.8.E4N32   1990
530.4'11—dc20   89-17516
                                                     CIP

Printed in the United States of America
89 90 91 92   9 8 7 6 5 4 3 2 1

Preface

This book contains contributions presented at the first International Symposium on Nanostructure Physics and Fabrication. The symposium was attended by 160 researchers from around the world, who met in College Station, Texas, on the campus of Texas A&M University from March 13 to 15, 1989. The purpose of the symposium was to bring together an interdisciplinary group of specialists in nanometer scale fabrication, physics of mesoscopic systems, electronic transport, and materials scientists and technologists to discuss the current status of nanometer scale electronic structures.

The symposium was organized into eleven sessions of 28 oral presentations that spanned three days, and was accompanied by two lively late afternoon poster sessions in which approximately forty technical posters were presented. These sessions addressed topics on conceptual origins of nanostructures, lateral periodicity and confinement, nanoconstruction and phase coherent effects, quantum devices and transistors, equilibrium and nonequilibrium response in nanoelectronic structures, quantum ballistic transport, quantum dots, ballistic transport at quantum point contacts, and quantum constriction and narrow wires.

In order to convey the many important ideas and the most recent results of the symposium's contributors, we have organized this book into seven chapters. The first chapter represents a condensed overview in which we attempt to unify and summarize briefly the various individual contributions that follow in the remaining six chapters. Because the field is very new, we have also provided in the first chapter some introductory comments and background material that has led to the formation of the field. The book, therefore, provides the most up-to-date view of a very active and growing subject that is gaining wide interest.

The book's origin and contents are for the most part a result of the symposium's success, which in turn was a result of the efforts and many fine suggestions put forward by the Organizing and Program Committee and the Advisory Committee. As always, in the background of any successful conference is a coterie of students, research associates, and faculty colleagues who see that things get done. Our coterie consisted of the following individuals, whom we owe much gratitude: C. C. Andrews, K. G. Balke, M. G. Blain, C. J. Brumlik, R. N. Burns, Z. Cai, F. Cheng, J. F. Chepin, B. H. Chu, D. P. Dave, J. J. Garris, A. Roy-Ghatak, W. B. Kinard, W. R. Klima, P. S. Kobiela, J. Lei, F. Li, Y. X. Liu, J. R. McBride, E. C. Palm, M. A. Park, R. Pathak, B. Radle, L. R. Richards, G. Spencer, T. Szafranski, W. Szott, M. Tierney, R. Tsumura, A. L. Vandervort, C. Wang, and W. K. Yue. We are indebted to Professor Glenn Agnolet for his many helpful suggestions, to Professors Wayne M. Saslow, Chia-Ren Hu, Joseph H. Ross, and Dr. John N. Randall for their help in judging the posters, and especially to the Publications Secretary, Phil R. Reagan, for his invaluable assistance in organizing the conference proceedings. Each of the referees, who helped review the contributions, deserve much praise from us. Our greatest thanks and appreciation, however, goes to the Symposium Secretary, Mary J. Miller, who tackled innumerable details with great proficiency. Finally, financial support from our sponsors listed on a nearby page is gratefully acknowledged and in particular we thank Dr. Larry R. Cooper for his special interests in helping us secure the needed support.

Mark A. Reed
Dallas, Texas
May 1989

Wiley P. Kirk
College Station, Texas
May 1989

Sponsors

Office of Naval Research
Air Force Office of Scientific Research
Defense Advanced Research Projects Agency
U.S. Army Research Office
American Vacuum Society, Texas Chapter
Texas Instruments, Inc.
Texas A&M University
    Physics Department
    Electrical Engineering Department
    College of Science
    Texas Engineering Experiment Station

Industrial Sponsors and Exhibitors

EKC Technology
Grant Associates
MKS
Group Five
Kurt J. Lesker
Balzers

TABLE OF CONTENTS

Contents

Chapter 4 - Quantum Devices and Transistors

Contents

Chapter 5 - Equilibrium and Nonequilibrium Response in Nanoelectronic Structures

Contents

Chapter 6 - Quantum Wires and Ballistic Point Contacts

Contents

Chapter 7 - Related Fabrication and Phenomena

Contents

Chapter 1

Overview and Background

"At any rate, it seems that the laws of physics present no barrier to reducing the size of computers until bits are the size of atoms, and quantum behavior holds dominant sway." Richard P. Feynman (from "Quantum Mechanical Computers", Optics News, pp. 11-20, February 1985, published by the Optical Society of America.)

OVERVIEW AND BACKGROUND

Mark A. Reed

Texas Instruments Incorporated
Dallas, Texas 75265

Wiley P. Kirk

Texas A&M University
College Station, Texas 77843

1. INTRODUCTION

In 1957 J. R. Schrieffer [1] suggested that the narrow confinement potential of an inversion layer may lead to the observation of non-classical electron transport behavior. This was unambiguously demonstrated in 1966 [2,3] by low-temperature magnetotransport, and initiated an intense effort in the exploration of 2-dimensional electrons gas (2DEG) systems. The compelling attraction of the field was and is the ability to observe macroscopic quantum size effects and (to a certain extent) control system parameters that determined these effects [4]. Until recently, only in areas of low-temperature physics has there been any opportunity to study and gain experience with macroscopic quantum mechanical effects. Unfortunately, the number of low-temperature systems available for study has been few and mostly limited to liquid and solid states of helium and a finite number of a few types of superconductors. Moreover, these low-temperatures systems always incorporate rather complex many-body effects that make them exceedingly difficult to control and understand on a first principle basis. Now, however, we are rapidly approaching a level of solid-state structural fabrication in which the energy and length scales are such that macroscopic quantum effects are manifest. We have, for the first time, reached a stage where it is possible to observe and perhaps even control large scale quantum effects in a variety of materials and states of condensed matter.

The advent of ultra-thin epitaxial film growth techniques ushered in the era of reduced dimensionality physics, and it became clear that heterojunction interfaces with large energy band discontinuities were a reality. These structures are inherently two-dimensional; thus investigations concentrated on heterostructures, where quantum effects are due to confinement in the epitaxial direction; i.e., quantum wells (more properly, "quantum planes") which are 2-dimensional systems [5]. The ability to vary composition, band offset, periodicity, and other variables results in nearly limitless possibilities in creating structures for physical exploration as well as for electron and optical device applications.

Within the last few years, advances in microfabrication technology have allowed laboratories around the world to impose additional lateral dimensions of quantum confinement on 2-dimensional systems with length scales approaching those of epitaxial lengths in the growth direction. The achievement of quantum wires [6,7,8] and quantum dots [9] have demonstrated that electronic systems with different dimensionality are now available to the experimentalist. The quantum limit of electronic transport has been demonstrated [10,11], leading to a fundamentally distinct approach to the understanding of mesoscopic electronic systems.

Yet as late as 1986, "low dimensional structures" implicitly meant 2DEGs or quantum wells. At the 1986 International Conference on the Physics of Semiconductors in Stockholm, Sweden, there were less than 10 papers (out of a total of 405 papers) on fabricated nanostructures. However, these few papers framed the beginnings of a new era in semiconductor transport. One of the major motivational reasons for examining small electronic systems was the observation of "universal conductance fluctuations" and weak localization effects in silicon [12] and GaAs [13] nanostructures. These works demonstrated that coherent and random quantum interference, respectively, are observable in few-electron systems, and perhaps could even be dominant in appropriate structures. However, fabrication processes and tolerances were not sufficiently well developed to create structures that would exhibit large quantum interference or quantum size effects; the observations were at most small, noisy perturbations.

A conceptually more pleasing investigation at the time, at least to transport physicists, was the observation of Aharonov-Bohm oscillations in small metallic ring structures [14]. These observations appeared to be the first clear observation of quantum interference of the electron wavefunction in a nanofabricated structure. At the time, there appeared to be little connection between this esoteric effect and conductance fluctuation work, except for their embodiment in small electronic systems.

During this time, a renaissance was occurring in vertical electronic transport through multilayer heterostructures. Large quantum tunneling effects were achieved in resonant tunneling structures [15] nearly a decade after the initial (rather disappointing) investigations by Chang et al. [16]. The importance of this renaissance was not the specific realization of the tunneling structure, but a realization that artificially-structured quantum states could indeed show nonclassical electronic transport. Around the same time, the achievement of ballistic transport (i.e., electronic transport in which carriers (electrons) traverse without undergoing a scattering event) in vertical hot-electron structures was reported [17,18].

The turning point in the understanding of nanometer scale electronic transport was the development of reliable semiconductor fabrication techniques on the nanometer scale, an example of which is the fabrication of semiconductor quantum wires [7], which allowed the fabrication of electronic systems in a totally new regime. The realization of ballistic structures had been done in the epitaxial dimension, but not in the lateral dimensions due to the small dimensional scales (tens of nanometers) involved. However, the advancement of electron beam lithography and the incorporation of this dimensional scale onto high mobility 2DEGs allowed one, for the first time, to create structures that were *smaller* than the relevant length scales (elastic, inelastic, phase-breaking, etc). In this regime, quantum transport becomes dominant and the wave nature of the electron becomes apparent. An elegant example of the behavior of electrons on this length scale was the creation of electron "waveguides" [8], which are surprisingly akin to their microwave counterparts that are nearly six orders of magnitude larger.

If structures are created where the interference of electron waves are a dominant effect, how is the system measured? In such systems, the "contacts" to the system are no longer ideal - they are now by definition part of the entire (interacting and interfering) electron wave system. Thus, electrical leads become intractably invasive [19]. This topic has spawned an active area of current research, especially with regard to measurements in the quantum limit.

An obvious area of interest is the limit of electrical conduction; i.e., what is the relevant physics when the number of carriers (or current carrying channels) in the system are countably small. Though this limit was theoretically explored by Landauer over 30 years ago, the realization of such systems has only now been possible. The recent discovery of quantized conductance in ballistic "point contacts" [10,11] , which is predicted by the Landauer formalism, has opened up a new era of nanostructure physics where one can fabricate nanostructures whose behavior is dominated by quantum interference effects. This new capability has enthused the experimentalist and theorist alike, an excitement akin to the advent of quantum well technology, with limitless possibilities for physical exploration and device technology on the nanoscale. It is within this context that the International Symposium on Nanostructure Physics and Fabrication was held.

2. CONCEPTUAL ORIGINS OF NANOSTRUCTURES

Within the last few years, the work of Rolf Landauer (nearly 20 years ago) on the limits of transport has undergone a renaissance with the realization of systems with countably few electronic conducting channels. The basic tenet of his work has been that electronic transport is fundamentally and simply described by summing up the conductances of all possible channels, each having a well defined quantum mechanical transmission coefficient. A consequence of this formalism is that conductance will be quantized, in units of e^2/h per spin, in the limit of a few ballistic electron channels. The experimental observations of this effect [10,11] and the regimes of applicability and approaches to conductance theory are the major topics of the keynote address given by Rolf Landauer (R. Landauer, Chapter 2). His remarks are an admittedly personal perspective of the strengths and weaknesses of existing conductance theories, and the lack of attention given to reservoir connections, invasive probes, and internal voltage distributions. He concludes by addressing the criteria that practical electron devices based upon these effects must satisfy.

An alternative view of the mesoscopic world, especially the interface between the classical and quantum mechanical worlds, is the thesis of the companion paper by Anthony Leggett (A. Leggett, Chapter 2). The fabrication of systems with dimension on the order of the electron phase-coherence length requires a better understanding of the exact meaning of phase coherence of electrons on this dimensional scale. The author also addresses a key weakness of existing theories of nanoscale electron transport, that of the invasiveness of external contacts to the quantum mechanical system. The limit in which contacts can be treated as classical objects is addressed, which is important to experimentalist and theorist alike.

The exploration of quantum effects on the nanoscale is unquestionably an experimental science. A variety of experiments over the last decade have shown different degrees of quantum interference effects due to quantum mechanical phase coherence over the sample. The following paper by Richard Webb (R. A. Webb, Chapter 2) discusses some of the measurements that reveal the richness of phenomena in this regime, demonstrating the interplay between Aharonov-Bohm geometry effects, random quantum interference, and the complexity of nonlocal effects. It is elucidating to compare, when one considers the energy scales involved, the quantum interference effects observed in these small metallic 1D structures with the more recent confined 2DEG semiconductor structures.

3. LATERAL PERIODICITY AND CONFINEMENT

Early attempts to produce quantum size effects in nanostructures arose from the imposition of lateral confinement onto a 2DEG. To improve experimental signal-to-noise ratio, periodic arrays were often investigated. A pioneer in the creation of these arrays is Hank Smith, who discusses the technique of X-ray nanolithography (H. Smith *et al.*, Chapter 3). The technique has produced periodic sub-50 nm metal lines, with demonstrated wide process latitude. Details of the method and comparison with other parallel processing techniques are given. The imposition of additional lateral confinement onto a 2DEG by defining periodic gate structures using this technique is also discussed.

Optical holographic lithography is a similar fabrication technique to define laterally periodic structures (J. Kotthaus *et al.*, Chapter 3). The infrared excitations and magnetotransport of these structures, defining a grid on the surface of a silicon MOSFET 2DEG, have been investigated in detail by the Hamburg group. The advantage of the gated-2DEG configuration is the ability to tune the degree of confinement, allowing these workers to examine the transition from a 2DEG to a 1DEG (in the case of parallel lines) or to a 0DEG (in the case of a periodic grid).

The extension to lateral arrays of 0D structures, or quantum dots, has received considerable attention due to their intrinsic interest, aside from the technological fabrication challenge they present. Fabrication by electron-beam lithography, etching, and studies of regrowth techniques for heterojunction confinement show promise for future quantum effect studies in this system (S. Beaumont, Chapter 3). In addition, Raman spectroscopy of similar structures is proving to be a unique analytical tool for the investigation of surface damage by this fabrication scheme (M. Watt *et al.*, Chapter 3).

An alternative approach to studying arrays of quantum dots is to impose a grid gate potential on a 2DEG and probe the density of states in the dots by capacitive techniques. This system provides a laboratory to study level crossing as the magnetic length of the electron system is varied (W. Hansen *et al.*, Chapter 3).

Other papers in the section on Lateral Periodicity and Confinement (Chapter 3) include the calculated effects of inhomogeneous 2DEGs on lateral confinement of nanostructures (J. Davies); optical emission and inelastic light scattering from a 1DEG (G. Danan *et al.*); the effect of quantum phase coherence on the photoluminescence spectra of quantum dot arrays (S. Bandyopadhyay *et al.*); the calculated effects of inhomogeneous dopant distributions on 2DEGs in lateral confinement nanostructures (J. Nixon *et al.*); an electrostatically laterally-confined resonant tunneling diode (W. Kinard *et al.*); a calculation of the intrinsic time response of a quantum dot (G. Neofotistos *et al.*); and finally calculations on the collective electronic (inter-subband plasma) modes of quasi-1D wire structures (S. Ulloa *et al.*).

4. QUANTUM DEVICES AND TRANSISTORS

Besides the intrinsic interest in nanostructures, there is speculation that quantum size and/or interference effects can be utilized in future electronic devices. An introduction to the variety of effects and some possible applications is given by Dick van der Marel (D. van der Marel, Chapter 4). In systems so small that the electron states take on a discrete set of energies (quantum dots), it is possible to conceive of devices that utilize transmission through these atomic-like states as a basis for quantum effect devices. The physics and magnetic interactions of these structures are theoretically investigated in this paper.

Devices that utilize electron wavefunction quantum interference, where a third terminal affects the interference of electrons on the output terminals, have also been conceived. In the embodiment presented here, as a T-shaped electron waveguide (F. Sols et al., Chapter 4), the gate modifies the effective length of the quantum system, very analogous to the mechanical stub tuner in a microwave circuit. An accompanying paper (D. Miller et al, Chapter 4) presents experimental observations of conductance oscillations in such a structure.

The eventual application of quantum effects in nanostructures to electronic devices is a prime motivator for current nanoelectronic research. A number of devices have been proposed, though few have been realized. One such device is the quasi-one-dimensional quantum wire transistor (Hiramoto et al., Chapter 4). These authors discuss the fabrication of quasi-1D conduction channels in GaAs/AlGaAs 2DEGs by focused ion beams. Low-temperature transport measurements in these structures imply the quasi-1D nature of the electrons.

Other papers in the section on Quantum Devices and Transistors (Chapter 4) include quantum device concepts based on Aharonov-Bohm effects and a "diffraction transistor" (S. Bandyopadhyay et al.); a device concept and the initial experimental results for a pseudomorphic unipolar resonant tunneling transistor (U. Reddy et al.); overshoot saturation in ultra-short MESFETs (J. Ryan et al.); device enhancement by wavefunction tailoring (S. Bhobe et al.); the use of the InP system for resonant tunneling millimeter-wave power diodes (I. Mehdi et al.); and effective quantization due to a field/phonon interaction in the high-field limit (R. Bertoncini et al.).

5. EQUILIBRIUM AND NONEQUILIBRIUM RESPONSE IN NANOELECTRONIC STRUCTURES

The investigation of nanoscale electronic transport is still in its infancy; hence the simpler near-equilibrium linear response is, for the most part, dominant. In 2D systems, the influence of applications toward devices has extended this study to far-from-equilibrium transport, and an understanding of the important processes involved is beginning to emerge. Investigations of nonequilibrium transport in heterojunction bipolar transistor (HBT) structures (Levi *et al.*, Chapter 5) where the mean free path is greater than the emitter-collector spacing sheds light on the elastic and dissipative processes involved.

The ability to realize and control ballistic carriers in nanostructures is of fundamental importance for future device applications. Ballistic vertical transport in epitaxial hot-electron structures is summarized by one of the pioneers in this field (M. Heiblum, Chapter 5), who demonstrates that these structures can be constructed as sensitive electron spectrometers. The extension of this technique with lateral dimensioning may provide a sensitive spectrometer for nanoscale electron devices.

The behavior of nanostructures driven far from equilibrium has received little attention, though it is just under these conditions that marked quantum interference effects exist. In this regime, elementary quantum mechanics does not appear to adequately describe the system, and an attempt at a quantum kinetic theory is presented here (W. Frensley, Chapter 5). This approach yields physical device potentials for realistic values of the inelastic phonon scattering.

An alternative approach to a quantum transport formalism is to express electron densities (per unit energy) in a form that resembles the multiprobe Landauer formalism (S. Datta, Chapter 5) under conditions of local thermal equilibrium. This approach has the advantage of yielding a physically measurable quantity to test the validity of the model.

Algorithms to numerically model quantum transport in laterally defined nanostructures are just starting to be developed (John Barker, Chapter 5). Older approaches to the problem involved techniques such as Green's function and finite-difference 2D Schrödinger equation methods. These techniques are compared with a new semi-analytical coupled-mode technique, which when applied to quantum waveguides explains why the experimentally observed interference effects are weak. Such techniques could help in the design of quantum mechanical wavefunction interference structures.

A fundamental question in nanoelectronic structures is the effect of resident charge on the transport processes of a system. This topic has been one of popular controversy [20,21] in resonant tunneling structures. An attempt at a definitive answer was presented by Laurence Eaves, who presented results of structures tailored to enhance "bistable" effects, as well as ballistic effects in large structures (M. Leadbeater *et al.*, Chapter 5).

Other papers in the section on Equilibrium and Nonequilibrium Response in Nanoelectronic Structures (Chapter 5) include a quantum transport equation approach to calculating nonequilibrium linear screening (B. Hu *et al.*); the use of a finite element method for two-dimensional potentials to solve for non-zero current carrying wavefunctions (C. Lent *et al.*); a study of the temporal characteristics of resonant tunneling diodes by a time-dependent

Schrödinger equation approach (Diff *et al.*); a simplified approach for analyzing quantum size scale structures (W. Lui *et al.*); influence of traps on high field transport in SiO$_2$ (R. Kamocsai *et al.*); an attempt to clarify the concept of tunneling times (J. Støvneng *et al.*); and a statistical approach to viewing variable-range hopping conduction (R. Serota *et al.*).

6. QUANTUM WIRES AND BALLISTIC POINT CONTACTS

The relation of measured global electronic properties and the relation to transmission coefficients [19], especially as applied to small electron systems, has recently been used to address a wide range of problems in nanostructures. A further extension of this formalism to observations of the quantum Hall effect in small systems is presented by Markus Büttiker (M. Büttiker, Chapter 6). Transport along edge states provide a simple and elegant picture with which to explain the various anomalous magnetotransport phenomena observed in nanostructures.

With the realization that (appropriate) electron systems are coupled to external leads, it is of fundamental importance to ask how invasive leads affect the measurement of a system in which the conductance is quantized. This question is investigated by a pioneer in the physics of electron waveguides (G. Timp *et al.*, Chapter 6), with some intriguing results for the quantum generalization of series conductances and multi-terminal measurements.

The realization of quantum transport in the limit of ballistic transport through a countable number of channels [10,11] has initiated intensive investigation into the phenomena and physics of nanoconstrictions of 2DEGs. Two of the initial investigators of this exciting new phenomena review this subject and its classical analogies, and propose new investigations into collimation of the injected electrons and applications toward hot-electron devices (H. van Houten and C. Beenakker, Chapter 6).

In addition to the intrinsic interest in the physics of ballistic point contacts, it has become apparent that these structures are excellent electron filters. In the presence of a quantizing magnetic field, they can be used to either inject into or detect specific magnetic edge states (B. van Wees *et al.*, Chapter 6). A fascinating use of these contacts is to probe magnetically-defined 0D states confined between the contacts.

The application of the Landauer formalism to conductance quantization, as well as to the related phenomena of quenching of the Hall effect in restricted geometries, has been successful in explaining the observed phenomena (A.D. Stone *et al.*, Chapter 6). Detailed analysis yields the general principle that, in ballistic systems, the device geometry can alter and in principle control the electron momentum distribution. Variations upon and elaborations to this approach is provided in a paper by another of the pioneers in understanding quantum transport (Y. Imry, Chapter 6). Recent experimental results that demonstrate these principles appear in a paper by Christopher Ford (C. J.

B. Ford, Chapter 6).

Other papers in the section on Quantum Wires and Ballistic Point Contacts include breakdown of the Hall effect for micron size constrictions (A. Sachrajda *et al.*); a new and novel technique that can form buried metal structures (wires) of nanometer dimensions by laterally-defined mesoepitaxy (S. Berger *et al.*); the prediction of double-frequency components in electrostatic Aharonov-Bohm oscillations in 1D rings (M. Cahay *et al.*); and an alternative view of "universal conductance fluctuations" in disordered mesoscopic metal systems (A. Pruisken and Z. Wang).

7. RELATED FABRICATION AND PHENOMENA

The present understanding of quantum transport in nanostructures had its origins in conductance fluctuation and localization measurements. Concurrent with this development has been ever improving techniques to fabricate nanostructures. Alec Broers, a pioneer in the fabrication of these structures, describes the present state-of-the-art in electron-beam lithography (at 350 KeV) and the associated analysis techniques, to determine ultimate lithography limits of nanostructure fabrication (A. Broers, Chapter 7). He has identified a basic electron-beam resist (PMMA) resolution limit at 15 nm due to the electron-polymer molecule interaction mechanisms, which may be difficult to surpass.

An alternative technique for producing quantum devices is precise electrical damage by low energy ion-beam implantation. This technique (A. Scherer, Chapter 7) can produce lateral confinement of 2DEGs with minimal lateral spread of the ion-beam induced damage. This approach has the advantage of simple *in situ* diagnostics. The technique has demonstrated complex self-aligned gated structures with dimensions below 100 nm.

Other papers in this chapter (Related Fabrication and Phenomena, Chapter 7) related to nanofabrication include the fabrication of nanometer scale features with a scanning tunneling microscope (STM) (E. Ehrichs *et al.*) and a metal-chalcogenide high resolution resist (A. Owen *et al.*).

It is a challenge to recognize, at a sufficiently early stage, related phenomena and technologies that may become important for future nanoelectronic research. This overview of course cannot provide an exhaustive list, but some topics are presented here to illustrate the fields of which a productive researcher should be aware. An example is the role of defects in small devices and device contacts. Early work in the investigation of small Si MOSFET wires revealed the importance of single defects on the electron transport through the wire [22]. This phenomena has been rediscovered in metal nanocontacts, along with the intriguing ability to "nanomachine" the structures by electromigration (D. Ralph *et al.*, Chapter 7). Other papers, which span related phenomena in Chapter 7 include electrochemical techniques to produce nanostructures (C. Martin *et al.*); mesoscopic spin glass systems that exhibit an absence of conductance fluctuations (van Haesendonck *et al.*); the nature of the superconducting transition in thin metal wires (N. Giordano); the concept of population inversion in superconducting quantum wells (L. Reggiani *et al.*); photoreflectance of GaAs structures (A. Badakhshan *et al.*); space charge effects in effective-mass superlattice structures (M. Cahay *et al.*); and 1/f noise as applied to mesoscopic electronic systesm (G. Hu *et al.*).

8. SUMMARY

It is a difficult, if not impossible, task to predict research that will have a future impact upon physics and semiconductor technology. However, the trend toward reduced length scales in electronics compels us to understand electron transport physics and electronics on the quantum scale. Equally important, the combination of new technologies and conceptual advancements in nanostructures has opened an exciting new frontier for us to explore.

REFERENCES

[1] J. R. Schrieffer, in "Semiconductor Surface Physics" (R. H. Kingston, ed.), p. 55. University of Pennsylvania Press, Philadelphia (1957).

[2] A. B. Fowler, F. F. Fang, W. E. Howard, and P. J. Stiles, Phys. Rev. Lett. **16**, 901 (1966).

[3] A. B. Fowler, F. F. Fang, W. E. Howard, and P. J. Stiles, J. Phys. Soc. Japan **21**, Suppl., 331 (1966).

[4] T. Ando, A. B. Fowler, and F. Stern, Rev. Mod. Phys. **54**, 437 (1982).

[5] R. Dingle, A. C. Gossard, and W. Wiegmann, Phys. Rev. Lett. **33**, 827 (1974).

[6] T. J. Thornton, M. Pepper, H. Ahmed, D. Andrews, and G. J. Davies, Phys. Rev. Lett. **56**, 1198 (1986).

[7] H. van Houten, B. J. van Wees, M. G. J. Heijman, and J. P. Andre, Appl. Phys. Lett. **49**, 1781 (1986).

[8] G. Timp, A. M. Chang, P. Mankiewich, R. Behringer, J. E. Cunningham, T. Y. Chang, and R. E. Howard, Phys. Rev. Lett. **59**, 732 (1987).

[9] M. A. Reed, J. N. Randall, R. J. Aggarwal, R. J. Matyi, T. M. Moore, and A. E. Wetsel, Phys. Rev. Lett. **60**, 535 (1988).

[10] B. J. van Wees, H. van Houten, C. W. J. Beenakker, J. G. Williamson, L. P. Kouwenhoven, D. van der Marel, and C. T. Foxon, Phys. Rev. Lett. **60**, 848 (1988).

[11] D. A. Wharam, T. J. Thornton, R. Newbury, M. Pepper, H. Ahmed, J. E. F. Frost, D. G. Hasko, D. C. Peacock, D. A. Ritchie, and G. A. C. Jones, J. Phys. C **21**, L209 (1988).

[12] W. J. Skocpol, P. M. Mankiewich, R. E. Howard, L. D. Jackel, and D. M. Tennant, "The 18th International Conference on the Physics of Semiconductors" (O. Engström, ed.), p. 1491. World Scientific Publishing Co. Pte. Ltd., Singapore (1987).

[13] T. J. Thornton, M. Pepper, G. J. Davies, and D. Andrews, "The 18th International Conference on the Physics of Semiconductors" (O. Engström, ed.), p. 1503. World Scientific Publishing Co. Pte. Ltd., Singapore (1987).

[14] R. A. Webb, S. Washburn, C. P. Umbach, and R. B. Laibowitz, Phys. Rev. Lett. **54**, 2696 (1985).

[15] T. C. L. G. Sollner, W. D. Goodhue, P. E. Tannenwald, C. D. Parker, and D. D. Peck, Appl. Phys. Lett. **43**, 588 (1983).

[16] L. L. Chang, L. Esaki, and R. Tsu, Appl. Phys. Lett. **24**, 593 (1974).

[17] M. Heiblum, M. I. Nathan, D. C. Thomas, and C. M. Knoedler, Phys. Rev. Lett. **55**, 2200 (1985).

[18] A. F. J. Levi, J. R. Hayes, P. M. Platzman, and W. Weigmann, Phys. Rev. Lett. **55**, 2071 (1985).

[19] M. Büttiker, Phys. Rev. Lett. **57**, 1761 (1986).

[20] V. J. Goldman, D. C. Tsui, and J. E. Cunningham, Phys. Rev. Lett. **58**, 1256 (1987).

[21] T. C. L. G. Sollner, Phys. Rev. Lett. **59**, 1622 (1987).

[22] K. S. Ralls, W. J. Skocpol, L. D. Jackel, R. E. Howard, L. A. Fetter, R. W. Epworth, and D. M. Tennant, Phys. Rev. Lett. **52**, 228 (1984).

Chapter 2

Conceptual Origins of Nanostructures

NANOSTRUCTURE PHYSICS: FASHION OR DEPTH?

Rolf Landauer

IBM Research Division
T. J. Watson Research Center
Yorktown Heights, New York 10598

I. INTRODUCTION

In a few pages I cannot give a broad review of Nanostructure Physics and Fabrication. This is a collection of observations; a relatively personal perspective. This volume represents a new and expanding field, with a diversity illustrated by the papers that follow. Rather than repeat the most common ingredients of these papers, I will concentrate on some questions where the prevailing literature lacks clarity. That, of course, will give my remarks a controversial flavor.

Science has periods and fields in which our efforts are well channeled. The channels are defined by conceptual tools, theoretical methods, experimental techniques, and a technology which allows us to look at a certain range of phenomenon. Occasionally new channels emerge, but soon the new channel becomes well defined and starts to act as a constraint. The physics we discuss here is a result of the technology of the transistor. The continued drive toward miniaturization allows us, indeed compels us, to look at small samples. Electron beam photoresist techniques, field effect structures, epitaxial deposition, III-V compounds, and heterojunctions have been part of the arsenal that made all the physics in this volume possible. Our new world may, in turn, lead to new devices. But we will urge caution in that regard. On the other hand, there is little question that with continuing miniaturization some of our new effect will appear as secondary features in conventional devices, and deserve understanding on that basis, quite aside from their intrinsic fascination.

17

For many decades quantum mechanics has manifested itself, in a detailed way, in atoms and molecules. The energy levels of the hydrogen atom correspond to stationary wave patterns in that potential. But larger structures, built in the laboratory, such as transistors, manifested quantum mechanics only indirectly. The Esaki diode went beyond that, at an early stage. Further manifestations came with the two-dimensional electron gas(1) and after that molecular beam epitaxial superlattices and resonant tunneling(2). In recent years, however, our ability to play with electron waves in laboratory structures has widened greatly. We can measure the resistance due to the scattering of electrons when they have to turn through a 90° bend in a wire(3). The quantized resistance of small apertures(4,5), was another fascinating discovery to which we return in Sec. 4. A number of experiments demonstrate that measurements made via a probe on a conducting sample reflect the whole history of the electron waves bouncing in and out of the probe. For example, the voltage drop measured between two voltage probes depends on the Aharanov-Bohm flux through a loop attached to a part of the conductor outside of the portion between the probes (6). New insights into the integral quantum Hall effect have been gained; in a supposedly settled area that we now finally really understand (7). There is energy level spectroscopy on small two-dimensional dots, much as on atoms (8). At an intersection between wires we may have a bound state (9). (The existing theories of this bound state are not self-consistent, and do not allow for the electrostatic interaction of the bound state with its geometrically complicated environment.)

The device physicist has long understood that nominally identical samples are not really identical; indeed, this was known in antiquity to the buyer of a jar of grain. The commotion about universal fluctuations has now taught our more fundamentally oriented colleagues the same fact. Unfortunately, while they now emphasize that small samples do show the "fingerprint of the sample," they still lean heavily on techniques which characterize the fluctuations over an ensemble. With rare exceptions, fundamentally oriented theoreticians are still not really ready to ask about a *specific* sample. Where, near a grain boundary does the voltage drop take place (10)? In contrast, the device community has always asked: Where in space does the voltage drop occur? Without that we cannot discuss a transistor, or even a vacuum diode. Such questions recur, forcefully, in resonant tunneling devices in connection with the space charge stored between the tunneling barriers. Many theoretical discussions in our field, unfortunately, use a formulation which emphasizes events at terminating reservoirs, and does not permit us to discuss spatial variations. We return to this, in detail, in Sec. 3.

II. APPROACHES TO CONDUCTANCE THEORY

Quantum mechanical conductance calculations fall into three classes, representing different historical periods. There are many variations on the basic themes, and a simple classification does injustice to some contributions. In the first three decades of quantum mechanics models adapted from the kinetic theory of gases invoked quantum theory through the use of Fermi-Dirac statistics, through the use of band structure, and most particularly through the use of quantum mechanical scattering cross-sections. The 1950's saw a number of efforts to be more consistently quantum mechanical. Kubo's work (11) is most widely known. There is a central problem with most of these more totally quantum mechanical formulations. They use the quantum mechanics of conservative systems, but then go on and try to explain dissipative effects. They also need to invoke the quantum mechanics of closed systems, i.e. systems with a well defined number of particles and a clearly specified Hamiltonian for all possible particle configurations. But a resistor has leads, and particles must be able to go in and out of those. All of this requires cheating, and we have become very clever at getting reasonable (and sometimes unreasonable) answers. The literature contains other critiques of linear response theory, we cite only two (12).

The third method for calculating conductance explicitly allows for the entrance and departure of carriers. It calculates resistance for the whole sample regarded as a single source of scattering for electrons incident on its boundaries. If we are told how electrons entering the sample eventually leave it, we can calculate resistance from that. This approach is particularly suited to small samples. If the sample provides only elastic scattering, then the dissipation occurs in the reservoirs feeding current to the sample. General reviews of the method are contained in Refs. (13-16) and the generalization to more than two terminals is discussed in Ref. (17). There have been attempts to derive similar results, giving conductance in terms of transmission (18-22), from linear response theory. We will later comment on some of these in more detail. Here, we cite Sorbello (23): "In retrospect, attempts to rigorously derive the Landauer formula from the Kubo approach are less a test of the Landauer formula than of the Kubo approach." We stress that the treatment of a sample in terms of its scattering behavior is a *viewpoint*, not a particular formula. While originally written down for a one-dimensional elastic scatterer, Fermi statistics, and no spin effects, the viewpoint has much broader applicability (13, 14).

Finally, we allude to a fourth possible approach, which has not yet been developed in a way which makes *broad* contact with nanostructures. We can represent dissipative effects by coupling the motion of the degree of freedom of interest, to many other degrees of freedom, to which we can pass energy. This is the approach used in macroscopic quantum tunneling. It can, in principle, also be applied to nanostructure problems (24). It could, in fact, be invoked to give a more formal description of the reservoirs within the formulation which uses the overall scattering behavior of the sample.

III. WHICH VERSION OF THE CONDUCTANCE FORMULA IS RIGHT?

The literature contains a number of alternative forms for conductance expressed as a function of transmissive behavior. There are legitimate reasons for this variety. That does not mean that *all* of the expressions in the literature have a reasonable domain of validity. First of all as stressed in Ref. (13) and (25), the transmissive behavior tells us what will happen, once the incident flux is specified. The incident flux, however, depends on the rest of the circuit. In general it will depend in a self-consistent way on the sample and on the remaining circuit. After all, carriers incident on the sample may have recently come from there. Thus, for example, the "reservoir" so beloved by theoreticians, is unlikely to occur in a real experiment. Current is likely to be fed into a sample from a conductor which has a velocity distribution resembling the shifted Fermi sphere typical of conductors with scattering. Thus, there will be a greater occupation probability for carriers parallel to the direction of carrier flow, than for carriers moving almost at right angles to that. That is in contrast to a "reservoir" in which all of the carriers *coming out of the reservoir* are in equilibrium with each other.

There are additional reasons for the variety of conductance expressions. We will not, here, repeat the discussion in Ref. (26), and only allude to some of its principal points. We can measure the electrochemical potential in a conductor through small reservoirs which come into equilibrium with the conductor. But we can also measure the electrostatic potential at the surface, by purely capacitive techniques. As argued in Ref. (26) the small signal linear voltage drop can be measured in such a way that the perturbation of the sample, through the measuring circuit, is second order in the measured voltage drop. Even if we do measure electrochemical potentials via carrier equilibration, that can be done through loosely coupled probes, e.g. an STM tip, which perturbs the kinetics of the sample minimally.

Normally, in physics, we look for minimally invasive measurement techniques. The prevailing technique, in this field, has been to measure potentials via conductively attached leads which resemble the sample. Unfortunately, as a result, theoreticians have tended to assume that it *must* be that way, despite the *existing published* demonstrations of alternative techniques. For example, Ref. (27) tells us: "...it is unreasonable to suppose that the voltage drop across some region of a small sample can be measured without the presence of the voltage probes strongly affecting the results of the measurement."

Two versions of the conductance formula in terms of transmissive behavior have been particularly prevalent. Both of these are specified in terms of the matrix T_{ij} which specifies the *probability* that a wave arriving in channel j to the left of the sample is transmitted into the outgoing channel i on the right. For the conductance measured by a two terminal measurement, where the voltages are measured well inside the terminating reservoirs, Ref. (16) tells us

$$G = \frac{2e^2}{h} \sum_{i,j} T_{ij} = \frac{2e^2}{h} \, \mathrm{Tr}(tt^{\dagger}), \tag{3.1}$$

where t is the complex transmission matrix specifying outgoing waves in terms of arriving waves. Eq. 3.1 assumes a two-fold spin degeneracy, and requires summing only over spatially distinguishable channels. The other common expression (25) is

$$G = \frac{2e^2}{\pi\hbar} \frac{(\Sigma_i T_i)(\Sigma_i v_i^{-1})}{\Sigma_i v_i^{-1}(1 + R_i - T_i)}. \tag{3.2}$$

This again assumes a two-fold spin degeneracy. i refers to the channel number, v_i is the longitudinal channel velocity, and $T_i = \sum_j T_{ij}$.

Eq. 3.1 represents a sum over channels in an ideal conductor which separates the sample from the reservoir; the channels are *not* those of the reservoir. Now a calculation for the voltage drop between reservoirs can only ignore the interface between the reservoir and the adjacent ideal conductor, if that interface causes no additional scattering. This was emphasized in (14) and (15) but does not seem to be generally appreciated. Eq. 3.2 assumes that the voltage is measured between ideal conductors which feed the current into the sample and which are in turn attached to reservoirs. Eq. 3.2 also assumes spatial averaging to eliminate quantum mechanical interferences.

Eq. 3.1 and its multi-reservoir generalization (17) are concerned with events occurring *strictly* at reservoirs. By contrast Eq. 3.2 is concerned with the voltage distribution along the path between reservoirs. If we ask a question of the latter sort, we cannot ignore the charge pile-up in the circuit, and Poisson's equation. The vast majority of papers in this field, which simply ignore Poisson's equation, become irrelevant. Furthermore, any measurement in which we make some attempt to make correction for a series resistance between probes is, by definition, concerned with the internal voltage distribution. But that is exactly what is done in Refs. (4) and (5). Nevertheless, the theoretical attempts to interpret these measurements, discussed in Sec. 4, seem to focus almost exclusively on Eq. 3.1.

The distinction between Eqs. 3.1 and 3.2 becomes unimportant for small transmission probabilities. In ballistic samples, however, we are not necessarily in that regime.

Eq. 3.2 arises from a viewpoint in which we follow the motion of carriers through the sample, as they are scattered, at first ignoring the space charges which pile up as a result. Then we allow for self-consistent screening, letting the initial transport charge pile-up be screened self-consistently the way any imbedded charge, e.g. an interstitial hydrogen ion, is screened. The screening charge is not associated with current flow, and does not change the initially resulting current flow pattern. Such a concern with the spatial variation of the potential drop will be essential for many problems. This includes electromigration (23), includes the nonlinear behavior of a sample in which the scattering of electrons depends on the applied potential, and also includes the high-frequency behavior. Unfortunately, there are many discussions of the high frequency case which analyze the acceleration of electrons in a time-dependent, but spatially uniform, field. They do not allow for the fact that capacitive effects can allow current flow past the hard places, where scattering is most intense.

IV. QUANTIZED CONSTRICTION RESISTANCES

Refs.(4) and (5) measure the conductance of a narrow constriction separating wider good conductors. The situation is symbolized in Fig. 1 which is closer to the actual setup of Ref. (5) than to Ref. (4). As the constriction is widened the conductance varies in a step-like fashion, with plateaus at $G = 2ne^2/h$, for integral n. A number of theories have been presented to account for this behavior (28-34). We do not list additional papers which make no attempt to invoke the actual geometry of a constriction, or papers that are not based on a calculation of transmission probabilities.

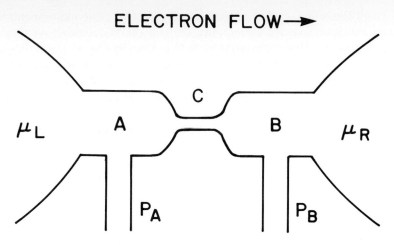

ELECTRON FLOW⟶

Fig. 1. Two ballistic conductors with probes P_A *and* P_B *separated by a narrow constriction C. Current connects to reservoirs at far left and right.*

The plateaus and steps can be understood most easily by considering a long uniform constriction ending in tapered structures [see Fig. 6 of Ref. (14)] which, like an acoustic horn, permit waves in the constriction to emerge into the wider leads, without reflection. The uniform section of the constriction is taken to be long enough so that evanescent wave functions, excited at one end by carriers incident from the wider region, are attenuated strongly before reaching the other end. The steps occur when we widen the constriction enough to permit one more transverse mode to propagate through the constriction without attention. The transmitted modes each contribute to the current in proportion to their longitudinal (along the constriction) density of states and their longitudinal velocity. In the usual fashion, these factors cancel. Thus, the conductance is simply proportional to the number of transmitted modes. The cited theories (28-34) calculate transmissive behavior between A and B, in Fig. 1, invoking a variety of geometrical assumptions, and a variety of analytical techniques. The validity of these calculations is not in question; they help us to understand the results of Ref. (4) and Ref. (5). Nevertheless, the collection, as such, is burdened by a serious conceptual error, admittedly of minor quantitative significance, but unfortunate in the *unqualified* way it appears in *all* of the cited publications. (A minor example of the constraints, discussed in our introduction, which can get established in a new field.) Refs. (28-34) all invoke Eq. 3.1 rather than Eq. 3.2, and none of them distinguish between the reservoirs shown in Fig. 1 and the wider leads connected to the probes. The experiments (4,5) both invoke series resistance corrections; they do *not* characterize events *strictly* between reservoirs. But the confinement to events strictly at reservoirs was a requirement for Eq. 3.1, stressed in Sec. 3. Ref. (5), more clearly than Ref. (4), resembles the geometry shown in Fig. 1 with its voltage probes. But Eq. 3.1 is *not* a four probe formula, it applies only and strictly to voltages measured between the current source reservoirs.

In our critique of Refs. (28-34) we must introduce some qualifications. First of all, at the time this is written, only some have appeared in print in their final form, the others may still be subject to change. Additionally, Refs. (28,30,31) and (32) are less ambitious than the others, and do not attempt to model transmission between wide regions of well defined and limited width. Instead, they analyze transmission through a constriction between regions of unlimited width. As will be seen from our subsequent discussion, this eliminates the distinction between the wide leads and the reservoirs shown in Fig. 1, and the calculations of Refs. (28,30,31) are not actually in error.

The emphasis in Refs. (28-34) on Eq. 3.1 rather than Eq. 3.2 is implicit in some of these papers. It is, however, most explicit in Ref. (33), which states: "...sheds light on questions concerning the physically relevant version of the Landauer formula (relating conductance to the scattering matrix of the disordered conductor) for *two-probe* measurements, and it supports the view that in this case the dimensionless conductance, $g = G/(e^2/h)$, is best described by the formula $g = \mathrm{Tr}(tt^\dagger)$,..."

Now let us analyze the voltage distribution, in the situation of Fig. 1, using the approach leading to Eq. 3.2. We will not actually employ Eq. 3.2, that is done in Ref. (26). We will ask about the actual self-consistent potential in the wide leads, without reference to the kinetics of the probes. If we take the literature on probes (17, 22) seriously, we would expect results which are far more complex than those actually observed. We might expect "universal fluctuations" which would cause the quantized resistance to wash out. Our treatment, neglecting probes, is therefore approximate. Nevertheless, we maintain that the potential in the wide conductors, where the probes are located, is far more relevant than that in the adjacent reservoirs addressed by Eq. 3.1. As in all applications of Eq. 3.2, our expressions assume spatial averaging to remove off-diagonal elements between channels. We treat the diffusion of noninteracting carriers; allowing later for self-consistent screening will not change the drop in electrochemical potential. In lead A the electrochemical potential, reflecting carrier density there, will be below the value μ_L deep inside the left-hand reservoir. That reservoir has an equilibrium distribution, given by μ_L ; this also characterizes carriers arriving at A from the left. The carriers arriving at A from the right will, in part, have come through the constriction and depend on

μ_R . The current j, through A, is due to an imbalance between left moving and right moving carriers, and has the magnitude $n_d W_e v_F < \cos \theta >$. Here, W is the width of lead A, v_F the Fermi velocity, and n_d the deficit in carrier density of the carriers from the right, relative to those from the left. θ is the angle of the carriers relative to the direction of current flow. $< \cos \theta >$ is an average for the mix of channels in the flow. The electrochemical potential in A is

$$\mu_A = \mu_L - \frac{d\mu}{dn} n_d = (j/We\ v_F < \cos \theta >) \frac{d\mu}{dn}, \qquad (4.1)$$

with a similar expression for μ_B. For simplicity, we use the same W, v_F, $< \cos \theta >$, $d\mu/dn$ in B as in A. Using $(d\mu/dn) = v_F/2k$, where k is the Fermi wave number, and $kW/\pi = N_W$ as the number of transverse channels, we find

$$\Delta R = j^{-1}[(\mu_L - \mu_R) - (\mu_A - \mu_B)] = h/\pi e^2 N_W < \cos \theta > . \qquad (4.2)$$

ΔR is the resistance between reservoirs minus that between A and B. If ΔR is small compared to the total resistance, then the conductance between A and B is $G = G_R(1 + G_R\Delta R)$, where G_R is the resistance between reservoirs, given by Eq. 3.1. With Eq. (4.2),

$$G/G_R = 1 + \frac{2T_r(tt^\dagger)}{\pi N_W < \cos \theta >} . \qquad (4.3)$$

In a quantized plateaus $Tr(tt^\dagger) = N_C$ where N_C is the channel count in the constriction. Thus, G/G_R in Eq. 4.3 becomes $1 + \dfrac{2N_C}{\pi N_W} < \cos \theta >^{-1}$. If the constriction ends in slowly tapered "horns," then the transverse wave numbers are reduced in passage out through the horn, and $< \cos \theta >$ is close to 1. The conductances evaluated in Refs. (29,33, 34) need the correction in Eq. 4.3.

Note that for the density of states in Eq. 4.1 we used the Fermi-Thomas expression $v_F/2k$, rather than one reflecting the precise geometry. The validity of this, and the reason for this, are discussed in detail in (26). Here, we simply allude to the fact that even a little unintended scattering in the wide leads, or irregularity in their width, will smear out the singularities present in Eq. 3.2 when one of the v_i in Eq. 3.2 happens to be exceptionally small.

The fact that G exceeds G_R is real; it represents the fact that the wide leads are not perfect reservoirs. The carriers, there, are not in equilibrium, but to a small extent reflect the electrochemical potential of the far away reservoir, on the other side. The wider the conductor at A and B is in comparison to the constriction, the closer the correction factor of (4.3) is to unity. It is a measure of the quality of the wide leads as approximate reservoirs.

Note that Eq. 4.3, which measures the quality of the leads as a reservoir for the current in the constriction, did not need to invoke elastic or inelastic scattering. As we go further into a reservoir the electronic motion must approach that characteristic of thermal equilibrium appropriate to the carriers emerging from the reservoir into the conducting sample. This can be achieved through simple geometrical dilution. As we go further into the reservoir, a higher proportion of the carriers present at a particular point originated far inside of the reservoir, rather than from the sample. Elastic or inelastic scattering in a wide lead does not cause an approach to an equilibrium distribution as we go further into the lead. The geometrical dilution is essential.

Ref. (32) evaluates the conductance of an interface between a half-space of unlimited width and a constricted conductor of *unlimited length*. The conductance expressions in Ref. (32) implicitly assign an electrochemical potential to the narrow region which characterizes only the electrons moving toward the interface. Ref. (32) does not point out that this is really the potential in an adjacent reservoir, and that the reservoir must be connected to the constriction in a particular way.

V. DEVICES

This field had its source in our ability to make small structures. Can we go on and predict that our new and fascinating effects will give rise to promising new devices? It is clear, of course, that as devices continue to become smaller the effects discussed here, or their close relatives, will turn up as secondary effects in more or less conventional transistor structures. But can we expect devices in which quantum coherence is a key ingredient? That is, in part, a matter of definition. The tunnel diode which appeared three decades ago, and has seen some practical use, depends on quantum mechanically coherent tunneling in its key step. The ballistic transistors discussed by Heiblum (35) can be considered examples of ballistic transport. Alternatively, however, they can be considered as a revival of the hot electron structures proposed decades ago. If we emphasize devices which do not just have one small dimension, but two or three, and devices in which we control quantum mechanical interference or resonance, then we will advise caution.

First a general theme. The enthusiasm of basic scientists has generated countless device inventions including tunnel diodes, Gunn devices, ferrite core logic, Josephson junctions, parametric excitation, optical bistability, devices switching at a molecular level, and a great many more. Some of these have found a limited or temporary niche. But the main thrust, in many of these proposals, has been toward the logic stages of the computer. The only devices, there, which have ever made very serious progress are simple switches. The relay, the vacuum tube, and the transistor are the key examples. They can be pushed toward one of two extreme states, like an open or shut door. They readily restandardize the signal toward the intended values, at every stage, by pushing it, for example, toward ground or B+. For more on this theme see Ref. (36) and (37). And the transistor continues to advance: 0.1 micron silicon field effect devices are here, and show a raw inverter delay of 13 psec. (38).

Let us briefly discuss resonant tunneling devices, represented in this volume by Frensley (39) and others. These need be thin in only one direction, and are less delicate than some other proposed devices. Non-resonant tunneling devices, in the form of Esaki diodes and Josephson junctions are fairly hardy structures. Resonant tunneling devices can come as two-terminal devices, or else be supplied with an additional degree of freedom to provide control. For an optimistic assessment see Ref. (40). Tunneling, however, depends exponentially on the parameters of the tunneling barrier. That will be hard to control. With a great deal of work the tunneling behavior of Josephson junctions was brought to a remarkable degree of control (41). But nature, there, gave us special help. Oxide formation tends to slow down as the oxide grows; that helps thickness control. Furthermore, oxidation involves electron transfer through the oxide; the growth process is directly related to the device kinetics. Molecular beam epitaxy may not be equally kind. Furthermore, in a resonant tunneling structure the total transmission depends delicately on the *relative* transmission of the *two* barriers, and typically these are not fabricated simultaneously. That puts particularly rigid demands on control. Resonant tunneling depends not only on control of the barriers, but also on the thickness of the well. All that adds up to a challenging set of specifications! And, to the best of my knowledge, none of the existing proposals have the desirable simplicity and parameter tolerance of a gate that can simply be slammed open or shut, with insensitivity to the exact value of the controlling inputs in *both* input states.

There are a number of proposals, or demonstrations, aimed at the use of controlled quantum interference effects in mesoscopic samples (42-48). What are the problems? Some are obvious, and do not characterize all of these devices. For example: effects which appear only at milli-Kelvins are not good device candidates. Some mesoscopic devices tend to have resistances of order h/e^2, and that is high compared to reasonable transmission line impedances. Totally quantum mechanically coherent computers have no manufacturing tolerances (49), some dissipation is needed. Devices which can only be studied with a lock-in amplifier are not useful for a practical system (50). And some of the mesoscopic device proposals have not asked: How does the inevitable voltage swing at the output affect the device's operation?

The study of mesoscopic devices has emphasized "universal" fluctuations and "fingerprint of the sample;" these are not characteristic of what we would like to see in a device. Self-averaging over many independently acting regions in a device is a good way to achieve a small statistical spread in device characteristics. We give that up by going to small and coherent samples. The lack of locality, i.e. the fact that a resistance between two connections depends on events outside that range, does not make circuit design easy!

But finally, of course, small devices are more delicate! (And if they are not small, why bother?) Why do we want smaller devices? So that we can make many more of them, within a given system. That means it won't be enough to have the smaller devices as reliable as the larger ones. They have to be *more* reliable. We have not discussed single electron tunneling devices as a separate category (51). But much of what has been said applies there, too.

Pessimism is, unfortunately, unpopular. But excessive optimism can only cause a premature and excessive counter-reaction when the optimistic promises are not speedily fulfilled.

REFERENCES

1. Fowler, A. B., Fang, F.F., Howard, W. E., and Stiles, P. J. (1966). Phys. Rev. Lett. **16**, 901.
2. Chang, L. L., Esaki, L., and Tsu, R. (1974). Appl. Phys. Lett. **24**, 593.
3. Timp, G., Baranger, H. U., de Vegvar, P., Cunningham, J. E., Howard, R. E., Behringer, R., and Mankiewich, P. M. (1988). Phys. Rev. Lett. **60**, 2081.
4. van Wees, B. J., van Houten, H., Beenakker, C. W. J., Williamson, J. G., Kouwenhoven, L. P., van der Marel, D., and Foxon, C. T. (1988). Phys. Rev. Lett. **60**, 848.
5. Wharam, D. A., Thornton, T. J., Newbury, R., Pepper, M., Ahmed, H., Frost, J. E. F., Hasko, D. G., Peacock, D. C., Ritchie, D. A., and Jones, G. A. C. (1988). J. Phys. C **21**, L209.
6. Umbach, C. P., Santhanam, P., van Haesendonck, C, and Webb, R. A. (1987). Appl. Phys. Lett. **50**, 1289.
7. Büttiker, M. (1989). Phys. Rev. Lett. **62**, 229; Streda, P., Kucera, J., and MacDonald, A. H. (1989). Phys. Rev. Lett. **62**, 230; Jain, J. K. and Kivelson, S. A. (1989). Phys. Rev. Lett. **62**, 231; Haug, R. J., MacDonald, A. H., Streda, P. and von Klitzing, K. (1988). Phys. Rev. Lett. **61** 2797; Washburn, S., Fowler, A. B., Schmid, H., and Kern, D. (1988). Phys. Rev. Lett. **61**, 2801.
8. Hansen, W., Smith, T. P., III, Lee, K. Y., Brum, J.A., Knoedler, C. M., Hong, J. M., and Kern, D. P., "Zeeman Bifurcation of Quantum Dot Spectra," to be published.
9. Schult, R. L., Ravenhall, D. G., and Wyld, H. W. (1988). "Quantum Bound States in a Classically Unbound System of Crossed Wires," preprint.
10. Kirtley, J. R., Washburn, S., and Brady, M. J. (1988). IBM J. Res. Develop. **32**, 414; Kirtley, J. R., Washburn, S., and Brady, M. J. (1988). Phys. Rev. Lett. **60**, 1546.

11. Kubo, R. (1986). Science **233**, 330.
12. van Vliet, C. M. (1988). J. Stat. Phys. **53**, 49; van Kampen, N. G. (1971). Physica Norvegica, **5**, 279.
13. Landauer, R. (1985). In *Localization, Interaction and Transport Phenomena* (B. Kramer, G. Bergmann, and Y. Bruynseraede, eds.) Vol. 61, p. 38. Springer, Heidelberg.
14. Landauer, R. (1987). Z. Phys. B **68**, 217.
15. Landauer, R. (1988). IBM J. Res. Develop. **32**, 306.
16. Imry, Y. (1986). In *Directions in Condensed Matter Physics*, (G. Grinstein, G. Mazenko, eds.) p. 101. World Scientific, Singapore.
17. Büttiker, M. (1988). IBM J. Res. Develop. **32**, 317.
18. Economou, E. N., and Soukoulis, C. M. (1981). Phys. Rev. Lett. **46**, 618.
19. Fisher, D. S., and Lee, P. A. (1981). Phys. Rev. **B23**, 6851.
20. Langreth, D. C. and Abrahams, E. (1981). Phys. Rev. **B24**, 2978.
21. Thouless, D. J. (1981). Phys. Rev. Lett. **47**, 972.
22. Stone, A. D., and Szafer, A. (1988). IBM J. Res. Develop. **32**, 384.
23. Sorbello, R. S., Phys. Rev. B., to be published.
24. Fu, Y. (1989). Abstract for paper P14 submitted to this conference.
25. Büttiker, M., Imry, Y., Landauer, R., and Pinhas, S. (1985). Phys. Rev. **B31**, 6207.
26. Landauer, R., "Conductance determined by transmission: probes and quantized constriction resistance," (1989). Preprint.
27. Stone, A. D. (1988). In *Physics and Technology of Submicron Structures,* (H. Heinrich, G. Bauer and F. Kuchar, eds.) Vol. 83, p. 108. Springer, Heidelberg.
28. Glazman, L. I., Lesovick, G. B., Khmel'nitskii, D. E., and Shekhter, R. E. (1988). Pis'ma Zh. Eksp. Teor. Fiz. **48**, 218. [Trans. JETP Lett. (1988) **48**, 238.]
29. Escapa, L., and Garcia, N. (1988). J. Phys.:Cond. Matt. to be published; Garcia, N., and Escapa, L., Appl. Phys. Lett. to be published.
30. Haanappel, E. G. and van der Marel, D. (1989). Phys. Rev. B, to be published.
31. Kirczenow, G. (1988). Solid State Comm. **68**, 715.
32. Kirczenow, G. J. (1989). J. Phys.:Cond. Matt. **1**, 305.
33. Szafer, A. and Stone, A. D. (1989). Phys. Rev. Lett. **62**, 300.
34. Avishai, Y., and Band, Y. B. (1988). "Ballistic Electronic Conductance of an Orifice and of a Slit." Preprint.
35. Heiblum, M., this volume.
36. Landauer, R. (1984). In *Chemically-Based Computer Designs*, (F. E. Yates, ed.) p. 86. Report CIME TR/84/1, Univ. California, Los Angeles.
37. Keyes, R. W. (1985). Science **230**, 138.
38. Sai-Halasz, G. A., Wordeman, M. R., Kern, D. P., Rishton, E., Ganin, E., Ng, H. Y., and Moy, D. (1988). IEEE Electron Device Lett. **EDL-9**, 633.
39. Frensley, W. R., this volume.
40. Bate, R. T. (1988). Sci. Am. **258**, 96.
41. Greiner, J. H., and 11 other authors (1980). IBM J. Res. Develop. **24**, 195.
42. Sols, F., Macucci, M., Ravaioli, U., and Hess, K. (1989). Appl. Phys. Lett. **54**, 350.
43. Datta, S., Melloch, M. R., Bandyopadyay, S., and Lundstrom, M. S. (1986). Appl. Phys. Lett. **48**, 487.
44. Exner, P., Seba, P., and Stovicek, P. (1988). J. Phys. A **21**, 4009.
45. deVegvar, P., Timp, G., Mankiewich, P. M., Behringer, R., and Cunningham, J. C., to be published.
46. Bandyopadyay, S., and Porod, W. (1988). Appl. Phys. Lett. **53**, 2323.
47. Washburn, S., Schmid, H., Kern, D., and Webb, R. A. (1987). Phys. Rev. Lett. **59**, 1791.

48. Fowler, A. B., U.S. Patent 4,550,330, Oct. 29, 1985.
49. Landauer, R. (1988). Nature **335**, 779.
50. Fowler, A. B. (1988). Talk at AFOSR Workshop on Quantum Devices, Atlanta, September 15-16.
51. Likharev, K. K. (1988). IBM J. Res. Develop. **32**, 144.

QUANTUM AND CLASSICAL CONCEPTS AT THE ONE-ELECTRON LEVEL[*]

A. J. Leggett

Department of Physics
University of Illinois at Urbana-Champaign
Urbana, IL 61801

The organizers have thoughtfully subtitled this session "Past, Present and Future." This contribution will refer almost exclusively to the future, which is a euphemistic way of saying that many of the ideas I shall float in it are distinctly quarter-baked. Some of them may well be wrong; if so, I hope they are, at least, not wrong for trivial reasons, and that correction of them will lead to some interesting physics.

In this talk I want to discuss three questions which are at first sight unrelated:

(1) What exactly is meant by a "dephasing" collision in the theory of quantum interference in mesoscopic systems?

(2) What are the theoretical limitations imposed by quantum mechanics on the measurement of resistance?

(3) When and why is one allowed to treat the leads to one's "system" in classical terms?

What these questions have in common is that they all probe what one might call the interface between quantum and classical physics, and in particular force one to justify the precise extent to which one is going to describe in classical terms a system which one knows at a fundamental level to be quantum-mechanical.

---

[*] This work was supported by the NSF under grant no. DMR86-12860.

1. WHAT IS A "DEPHASING" COLLISION?

In the theory of quantum interference in mesoscopic systems it is conventional to distinguish between "phase-preserving" and "phase-randomizing" or "dephasing" collisions (see e.g. ref. (1)). The usual view is that in collisions of the former type it is essential to keep track of the phase of the single-electron wave function, even though it may undergo complicated and perhaps at first sight "random" changes; by contrast, once the electron has made a "dephasing" collision all phase memory is washed out. Often, in calculations, one introduces a characteristic length L_ϕ over which the electron can diffuse before making a dephasing collision; then characteristic quantum interference effects, such as the Aharonov-Bohm[2] effect in the magnetoresistance of mesoscopic rings, are found to disappear exponentially when the relevant characteristic dimension (e.g. the ring circumference) becomes large compared to L_ϕ. It is conventionally assumed that collisions with static objects such as chemical impurities or the walls of the sample are phase-preserving, whereas processes involving emission or absorption of a phonon are dephasing. While there is no reason to suspect that the calculations performed on the basis of this assumption are in error, it is nevertheless desirable to try to understand the precise basis for the distinction. An additional motivation is that in some other cases, e.g. the experimentally interesting case of interaction with magnetic impurities which may or may not be aligned by an external field, the situation seems less clear-cut.

A first guess is that the distinction between phase-preserving and dephasing collisions is simply that between processes involving no energy transfer to the system and those involving finite transfer. Such an equivalence is often suggested, perhaps unintentionally, by the frequent use of the words "elastic" and "inelastic" in this context. However, this guess is certainly wrong, as is shown by a series of beautiful experiments[3] recently carried out in neutron interferometry; in these experiments it was shown quite unambiguously that phase coherence between two "branches" of the wave function was preserved even in the presence of finite energy transfer to one of the branches from the electromagnetic field in an r.f. cavity. In this case the crucial observation is that since the field is in a coherent state (in the usual sense of quantum optics) its final state is approximately identical whichever branch the neutron is in, so that we cannot determine the branch by inspection of the field.

This suggests the following revised guess: A "dephasing" collision is one in which the state of the "environment" is changed, a phase-preserving one is one in which it is not. Roughly speaking this is equivalent to saying that a phase-preserving collision is one in which the environment may be treated classically. In a recent preprint[4], Stern et al. advocate this point of view, illustrating it with a thought-experiment which I shall develop further below. In this thought-experiment an electron propagates between diametrically opposite points on a ring through which an Aharonov-Bohm (AB) flux is applied, and one looks for the effects of interference between the two possible paths. On one path, say the left one L, there is a localized spin. The interaction of the electron with the spin is chosen so that if the spin is initially in (say) the x-direction, then passage of the electron through the L path will rotate it into the -x direction, while, of course, if the electron passes through the R path the spin remains in the original state. Since the two orientations of the spin correspond to mutually orthogonal state vectors, inspection of the final state of the spin (i.e. a determination of whether it is in the positive or negative x-direction) will therefore allow one to determine which path the electron took. As expected, under these conditions, whether or not one in fact chooses to "inspect" the spin in this way, the interference pattern produced by the electron ensemble at the far end of the ring is totally destroyed. Stern et al. point out that this is exactly the result one would have obtained by assuming that the collision of the electron with the spin is completely "dephasing."

So far, so good. However, if the suggestion made in the first sentence of the last paragraph is to be applied in earnest to the classification of collision processes in mesoscopic systems, it raises at least two important questions. First, what exactly is to count as the "environment"? In the simple example above the question hardly arises, since there is only one degree of freedom (the motion of the electron around the ring) which is directly involved in the production of the interference pattern and one (the spin) which is not. In real life, however, one has a large number of coordinates, and the distinction is less obvious. For simplicity let us consider the artificial case of a single electron in an "AB" ring or cylinder (for the moment we consider it to be isolated), but now let the ring have finite transverse dimensions and consider also the interaction with phonons. Schematically, we may write the complete set of coordinates of the system (electron <u>plus</u> phonons) as (θ, x, y, R), where θ indicates the angle around the ring, x and y the other two coordinates of the electron, and R the collection of atomic (phonon) coordinates. Now, it is immediately obvious that there is an asymmetry between θ on the one hand and x, y, R on the other. This may be verified by writing out explicitly the partition function for the ring in a finite AB flux. This is proportional to the trace of the density matrix $\rho(\theta,\theta':x,x':y,y':R,R')$, with the proviso that θ and $\theta + 2n\pi$ are to be identified (no such proviso being made for the noncyclic variables x, y, R). The density matrix is in turn written as a path integral, with a weight factor depending on the "Euclidean" Lagrangian $L_E(\theta(\tau),x(\tau),y(\tau),R(\tau))$. In the Lagrangian only the term involving θ is coupled to the flux ϕ (through a contribution proportional to $\dot{\theta}\bullet\phi$): all other variables enter on an equal footing. Thus, at first sight at least, there is simply no conceptual basis for treating "elastic" (electron-impurity) collisions, which mix x and y with θ and with one another, any differently from "inelastic" (electron-phonon) ones, which mix x, y and R with θ and with one another.

The second problem is even more basic: The statement that interaction processes which change the state of the environment are equivalent to a complete loss of phase information on the system is true only to the extent that we implicitly exclude the possibility of recovering the missing information by appropriate measurements on the environment. To illustrate this point, let us suppose that in the thought-experiment of Stern et al., we inspect, after each electron of the ensemble has passed through, not the x-component but the z-component of the localized spin (subsequently resetting it again in the x-direction for the next electron, of course) and then keep the data only for those cases in which it was +1. Lo and behold, we recover the interference pattern!* Now, of course, it is not likely, in a realistic mesoscopic interference experiment, that we can inspect the "environmental" degrees of freedom directly in this way. However, it is not at all excluded that Nature will effectively do it for us whether we like it or not. Indeed, Nature is much cleverer at recovering the effects of phase coherence than she is often given credit for by practitioners of the quantum theory of measurement. In the literature of that area, one often gets the impression that the moment that the system of primary interest has interacted sufficiently with its "environment" that the off-diagonal elements of its density matrtix are suppressed to a value much less than unity, then the amplitude of all effects which depend on the associated phase coherence is correspondingly reduced. This is quite simply untrue: a spectacular counter-example is the familiar "spin-boson" problem, (cf. e.g. ref. (5)), where the interactions with the high-frequency parts of the "environment" (the oscillators), while very strongly suppressing the off-diagonal elements of the "spin" density matrix, do not reduce the <u>amplitude</u> (as distinct from the frequency) of the characteristic coherent oscillations. (See also ref. (6) for a further discussion of this point.) The lesson of this and similar examples is that it is simply not adequate, in discussing the generic effects of an "environment" on the system of prime interest, to focus only on the "immediate" (superficial) destruction of phase coherence: one must give Nature time to do her work, that is, in effect consider the full dynamics of the coupled system and environment, and in particular take careful account of the ratio of the relevant frequency scales.

I believe that it should be possible to investigate these questions quantitatively by a development of the model of Stern et al., in which in general one would have to allow for a large number of independent "spins" each producing a small phase shift. I hope to discuss this approach in detail elsewhere, and here, both for reasons of space and because the work is still at a preliminary stage, content myself with a few rather general remarks. (1) It seems necessary, in order to get quantitative results, to formulate the problem in terms of energy eigenvalues or some related quantities, rather than in terms of

---

* Provided that we chose the interaction Hamiltonian to be diagonal in σ_z, as Stern et al. do.

transmission probabilities, etc., as is usually done. In the experimentally realistic case of a ring with leads attached, what we actually measure is the magnetoresistance, but this can be related to the flux-dependence of the energy eigenvalues for a problem in which we impose as boundary condition a definite (nontrivial) phase relation between the probability amplitudes at the entry and exit points. In the simplest case, for a definite set of values of the $\{\sigma_Z\}$, this flux-dependence can be put in the form $f(\phi/\phi_0 - \delta\{\sigma_Z\})$, where f is periodic with period unity and $\delta(\sigma_Z)$ is an appropriate phase shift. (2) The simplest way of introducing the dynamics of the "environment" is to apply to the spin(s) (a) magnetic field(s) in some direction. The problem is substantially different depending on whether the "field" is applied parallel or perpendicular to the $\hat{z}$-axis (we assume that the interaction with the electron is diagonal in σ_Z); both situations appear to be experimentally relevant, in different contexts. (3) In the case of a field parallel to z, the situation is conceptually simple: crudely speaking we simply get an appropriate thermal mixture of the states we would have predicted for each particular value of the $\{\sigma_Z\}$. It is fairly clear that if the flux-dependence of the energies of these states (i.e. the quantity $\delta\{\sigma_Z\}$) is "random" then sufficiently high temperature will produce an exponential washing-out of the flux-dependence of the free energy, and hence presumably of the magnetoresistance. (4) The opposite case (field perpendicular to $\hat{z}$) is much more delicate, as can be seen even in the case of a single spin. It is tempting, but almost certainly wrong, to calculate the phase shifts δ_+, δ_- for the two possible values of spin, derive the corresponding energies $E_+(\phi)$, $E_-(\phi)$ and then diagonalize a 2×2 Hamiltonian matrix which has diagonal elements $E_+(\phi)$, $E_-(\phi)$ and off-diagonal elements a $\mathcal{H}$ ($\mathcal{H} =$ "field," $\alpha \sim 1 =$ overlap integral of $\psi_+(\phi)$ and $\psi_-(\phi)$). It is wrong, at least in the large-$\mathcal{H}$ limit, because the phase shifts themselves are not given a priori but must be recalculated in the presence of the field. This conclusion is qualitatively confirmed by a Born-Oppenheimer approach to the problem in this limit. The problem therefore becomes less trivial then it looks, even for one spin. (5) Finally, nothing in the model as developed so far appears to do anything to illuminate the difference, which appears to be well confirmed both experimentally[7] and theoretically[8], between the roles played, in a real-life mesoscopic interference experiment, by the transverse electron degrees of freedom and the phonon ones. The fundamental physical reason for this difference remains, at least to me at the time of writing, a mystery.

2. WHAT IS THE ACCURACY WITH WHICH WE CAN MEASURE A RESISTANCE?

The conventional definition of resistance, or rather its inverse, the conductance G, is

$$G = \lim_{v \to 0} I/V. \qquad (1)$$

where I is the current and V the "voltage" (i.e. what is read on a voltmeter: but see below). In this section I raise the question: Can we actually take the limit $V \to 0$ and get a well-defined value for G? It should be said at once that there is no difficulty in doing so (at least as far as the arguments presented below go) provided one is prepared to measure for an infinite time: the considerations below refer exclusively to the experimentally realistic condition of a possibly long but finite measurement time T.

Before we start, a comment is necessary on the basic notions of "current" and "voltage." The "current" I shall be discussing explicitly below is the convective current, that is, the motion of real electric charge; it does not include the so-called displacement current $\partial E/\partial t$. It may well be argued[9] that in real life one most commonly measures (e.g. by the deflection of a compass needle) a "current" which is the sum of charge and displacement currents. If so, this requires some reformulation of the statements below, but I believe does not change the general nature of the results. Similarly, the "voltage" referred to below is the "true" voltage, i.e. the electrostatic potential drop which would be seen by (say) a muon moving between the points in question; it is not the electrochemical potential which most (though not necessarily all) voltmeters measure. Again, I do not believe that taking this complication into account would qualitatively change the results: see below.

Let's start with a simple and prima facie plausible argument (in which I shall omit all numerical factors of order 1). What is likely to prevent us getting a well-defined result for the conductance in the limit $V \to 0$? The obvious candidate is thermal noise. To estimate the effects of this, we note that for a measuring time T the effective frequency band width is of order 1/T, and by the standard formulae of Johnson noise theory we have for the fluctuations of current and voltage respectively

$$(\delta I)^2 \sim <I_\omega^2> \Delta\omega \sim \frac{k\theta}{R}\Delta\omega \sim \frac{k\theta}{RT} \qquad (2a)$$

$$(\delta V)^2 \sim <V_\omega^2> \Delta\omega \sim Rk\theta\,\Delta\omega \sim Rk\theta/T \qquad (2b)$$

where θ is the absolute temperature and R the "true" value of the resistance, i.e. that which would be measured in an experiment lasting an infinite time. In writing the above estimates we implicitly assumed that the frequency-dependent resistance is effectively constant for $\omega \lesssim 1/T$: if T is "macroscopic" (say ~1 sec) this is certainly a good approximation. It is clear that $\delta I/<I>$ and $\delta V/<V>$ are of the same order, so the relative error ΔG in the measured conductance G can be estimated to be

$$\left(\frac{\Delta G}{G}\right)_{th} \sim \left(\frac{\delta I}{I}\right)^{1/2} \left(\frac{\delta V}{V}\right)^{1/2} \sim \left(\frac{k\theta}{IVT}\right)^{1/2} \sim \left(\frac{k\theta}{E_{diss}}\right)^{1/2} \tag{3}$$

where E_{diss} is the total Joule energy dissipated in the measurement. The result (3) is clearly plausible from an intuitive point of view. To get a rough idea of its significance, we take $V \sim 1$ nV, $R \sim 1$ kΩ, $T \sim 1$ sec, $\theta \sim 10$ mK; then $\Delta G/G \sim 10^{-2}$, so even under these rather extreme conditions the limitation on measurement is not too severe.

Now we ask: What is the effect of <u>quantum</u> noise? The most obvious guess is to replace the thermal energy $k\theta$ in the above formulae by the zero-point noise $\hbar\omega \sim \hbar\Delta\omega \sim \hbar/T$; and consideration of the quantum (Nyquist) formula for Johnson noise confirms this. Thus we obtain

$$\left(\frac{\Delta G}{G}\right)_{q.n.} \sim \left(\frac{\hbar/T}{E_{diss}}\right)^{1/2} \tag{4}$$

For the parameter values quoted above (and indeed for any realistic ones) this quantity is completely negligible compared to the uncertainty due to thermal noise, so we would prima facie conclude that quantum effects do not limit the accuracy of a measurement of resistance.

I shall now argue that this conclusion is <u>much too optimistic</u>. The argument rests on the following easily proved assertion: Consider a given piece of surface S such that the mean electron density n is constant on it, and define the operator

$$\hat{I}_S \equiv \int \hat{j}(r) \cdot dS \tag{5}$$

where $\hat{j}(r)$ is the usual electric current density operator. Moreover, consider two arbitrary points r_1 and r_2, and $\hat{V}_{12}$ to be the (operator) difference of the quantities $\hat{V}(r_1)$ and $\hat{V}(r_2)$, where

$$\hat{V}(r) \equiv \int \hat{\rho}(r')/(4\pi\epsilon_0|r-r'|) \, dr' \tag{6}$$

with $\hat{\rho}(r)$ the electric charge density operator. Then, directly from the commutation relation of $\hat{\rho}(r)$ and $\hat{j}(r)$ and the definition (6), it is easily shown that

$$[\hat{I}_s, \hat{V}_{12}] = \hbar \omega_p^2 (\Delta S/4\pi) \tag{7}$$

where $\omega_p \equiv (ne^2/m\varepsilon_0)^{1/2}$ is the plasma frequency corresponding to the surface, and ΔS is the difference of the (signed) solid angle subtended by it at $\underline{r}_1$ and $\underline{r}_2$. From (7) it follows immediately that the product of the uncertainties ΔI_S, ΔV_{12} in the values of I and V_{12} taken at equal time satisfies the inequality

$$\Delta I_S \cdot \Delta V_{12} \geq \frac{1}{2}\hbar\omega_p^2(\Delta S/4\pi). \tag{8}$$

We are, of course, interested not in the "equal-time" uncertainty but rather in the uncertainties in the quantities

$$\hat{I}_T \equiv \int_0^T \hat{I}(t)dt, \quad \hat{V}_T \equiv \int_0^T \hat{V}(t)dt. \tag{9}$$

A plausible, though not rigorous, way of estimating the commutator of these two quantities is to write the phenomenological equation

$$\hat{I}(t) \sim \hat{I}(0) \, \text{exp-}i\omega_p t \, \text{exp-}|t|/\tau \tag{10}$$

where τ is (of the order of) the relaxation time of the macroscopic electric current (which, except for ultrasmall samples, can usually be identified with the usual elastic scattering time at least to an order of magnitude). We then easily find

$$\Delta I_T \cdot \Delta V_T \sim \Delta I_S \cdot \Delta V_{12} \, (T/\omega_p^2 \tau). \tag{11}$$

Since the average value of I_T is simply $I \cdot T$ where I is the mean current, the result for $\Delta G/G$ is

$$(\Delta G/G)_{\text{q.n.}} \sim \left(\frac{\Delta I_T}{I_T}\right)^{1/2} \left(\frac{\Delta V_T}{V_T}\right)^{1/2} \sim \left(\frac{1}{2}\frac{\Delta S}{4\pi} \cdot \frac{\hbar/\tau}{E_{\text{diss}}}\right)^{1/2}. \tag{12}$$

The inequality (12) implies that for samples of any shape but a long thin wire (for which $\Delta S/4\pi \ll 1$) the quantum limitation on the measurement of resistance can easily be comparable to or even greater than the thermal limitation. Indeed, if we take the above set of parameters ($V \sim 1$ nV, etc.) and set t equal to the quite reasonable value of 10^{-13} secs, we find that for many samples $\Delta G/G$ is of order unity!

It would certainly be interesting to know whether anyone claims to have measured a resistance with an accuracy greater than that allowed by eqn. (12) and if so, to investigate the precise nature of the measurement and whether and how it circumvents the above argument. One obvious point to make in this connection is that most real-life voltmeters will have a response time τ_r which is itself long compared to τ; in fact, even if otherwise ideal, those which measure electrochemical potential μ will presumably respond to fluctuations in the "true" voltage V with a time-lag of at least the inelastic time τ_{in}, which may be much longer than τ. Crudely speaking we should expect the effect of this to be to replace, in the expression for fluctuations ΔG_m in the measured conductance G_m, the quantity $\hbar/\tau$ by $\hbar/\tau_r$. However, it is clearly a question, particularly if $\tau_r >> \tau_{in}$, whether we can now regard ΔG_m as a measure of the true accuracy of "voltage" measurement.

3. WHEN CAN WE DESCRIBE THE LEADS CLASSICALLY?

The last question I shall raise is the most frustrating of the three, but also the most important from a conceptual standpoint. It touches the very basis of our description of experiments at the macroscopic level. Since I have discussed it elsewhere[10], I shall keep the details to a minimum here.

To introduce the question, let me start with a quotation from Nils Bohr[11]: "The fundamentally new feature of the analysis of quantum phenomena is... the introduction of a fundamental distinction between the objects under investigation and the measuring apparatus. This is a direct consequence of the necessity of accounting for the function of the measuring instruments in purely classical terms, excluding in principle any regard to the quantum of action." [emphasis supplied]. In the context of mesoscopic systems it seems natural to regard the "measuring apparatus" as including the current and voltage leads.

Let us try to see how we could give a formal implementation of the procedure advocated by Bohr, starting from the conventional assumption that "in principle" both system and apparatus (leads) are described by quantum mechanics. Let q denote, schematically, the variables of the "system" S and x those of the "apparatus", A. In general the "universe" S + A will be described by some quantum-mechanical density matrix $\hat{\rho}_{S+A}(t)$, and its dynamics may be governed by a Hamiltonian which depends, inter alia, on some operator L which acts on the variables of A: $H = H(\Lambda)$. To obtain the usual description in which the apparatus is treated classically, two steps are necessary: (1) We (implicitly) assume that the density matrix $\hat{\rho}_{S+A}(t)$ may be expressed as a product $\hat{\rho}_S(t)\hat{\rho}_A(t)$, where $\hat{\rho}_S$ is an operator in the Hilbert space of S, etc. (2) We replace, in H, the operator L by its expectation value $\text{Tr } \hat{\rho}_A(t) L$.

I will now give an example of an experiment for which I believe that this procedure may be totally illegitimate. The example involves the so-called "Bloch oscillation" phenomenon predicted to occur in normal and superconducting tunnel junctions (see e.g. ref. (12)). If for definiteness we consider the normal case, then we may define the phenomenon for present purposes as follows: A fixed external current I is driven through a "black box" and the potential drop V(t) between two points in the box is measured. It is then predicted that V(t) has a component which oscillates <u>coherently</u> at the (circular) frequency of the transfer of single electrons through the box, namely $\nu_B = I/e$.

One variant of the standard derivation of this prediction proceeds schematically as follows (see e.g. ref. (13) for details of the argument). First, the external current I, <u>which is treated as a c-number</u>, is integrated to get the so-called "external charge" $Q_x(t)$, which is of course also then a c-number. In general, at least for certain types of circuit, $Q_x(t)$ can be identified as the charge contained within a certain definite region of space. The Hamiltonian is written as a function of Q_x, and it is shown that the energy eigenvalues, and hence other physical quantities such as the voltage drop mentioned above, are periodic as functions of Q_x with a period equal to the electron charge e. Since $Q_x(t)$ varies in time at a rate $dQ_x/dt \equiv I$, it immediately follows that the voltage drop will in general have a component which oscillates with frequency I/e, as stated above.

The problem with this argument is that if the quantity Q_x is indeed the total electric charge within a well-defined region of space, then the corresponding quantum-mechanical operator has eigenvalues <u>which are integral multiples of e</u>. (Needless to say, the <u>expectation value</u> $<Q_x(t)>$ need not be an integral multiple of e.) Now, suppose we take seriously the notion that the combined "universe" of system + leads may be represented by a wave function $\Psi(q:x)$ where q schematically denotes the system variables and x those of the leads (apparatus). (We carry out the analysis for simplicity for a pure state: the more realistic case of a mixture introduces no special difficulties.) Let y represent all apparatus variables <u>other</u> than Q_x. Then we can expand the "universe" wave function Ψ in the normalized eigenstates of Q_x, which we label:

$$\Psi(q,y,Q_x:t) = \sum_N a_N(t)\, \psi_N(q,y:t)|N> \tag{13}$$

where $\psi_N(q,y,r)$ is normalized to unity. If the expectation value of the quantity Q_x is to vary as It, as required, the quantities $a_N(t)$ must obviously satisfy the condition

$$\sum_N N|a_N(t)|^2 = It. \tag{14}$$

We will assume that this condition is guaranteed by some term(s) in the Hamiltonian which do not themselves depend on Q_x. Then it is easy to show that the individual ψ_N satisfy the equation

$$i\hbar \frac{d\psi_N}{dt} = \hat{H}(Ne)\psi_N(t) \tag{15}$$

where $\hat{H}(Ne)$ means $\hat{H}(\hat{Q}_x)$ evaluated for $\hat{Q}_x = Ne$. But we postulated that the periodicity of the energy levels $E_n(Q_x)$ in Q_x was precisely e, so $\hat{H}(Ne)$ is <u>independent</u> of N and the whole basis for the supposed "Bloch-oscillation" phenomenon goes up in smoke!

The above argument does <u>not</u> show that "Bloch oscillations" are necessarily spurious. However, it does make it clear that the most obvious way of obtaining them – that is, by replacing the <u>operator</u> Q_x by its expectation value $<Q_x(t)>$, which in general is not one of its eigenvalues – is in this case illegitimate. Generalizing, we can say that a sufficient condition for failure of the procedure described above to implement Bohr's demand is that the behavior of the system is controlled by some function $f(\Omega)$ of an apparatus variable Ω which is appreciably nonlinear over the characteristic spacing of the eigenvalues of the associated operator Ω: in such a case, barring pathologies, the replacement of Ω by its expectation value will give incorrect results.

This strongly suggests that there may be cases where Bohr's dictate can simply not be consistently met, and where we <u>must</u> keep the characteristically quantum correlations between the "system" and the "apparatus" at least up to a much more macroscopic level then we would naively have thought. This, in turn, suggests the perhaps alarming possibility that we may eventually run into a "quantum preparation paradox" which is in some sense the inverse of, and at least as severe as, the notorious "quantum measurement paradox."

However this may be, it is clear that mesoscopic systems are, almost by definition, precisely the frontier area between the microscopic world which obviously requires a quantum-mechanical description, and the macroscopic world where (or so we had hoped!) classical physics should be adequate. It is surely no accident that it is precisely here that we have run up against the above problems, all of which are concerned in one way or another with the interpretation of quantum mechanics and its "join" to our classical ideas. I believe that such problems are likely to loom increasingly large in the development of nanostructure physics over the next decade.

I am grateful to Drs. Stern, Aharonov and Imry for sending me their preprint. I should also like to acknowledge helpful discussions with Joe Imry, Rolf Landauer and, especially, Doug Stone, who saved me from at least one serious misconception. The responsibility for any remaining is of course entirely mine.

REFERENCES

1.   Y. Imry, in Directions in Condensed Matter Physics, ed. G. Grinstein and G. Mazenko, World Scientific, Singapore 1986.
2.   Y. Aharonov and D. Bohm, Phys. Rev. **115**, 485 (1959).
3.   G. Badurek, H. Rauch and J. Summhammer, Phys. Rev. Lett. **51**, 1015 (1983).
4.   A. Stern, Y. Aharonov and Y. Imry, preprint.

5. A. J. Leggett, S. Chakravarty, A. T. Dorsey, M. P. A. Fisher, A. Garg and W. Zwerger, Revs. Mod. Phys. **59**, 1 (1987).

6. A. J. Leggett, in Proc. Intl. Symp. on Foundations of Quantum Mechanics, ed. S. Kamefuchi et al., Physical Society of Japan, Tokyo 1984, p. 74.

7. S. Washburn and R. A. Webb, Adv. Phys. **35**, 375 (1986).

8. A. D. Stone, in Proc. 2nd Intl. Symp. on Foundations of Quantum Mechanics, ed. M. Namiki et al., Physical Society of Japan, Tokyo 1987.

9. R. Landauer, private communication.

10. A. J. Leggett, in Frontiers and Borderlines in Many-particle Physics, ed. R. A. Broglia and J. R. Schrieffer, North-Holland, Amsterdam, 1988.

11. N. Bohr, in Essays 1958-62 on Atomic Physics and Human Knowledge, Interscience, New York 1963.

12. K. K. Likharev and A. B. Zorin, J. Low Temp. Phys. **59**, 347 (1985).

13. E. Ben-Jacob, Y. Gefen, K. Mullen and Z. Schuss, in SQUID '85, ed. H. D. Hahlbohm and H. Lübbig, Walter de Gruyter, Berlin 1985, p. 203.

QUANTUM INTERFERENCE EFFECTS IN CONDENSED MATTER PHYSICS

Richard A. Webb

IBM Thomas J. Watson Research Center
P.O. Box 218 Yorktown Heights, New York 10598

Introduction

The transport properties of very small non-superconducting structures has recently gained increased attention with the experimental verification that samples can now be fabricated in the diffusive (1) and ballistic (2) limit where the measured resistance exhibits new contributions arising from complete quantum mechanical phase coherence over the entire sample. Perhaps the first surprising result is that this coherence is maintained in spite of strong elastic scattering (3) and is observable in samples containing 10^7 to 10^{10} electrons. In addition, these systems exhibit some interesting universal properties, but the details are extremely sensitive to the how the measurement is made. As will be shown, the non-local aspects associated with what is actually being measured is of central importance. Experimental work in this field has already provided a partial answer to one of the important questions in quantum mechanics "How does a quantum system evolve into the more familiar everyday classical system?". This transition can be studied by varying the temperature, or sample size, or by applying increasingly larger voltages across the phase coherent region. Surprisingly, quantum interference effects in insulators has also recently been demonstrated (4), but we will not be able to discuss this subject here.

43

The Aharonov-Bohm Effect in a Disordered Solid

One of the primary experimental signatures that quantum interference effects are playing an important role in determining the transport properties of any electrical circuit is that the resistance fluctuates as either the Fermi energy, or an externally applied magnetic field or voltage is altered. In a short metallic wire, these fluctuations are a random function of the externally changed variable, but in a doubly connected structure, such as metallic loop of wire, the electrical resistance is found to fluctuate periodically as shown schematically in Fig. 1. All of these phenomena are directly related to the Aharonov-Bohm effect. They have no classical analogs and are a signature that the phase of the electron wavefunction is being altered by the externally applied vector or scalar potentials and this in turn changes the measured electrical resistance.

As shown by Aharonov and Bohm in 1959 (5) for electrons traveling in a vacuum, there exist simple ways to smoothly alter the phases of wave-functions. Their basic observation was that the phase of an electron wave could be altered by placing the wave in contact with a vector potential A or a scalar potential V. They emphasized that this result would be true even if the waves traveled in a region where both the magnetic field and electric field were zero; no classical forces on the electron wave. The prediction is that the phase ϕ of the wave-function is wound up or down by the potentials according to $\Delta\phi = (e/\hbar)\int(Vdt - A \bullet ds)$, where ds and dt are elements of the path and time respectively. In a doubly connected geometry such as a ring, the electron wave can travel along both branches of the structure and interfere with itself when summed at the terminus. The resulting intensity of the recombined beam will oscillate periodically, if a magnetic flux is applied to the interior of the ring, with a flux period $\Delta\Phi$ = h/e. If a voltage difference is applied between the two arms of the ring, the intensity will again oscillate with a voltage period given by $\int Vdt$ = h/e, where the integral is over the time period that the electron spends in contact with the scalar potential V. The role of the vector potential has been demonstrated in vacuum (6) where the electrons suffer no collisions of any kind — elastic or otherwise, in metal whiskers where transport is deep in the ballistic regime (7) (again no collisions), and in long thin metal cylinders where a magnetoresistance oscillation of twice the Aharonov-Bohm frequency ($\Delta\Phi$ = h/2e) occurs near zero magnetic field because of the incipient localization in the metal which forms the cylinder (8). In the first two cases the absence of any scattering "guarantees" the occurrence of Aharonov-Bohm effects; the last example constitutes the original experimental proof that the these effects are not destroyed by elastic scattering. The observation of oscillations from the cylinders, whose length $L \sim 1\text{mm}$ $>> L_\phi$, is possible because the interference from time-reversed motion along paths around the cylinder does not average to zero(8), as the cylinder length exceeds L_ϕ. L_ϕ is the characteristic distance an electron can propagate before losing phase memory in it's wavefunction.

The simplest circuit analogue of the Aharonov-Bohm geometry is a loop of wire of small thickness and width fed by two leads. The possibility of observing h/e flux periodic effects from a loop is intimately connected with the states of the loop itself (in absence of leads). Theoretical analysis of one dimensional rings in the presence of elastic scattering clearly indicated these periodic effects would not be destroyed (9). Subsequent work on resistive rings with leads (10) revealed that, in contrast to the results from cylinders, the magnetoresistance would have flux period h/e. Experimental confirmation of these predictions is shown in the bottom part of Fig. 1 where the h/e resistance oscillations from a gold ring 0.8 μm in diameter at 30 mK is shown. Today these effects have been seen in many laboratories around the world. We see then, in this resistance measurement on a disordered metal loop, the direct signature of interference of the electron wave-function. There is another quantum mechanical contribution to the resistance of this loop. The periodic oscillations are superimposed on a randomly fluctuating background resistance. As the electrons diffuse through the wires which form the ring, their trajectories randomly enclose flux which pierces the metal. These random fluxes also contribute to Aharonov-Bohm interference, and this interference appears in the magnetoresistance as random fluctuations in R(H), see the top part of Fig. 1. The characteristic magnetic field increment to induce a fluctuation is much larger than the oscillation period. Naively speaking, the difference in field scale reflects the different areas enclosing flux: For the periodic oscillations all of the area encircled by the loop contributes, while for the random component, only the area covered by metal contributes. It has been shown (11) that the ratio of the field scales is approximately the ratio of these areas.

It has been proved that, in a two probe measurement, the amplitudes of the various terms are such that when the resistance is inverted to yield a conductance G, the fluctuations have RMS amplitude (12) $\Delta G \simeq e^2/h$. This conductance fluctuation amplitude is "universal" in several loose senses of the word. For instance, the precise factor relating ΔG to e^2/h is weakly dependent on geometry, independent of length (so long as phase coherence is maintained), and independent of the average resistance. *Any metal sample exhibits conductance fluctuations of about this amplitude.*

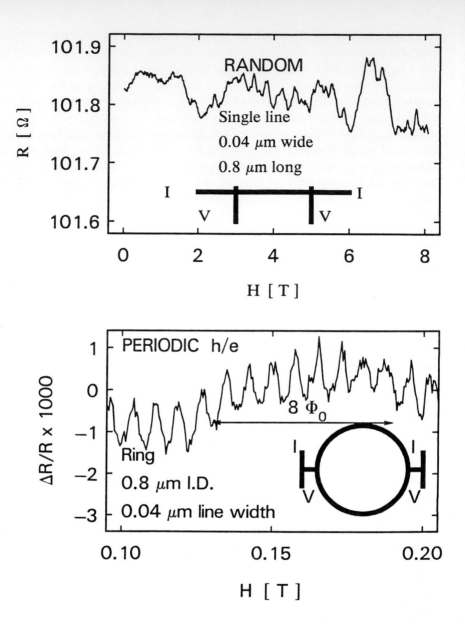

1). The resistance of a 100 Ω small wire (top) or 30 Ω small ring (bottom) as a function of an externally applied magnetic field H.

In the above discussion of the interference amplitude, the absolute phase change does not enter — only the relative phases of the two paths. This in turn implies that the interference is not necessarily destroyed as the magnetic field increases to rather strong values so that a large amount of flux pierces the metal in the device. The increasing of the magnetic field does not alter the character (average period, amplitude, ...) of the oscillations until the field reaches classically strong values where Landau orbits form and destroy the diffusive motion of the electrons. The oscillations persist to rather large magnetic field without any attenuation what-so-ever — in fact they have been observed in gold at H > 15T without attenuation. (Landau orbits will not form in this metal at low temperatures until B ≃ 200T.)

The Quantum Domain

One question frequently arises: "How does a quantum system evolve into the more familiar classical system?" Experiments on these small phase coherent systems have already provided a partial answer to some aspects of this question. In general, temperature, sample size, diffusion coefficient, and details of how the measurement is made play a crucial role in determining whether or not quantum interference effects will be a significant contribution to the electrical resistance in a disordered system. Of these, phase randomization due to elastic scattering and finite temperature effects has historically been thought to be the most devastating. As demonstrated in the cylinder and gold loop experiments, elastic scattering is not destructive. The question of thermal averaging arises because, in the systems discussed here containing 10^8 atoms, the average separation between electron energy levels is $\Delta E \simeq 10^{-8} \text{eV}$, which corresponds to a temperature of 10^{-4} K. When the thermal energy exceeds some characteristic energy of the system, the delicate quantum effects begin to disappear. The naive analogy to Shubnikov-de Haas oscillations leads one to expect that the oscillations will decay according to a Dingle factor $\Delta R \propto \exp(-\Delta E/k_B T)$. The surprising discovery from the resistance fluctuations is that the spacing between energy levels is not the characteristic energy scale and that the decay is much slower. The energy scale that governs the physics is related to the Thouless energy (13), $E_C = \hbar D/L_\phi^2$. Once $k_B T$ exceeds E_C, the magnitude of the quantum corrections to the resistance, ΔR, decays slowly as the square-root of the number of uncorrelated energy "bands", $\Delta R(T) \sim \sqrt{E_C/k_B T}$. For the gold ring displayed in Fig. 1, $E_C/k_B \simeq 0.03K$, but because of the weak algebraic averaging, the quantum interference effects are readily observable above 1

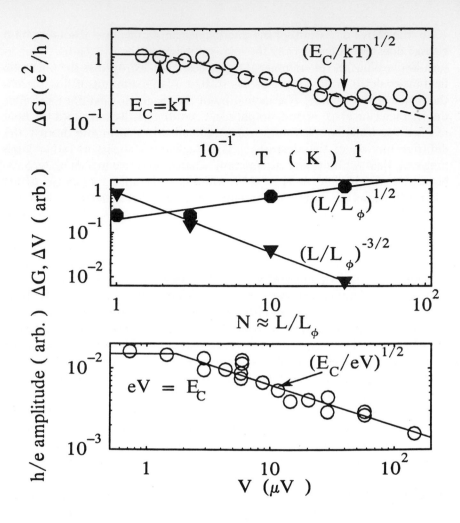

2). Three ways to destroy quantum interference effects in a disordered system by averaging away the signal. The top figure represents energy averaging. The middle figure displays the length dependence of both the rms value of the voltage and conductance fluctuations for a thin wire or series array of phase coherent rings as a function of N, the number of phase incoherent segments in the sample. The bottom part displays the voltage dependence of the decay of the h/e oscillations in a gold ring.

K as shown in Fig. 2a. In semiconductors where E_C can be $\sim$10K, quantum interference effects have been reported above 100 K (1).

This weak algebraic averaging of the quantum interference also appears in the size dependence with the characteristic length scale being L_ϕ. When the length of the sample is increased beyond $L = L_\phi$, the resistance fluctuations grow as, $\Delta R(L/L_\phi) \sim (L/L_\phi)^{1/2}$, but the average resistance grows linearly with L, thus the *relative* magnitude ($\Delta R/R$) of the quantum interference contributions decay as the square-root of the number of uncorrelated length segments. This means that the conductance fluctuations ΔG decay as $(L/L_\phi)^{-3/2}$. Both of these are shown in Fig. 2b where the magnitude of the h/e oscillations for N phase coherent rings in series is displayed as a function of N. For a single ring having half the perimeter larger than L_ϕ, the loss of phase coherence reduces the oscillations severely: the oscillations decrease exponentially as the path length L increases. The reason is simply that the probability of an electron reaching the terminus of the loop retaining phase coherence depends exponentially upon L/L_ϕ so that $\Delta R(L/L_\phi) \sim e^{-L/L_\phi}$.

Since all of these averaging mechanisms depend upon L_ϕ, the temperature dependence of L_ϕ determines the observability of quantum interference effects. Generally as the temperature increases, the probability of an electron absorbing and re-admitting a phonon increases. For narrow wires, the temperature dependence of the phase coherence length is usually $L_\phi \propto 1/T^p$, where p is between 0.33 and 1.0. Thus $L_\phi(T)$ is also a weak power law in T and not an exponential.

One obvious way to destroy quantum interference effects in a disordered conductor is to pass too much current through it. Joule heating effects are familiar from classical electronics, and an excessive amount of current will certainly raise the temperature of the sample. Perhaps not so apparent is the observation that currents too small to cause Joule heating are important. If the voltage drop across a phase coherent segment exceeds E_C/e, then averaging over incoherent energy regions will take place again, and the magnitude of the quantum interference effects will be reduced by $(E_C/eV)^{1/2}$ as shown in Fig. 2c.

Nonlocal Quantum Interference

There have been several unexpected experimental consequences of long range phase coherence in disordered systems. One of the most surprising is that the RMS amplitude of the voltage fluctuations measured as a function of magnetic field in a four terminal sample becomes independent of length (14) when the separation L between the voltage probes is less than L_ϕ. The

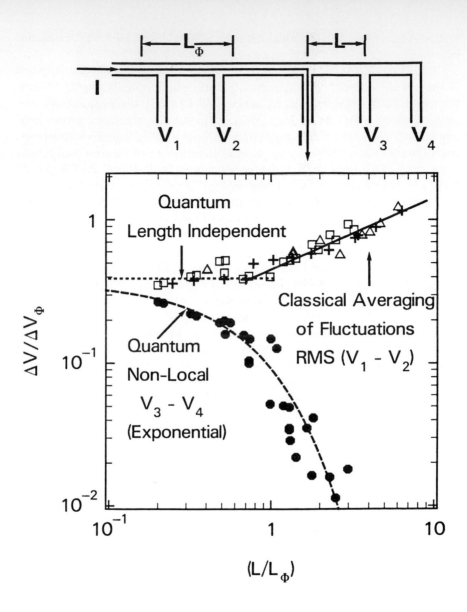

3). Length dependence of the average voltage fluctuation amplitude in a long line for local and non-local measurements.

top part of Fig. 3 schematically displays a typical multi-terminal sample configuration where the current can be injected and removed using any of the six leads. If the voltage difference V_1-V_2 is measured across the portion of the sample where a classical current exists, we would expect the average resistance to increases linearly with separation L. The average voltage difference between any two points along the sample is given by $\Delta V_{12} = E\,L$. The quantum interference contributions to the voltage fluctuations (measured by changing the magnetic field), however, behave much differently. Figure 3 displays the results of a recent experiment on an Sb wire, which has eight probes with separations $0.2 < L < 3.6\,\mu m$. The RMS values of the voltage fluctuations are plotted as a function of L. The length scale has been normalized to L_ϕ (1.1 μm), and the voltage scale has been normalized to $V_\phi \simeq IR_\phi^2 e^2/h$, where R_ϕ is the resistance when the voltage probes are exactly L_ϕ apart. The striking feature of these data is that as $L \to 0$ the RMS amplitude of the voltage fluctuations does *not* go to zero. The quantum interference contributions to the measured voltages give the same result when the probes are separated by 1 μm or 0 μm (zero average resistance) ! For large separations between voltage probes, $L/L_\phi > 1$, the voltage fluctuations grow just as expected from classical physics, $\Delta V \propto (L/L_\phi)^{1/2}$.

The explanation of this length independence is that the voltage probes do not define the length scale for the measurement. The effective length of the sample is really determined by the properties of electrons themselves. The electron is a wave whose phase coherence extends over a region L_ϕ long. This means that the electron waves exist in classically forbidden regions of the sample and in particular the waves can propagate coherently into the current and voltage probes a distance on the order of L_ϕ. The application of a magnetic field alters the interference over the entire phase coherent region. The microscopic details of the scattering impurities are different in each section of the sample as well as in the voltage leads. Thus a voltage difference is measured between leads 1 and 2, even when the two leads are attached to the same point on the sample, because of the propagation of electrons into the probes. This non-locality can be directly measured by using voltage probes V_3 and V_4 shown in the top part of Fig. 3. The average current between these two leads is always zero as required by current conservation but the potential difference V_3-V_4 is found to fluctuate randomly as the magnetic field is changed. The magnitude of the fluctuation decays exponentially as the distance L from the end of the classical current path increases as $\exp$-(L/L_ϕ) as shown by the dashed line fitted to the nonlocal data displayed in Fig. 3. Recent theoretical analysis (15) is consistent with the general shape of the curves in Fig.3. These data clearly demonstrate that

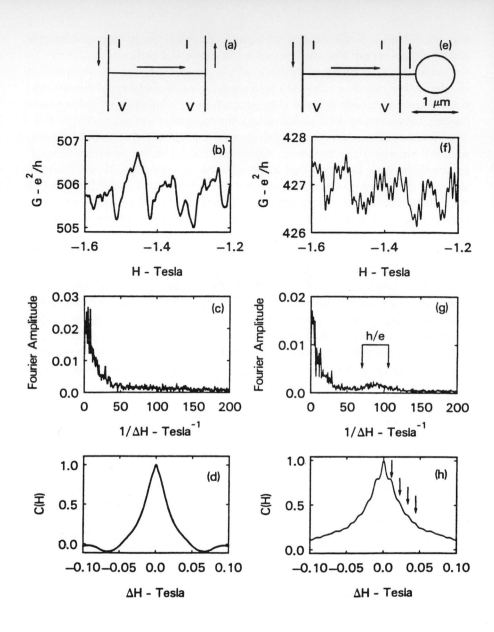

4). The magnetoresistance, Fourier transforms, and autocorrelation functions of two similar wires, one of which has a ring dangling outside of the classical current path.

there is a nonlocal relationship between E and j at any point in the sample due to quantum interference throughout a region L_ϕ long.

Perhaps the clearest experimental evidence of nonlocal behavior is shown in Fig. 4. Two identical four probe gold wires were fabricated, but on the second sample a small ring was attached 0.2 μm from the classical current path. Figure 4b and 4f displays the magnetic field dependence of the conductance of both wires. Both samples exhibit random conductance fluctuations with the same characteristic field scale, but in addition, the second sample has more high frequency "noise". Fourier transforms of both sets of data clearly show that this high frequency "noise" is in fact an h/e oscillation with the characteristic period equal to what is expected based on the diameter of the ring (16). Figures 4d and 4h display the autocorrelation function C(H) for the samples with both showing the same characteristic field scale, but the sample with the nonlocal attached ring also shows periodic h/e oscillations as indicated by the arrows. In order to observe this effect, some large fraction of the electrons had to encircle the ring coherently. The magnitude of the h/e oscillations decays exponentially with distance away from the classical current path, but the characteristic length scale of the decay is not L/w, as it would be in classical physics. Rather, it decays as $\Delta V \propto e^{-L/L_\phi}$. Thus the quantum interference measured in any small sample will have nonlocal contributions, and the distance between voltage probes does not really define the sample.

Concluding Remarks

I have only been able to review some of the recent discoveries associated with the physics of phase coherent systems. This sub-field of condensed matter physics is relatively new and we can expect more surprises in the near future. In addition the fundamental questions we are asking have direct relevance to the future of the electronics industry. How small can we make our circuits before something new happens? Or phrased another way, is there any possibility that quantum interference can be used to fabricate practical devices? We have already begun to better understand both what it means to make a measurement on a quantum system and how the classical domain evolves into the quantum domain but our understanding is still very incomplete. Only further research in this field will answer these questions.

References

1. See the reviews by A.G. Aronov and Yu.V. Sharvin, Rev. Mod. Phys. **59**, 755 (1987); Y. Imry, in *Directions in Condensed Matter Physics*, eds. G. Grinstein and E. Mazenko, (World Scientific, Singapore, 1986); S. Washburn and R.A. Webb, Adv. Phys. **35**, 375 (1986).

2. See G.L. Timp and M.L. Roukes, these prodeedings.

3. R. Landauer, Philo. Mag. **21**, 863 (1970).

4. Ya.B. Poyarkov, V.Ya. Kontarev, I.P. Krylov and Yu.V. Sharvin, JETP Lett. **44**, 373 (1986); and Z. Ovadyahu, F.P. Milliken, and R.A. Webb to be published.

5. Y. Aharonov and D. Bohm, Phys. Rev. **115**, 485 (1959).

6. A. Tonomura, N. Osakabe, T. Matsuda, T. Kawasaki, J. Endo, S. Yano, and H. Yamada, Phys. Rev. Lett. **56**, 792 (1986).

7. N.B. Brandt, D.B. Gitsu, V.A. Dolma, and Ya.G. Ponomarev, JETP, **65**, 515 (1987); and references cited therein.

8. D. Yu. Sharvin and Yu. V. Sharvin, Pis'ma Zh. Eksp. Teor. Fiz. **34**, 285 (1981) (JETP Lett. **34**, 272 (1981)).

9. M. Büttiker, Y. Imry, and R. Landauer, Phys. Lett. **96A**, 365 (1983).

10. Y. Gefen, Y. Imry, and M.Ya. Azbel, Phys. Rev. Lett. **52**, 129 (1984).

11. A.D. Stone, Phys. Rev. Lett. **54**, 2692 (1985).

12. B.L. Al'tshuler, JETP Lett. **41**, 641 (1985); B.L. Al'tshuler and B.I. Shklovskii, JETP, **64**, 127 (1986); P.A. Lee, A.D. Stone, and H. Fukuyama Phys. Rev. B **35**, 1039 (1987); P.A. Lee and A.D. Stone, Phys. Rev. Lett. **55**, 1622 (1985).

13. A.D. Stone and Y. Imry, Phys. Rev. Lett. **56** 189 (1986).

14. A. Benoit, C.P. Umbach, R.B. Laibowitz, and R.A. Webb, Phys. Rev. Lett. **58**, 2343 (1987); W.J. Skocpol, P.M. Mankiewich, R.E. Howard, L.D. Jackel, D.M. Tennant, and A.D. Stone, *ibid*, 2347 (1987).

15. H.U. Baranger, A.D. Stone, and D. diVincenzo, Phys. Rev. B **37**, 6521 (1988); C.L Kane, P.A. Lee, and D. diVincenzo, *ibid*. **38**, 2995 (1988); S. Hershfield and V. Ambegaokar, *ibid*. 7909 (1988).

16. C.P. Umbach, P. Santhanam, C. van Haesendonck, and R.A. Webb, Appl. Phys. Lett. **50**, 1289 (1987).

Chapter 3

Lateral Periodicity and Confinement

FABRICATION OF QUANTUM-EFFECT
ELECTRONIC DEVICES USING X-RAY NANOLITHOGRAPHY[1]

Henry I. Smith
K. Ismail
W. Chu
A. Yen
Y.C. Ku
M.L. Schattenburg[2]
D.A. Antoniadis

Department of Electrical Engineering and Computer Science
Massachusetts Institute of Technology
Cambridge, MA 02139

I. INTRODUCTION

It is well known that scanning-electron-beam lithography (SEBL) is capable of creating patterns of arbitrary geometry in resist with minimum linewidths down to about 10 nm. This is not to say that all SEBL systems can do this, only that with proper design of the electron optics, and under proper conditions of exposure, this can be done. Even relatively simple systems, such as scanning electron microscopes converted for SEBL, can be used to write pattern with linewidths of a few 10's of nanometers. Thus it would appear that for research in quantum-effect electronics SEBL is the preferred lithography method. Although a converted SEM can write fine patterns, applications often demand more than an SEM can deliver in terms of pattern precision, area, field butting, proximity-effect correction, alignment, etc. Once one decides to address these other issues, the cost of SEBL escalates rapidly. A commercially available electron beam system appropriate for research in the sub-100 nm domain costs about $2M. In addition, there are costs associated with an SEBL facility, its operation and maintenance. For these and several additional

[1]Supported by the Joint Services Electronics Program (Contract DAAL03-89-C-0001), the Air Force Office of Scientific Research (Grant AFOSR-88-0304) and the National Science Foundation (Grant ECS-8709806).
[2]Center for Space Research, Massachusetts Institute of Technology, Cambridge, MA 02139.

reasons we have taken a different approach to sub-100 nm lithography in the Submicron Structures Laboratory at MIT. We have instead pursued the development of x-ray nanolithography. In this paper we describe the essential features of x-ray nanolithography, exposure results, applications in the fabrication of quantum-effect electronic devices, and future development plans.

II. X-RAY NANOLITHOGRAPHY

Although our use of x-ray nanolithography is driven, to first order, by economic considerations, there are additional reasons for developing this technology, as shown in Fig. 1, which plots the time required to write a 1 cm- square field with dense patterns, as a function of minimum linewidth. We assume that any feature can be subdivided into square pixels, and that the minimum linewidth is 10 pixels wide. We further assume that exposure requires 100 particles, either electrons or photons, per pixel. The number of particles per pixel is kept fixed, irrespective of linewidth, in order to keep constant the probability that any pixel will be in error (i.e., improperly exposed due to statistical fluctuations). The choice of 100 particles per pixel is arbitrary. In fact, a more careful statistical analysis reveals that a larger number is required, especially at sub-100 nm linewidths (2). Figure 1 clearly shows that direct write SEBL is inappropriate for mass production of sub-100 nm electronic systems, whereas exposure times for x-ray nanolithography are reasonable, even for a bench-top "conventional" system based on electron bombardment of a water-cooled target.

Our in-house x-ray lithography system is very simple and low cost. It consists of a water-cooled carbon or copper target that is bombarded with electrons in a focal spot $\leq$ 1 mm diameter. Figure 2 shows the kind of high quality replication achieved, which is clearly compatible with the current needs of quantum-effect-electronics research. It is noteworthy that such results are consistently achieved without regard to the substrate atomic number, or the precise resist thickness, or the development time. This insensitivity is referred to as process latitude. In day-to-day operation in a research environment process latitude is a significant advantage. It is, of course, even more of an advantage in manufacturing (3).

Masks are made on Si_3N_4 or Si membranes, 1 μm thick, if the Cu_L x-ray at 1.33 nm is used; and on polyimide membranes if the C_K x-ray at 4.5 nm is used. The polyimide undergoes some distortion and hence cannot be used if multilevel alignment or placement precision is required. Membranes of Si_3N_4 or Si can be essentially free of distortion. Contrary to one's intuitions, such thin membranes are extremely strong, provided they are grown under dust-free conditions so as to be free of pinholes or defects. They do not break with ordinary handling and hence can be used repeatedly. Diamond or other forms of hard carbon would be excellent membrane material, but they are not readily available today.

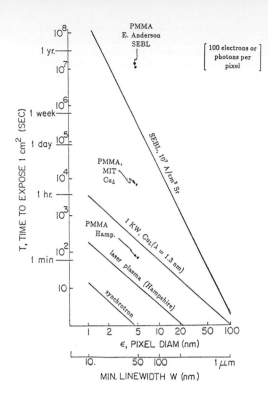

Fig. 1. Plot of the time to expose a 1 cm² area of dense patterns using scanning-electron-beam lithography (SEBL) and various forms of x-ray lithography, as a function of minimum linewidth. We assume a minimum linewidth is 10 pixels across and that 100 particles, either electrons or photons, are used to expose each pixel. A brightness of 10^7 A/cm² Sr corresponds to a thermal field emitter source. The isolated points correspond to experimental results, i.e., actual exposure of polymethyl methylacrylate (PMMA) with an SEBL system (1), and x-ray exposure of PMMA at MIT and Hamsphire Instruments.

Fig. 2. Scanning electron micrograph of a pattern exposed in PMMA by x-ray nanolithography using the C_K x-ray at 4.5 nm. The mask was held in contact with the substrate. A linewidth of 30 nm is readily achieved with large process latitude, and independent of substrate atomic number.

59

In the sub-100 nm linewidth domain, diffraction and photoelectron range are the factors that limit resolution (4,5). This is illustrated in Fig. 3. Most of our research to date has been done with the x-ray mask held in intimate contact with the substrate, in which case diffraction can be ignored and the C_K x-ray provides the highest resolution. Linewidths below 20 nm have been achieved with the C_K x-ray (6). Contact with the substrate does not appear to damage masks, presumably because membranes can deform around dust particles and surface asperities. However, mask-substrate contact is incompatible with multiple-mask alignment. In order to achieve sub-100 nm resolution simultaneously with alignment we have investigated several microgap techniques (5,7). Mask-substrate gaps of 4 μm over centimeter square areas appear to be practical. Such gaps, in turn, dictate the use of shorter wavelengths such as the Cu_L x-ray at 1.33 nm. Although Fig. 3 shows the resolution of Cu_L x-ray lithography is limited by an effective photoelectron range of 20 nm, experiments indicate that this figure may be too pessimistic. We are currently investigating the range in polymethyl methacrylate (PMMA) of photoelectrons generated by Cu_L x-rays.

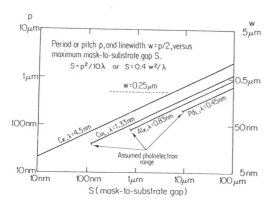

Fig. 3. Plot of the minimum linewidth versus mask-to-substrate gap, S, for four x-ray emission lines. The criterion $S = 0.4\ W^2/\lambda$ is conservative, designed to give larger than necessary process latitude. A more common criterion is $S = W^2/\lambda$. The plots are terminated at values corresponding to the estimated photoelectron range. A pi-phase shifting mask (8) allows one to use a gap about 4 times larger than predicted by this plot.

Figure 3 states that for replicating 50-nm-linewidths with the Cu_L x-ray, a mask-to-sample gap of 1 μm should be used. The criterion used to plot Fig. 3, $S = 0.4\ W^2/\lambda$, is quite conservative and hence a larger gap, e.g., 2 μm, is certainly permissible without significant sacrifice in process latitude. An approach we have taken to allow a further increase in gap is to offset the deleterious effects of diffraction by means of a pi-phase-shifting mask technique (8,9). The sharp vertical profiles of 100 nm lines in PMMA, obtained at a 4 μm gap with a pi-phase-shifting mask, shown in Fig. 4, lead us to be confident that at the same gap 50-nm linewidths are feasible with the Cu_L x-ray (5).

Fig. 4. Scanning electron micrograph of an x-ray exposure of 0.1 μm lines and spaces at a gap of 4 μm. The resist is 0.4 μm-thick PMMA. The mask was of the pi-phase shifting type (8).

Microgap X-Ray Exposure

Numerous investigators have shown that it is feasible to align masks to substrates with a precision ≤ 10 nm. We are currently trying to implement a scheme that simultaneously provides microgap control (at ~4 μm) and multiple mask alignment to < 10 nm (7).

III. QUANTUM-EFFECT ELECTRONICS

Figure 5 is a cross-sectional schematic of a grid-gate MODFET in GaAs/ AlGaAs (10). Figure 6 is a micrograph of the Ti/Au Schottky grid gate made via x-ray nanolithography and liftoff (4). X-ray masks with 200 nm-period gratings were fabricated using holographic lithography and liftoff (4). From such masks daughters masks were made which were either gratings or, by means of two sequential x-ray exposures at 90 deg. relative to one another, grids (i.e., periodic in two orthogonal directions). X-ray masks for nanolitho-graphy have also been made by SEBL (1) and by focused-ion-beam lithography (11).

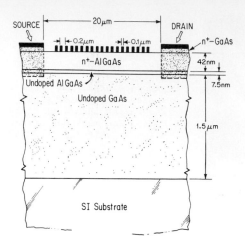

Fig. 5. Schematic of a grid-gate MODFET. The Schottky grid gate (shown as a black grating of 0.2 μm period between source and drain) produces a periodic potential modulation in the 2-D electron gas located at the interface between the undoped AlGaAs and GaAs.

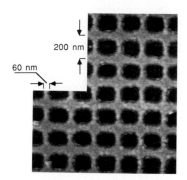

Fig. 6. Scanning electron micrograph of Ti/Au Schottky grid gate shown schematically in Fig. 5.

Figure 7 shows results of measurements with the grid-gate MODFET (10). We interpret the structure in the transconductance as indicative of electron back-diffraction from the two-dimensional periodic potential induced by the Schottky grid gate. Alternatively, one can say that the conduction band is broken up into minibands by the lateral surface superlattice (LSSL).

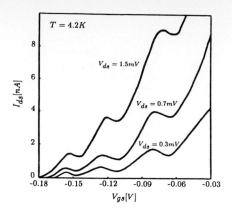

Fig. 7. Plot of the current from source to drain, I_{DS}, versus grid-gate voltage, V_{GS}, for several values of the source-to-drain voltage, V_{DS}. The regions of negative slope (negative transconductance) are indicative of electron back diffraction.

Using a grating-type mask, multiple parallel quasi-one-dimensional quantum wires were fabricated in AlGaAs/GaAs heterostructures, as depicted schematically in Fig. 8. Figure 9 shows results of measure-ments made on these structures (12). Because there are many (~100) parallel quantum wires, effects due to random impurities such as conductance fluctuations are averaged out, and the oscillations reflect the quasi-one-dimensional density-of-states. It is noteworthy that in certain regions the mobility is higher than for a corresponding 2-D electron gas, and in other regions it is lower. We believe this reflects a reduction of the scattering probability as a result of there being fewer states to scatter into in a 1-D wire when the Fermi energy is below the threshold for occupying the next subband. When the Fermi energy is positioned so that intersubband scattering can take place the mobility is correspondingly reduced.

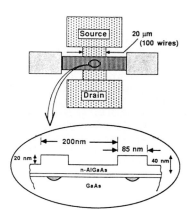

Fig. 8. Schematic top view and magnified cross section of the multiple parallel quantum wires. The ohmic contacts to the left and right are test contacts to examine transport perpendicular to the wires.

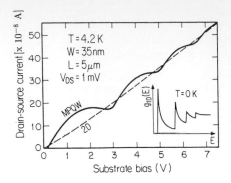

Fig. 9. Drain-source current (I_{DS}) as a function of substrate bias (V_{SUB}) for the multiple parallel quantum wires (solid line) and a scaled-down 2D channel (dashed line). The inset shows the theoretical density of states in a quasi-1-dimensional wire at 0 K.

We have also used x-ray nanolithography in our research on Si, to fabricate lateral-surface superlattices (13), quasi-one-dimensional quantum wires (14-16), and MOSFETs with channel lengths ~100 nm (17-20). In the most recent work with Si quantum wires, periodic conductance oscillations as a function of gate bias, at zero magnetic field, were observed at temperatures below 1K. These oscillations were attributed to a pinned charge density wave (16). In Si MOSFETs we were able to make the first demonstration of velocity overshoot (17,19). More recently, we have shown that hot electron effects are greatly reduced at channel lengths below ~150 nm.

IV. SUMMARY

In summary, we have shown that x-ray nanolithography can be used effectively to make quantum-effect electronic and short-channel devices. There are several advantages, notably low cost, simplicity, large process latitude and quick turn-around time once a mask is made. Mask making can be a difficult task but, in principle, should not be significantly more difficult than direct-write SEBL or focused-ion-beam lithography. A variety of other fabrication techniques can be brought to bear in fabricating a given x-ray mask: photolithography, holographic lithography, sidewall shadowing, etc. The latter, in fact, was the method used in the Si quantum-wire and short-channel MOSFET work (14-20). Lastly, x-ray lithography is compatible with high throughput manufacturing. Our current research in x-ray nanolithography is focused on carrying out multiple mask alignment (to < 10 nm) while maintaining a fixed gap of about 4 μm (7), and exploring additional means of circumventing the problem of diffraction, which becomes an important consideration at linewidth below 50 nm.

V. REFERENCES

1. E.H. Anderson, D. Kern, and H.I. Smith, Microel. Eng. 6, 541 (1987).

2. H.I. Smith, J. Vac. Sci. Technol. B 4, 148 (1986).

3. H.I. Smith, J. Vac. Sci. Technol. B 6, 346 (1988).

4. A.C. Warren, I. Plotnik, E.H. Anderson, M.L. Schattenburg, D.A. Antoniadis, and H.I. Smith, J. Vac. Sci. Technol. B 4, 365 (1986).

5. M.L. Schattenburg, I. Tanaka and H.I. Smith, Microel. Eng. 6, 273 (1987).

6. D.C. Flanders, Appl. Phys. Lett. 36, 93 (1980).

7. A. Moel, M.L. Schattenburg, J.M. Carter, and H.I. Smith, to be published J. Vac. Sci. Technol. (Nov/Dec 1989).

8. Y.-C. Ku, E.H. Anderson, M.L. Schattenburg, and H.I. Smith, J. Vac. Sci. Technol. B 6, 150 (1988).

9. M. Weiss, H. Oertel, and H.L. Huber, Microel. Eng. 6, 252 (1987).

10. K. Ismail, W. Chu, A. Yen, D.A. Antoniadis and H.I. Smith, Appl. Phys. Lett. 54, 460 (1988).

11. W. Chu, A.T. Yen, K. Ismail, M.I. Shepard, H.J. Lezec, C.R. Musil, J. Melngailis, Y.C. Ku, J.M. Carter, and H.I. Smith, to be published J. Vac. Sci. Technol. (Nov/Dec 1989).

12. K. Ismail, D.A. Antoniadis, and H.I. Smith, Appl. Phys. Lett. 54, 1130 (1989).

13. A.C. Warren, D.A. Antoniadis, H.I. Smith, and J. Melngailis IEEE Elect. Dev. Lett. EDL-6, 294 (1985).

14. J.H.F. Scott-Thomas, M.A. Kastner, D.A. Antoniadis, H.I. Smith, and S. Field, J. Vac. Sci. Technol. B, 6, 1841 (1988).

15. M.A. Kastner, S.B. Field, J.C. Licini, and S.L. Park, Phys. Rev. Lett. 60, 2535 (1988).

16. J.H.F. Scott-Thomas, S.B. Field, M.A. Kastner, H.I. Smith, and D.A. Antoniadis, Phys. Rev. Lett., 62, 583 (1989).

17. S.Y. Chou, D.A. Antoniadis, and H.I. Smith, IEEE Elect. Dev. Lett. EDL-6, 665 (1985).

18. G.G. Shahidi, D.A. Antoniadis, and H.I. Smith, IEEE Elect. Dev. Lett. EDL-9 (2), 94 (1988).

19. G.G. Shahidi, D.A. Antoniadis, and H.I. Smith, J. Vac. Sci. Technol. B 6, 137 (1988).

20. G.G. Shahidi, D.A. Antoniadis, and H.I. Smith, IEEE Elect. Dev. Lett. 9, 497 (1988).

TRANSPORT PROPERTIES AND INFRARED EXCITATIONS OF LATERALLY PERIODIC NANOSTRUCTURES

Jörg P. Kotthaus

Institut für Angewandte Physik, Universität Hamburg
Jungiusstrasse 11, D-2000 Hamburg 36, West Germany

I. INTRODUCTION

Quantum confinement of electrons at a semiconductor interface results in quasi-two-dimensional electron systems (2DES) in which the electronic motion is unbound within the interface plane but quantized perpendicularly to the interface. The unique electronic properties of such 2D electron systems have been widely investigated for more than two decades. With lithographic technologies it recently has become possible to further restrict the electronic motion at semiconductor interfaces in the lateral directions to dimensions in the range around 100 nm which are becoming comparable to the Fermi wavelength of electrons in a 2DES in semiconductor heterostructures. At low temperatures and for high quality device structures such as GaAs–AlGaAs heterojunctions other relevant electronic length scales such as the elastic mean free path $\ell_{el} = v_F \cdot \tau_{el}$ and the phase coherence length $\ell_\phi = \sqrt{D \cdot \tau_{in}}$, which describes the inelastic mean free path, can easily exceed 1 μm and thus can be much larger than lateral confinement dimensions W. Under these conditions lateral quantization phenomena have recently become observable in a variety of semiconductor nanostructures using various confinement schemes.

Here I want to summarize recent investigations carried out in our laboratory on laterally periodic field effect devices with periodicities in the range between 200 nm and 500 nm (1–6). In these devices electrostat-

67

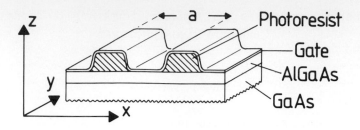

Fig. 1. Schematic cross section through a GaAs-AlGaAs heterojunction with a periodically microstructured gate.

ic confinement via field effect is used to create either 2D systems with strong periodic modulation of the electron density, or arrays of quasi-one-dimensional (1D) electron inversion channels, or even arrays of effectively zero-dimensional quantum dots. Using holographic lithography such structures have been fabricated on GaAs, InSb, and Si homogeneously on comparatively large areas of typically 10 mm² and thus can be used for investigations with infrared transmission spectroscopy as well as static or quasi-static transport experiments (see, e.g., 7).

In the following most of the discussion will be centered around a particular laterally periodic nanostructure, namely a GaAs-AlGaAs heterojunction with a periodically structured gate as schematically pictured in Fig.1. In fabricating these devices we start with a standard modulation doped heterojunction grown by molecular beam epitaxy and containing a single 2DES at the GaAs-AlGaAs interface which typically has a 4.2 K mobility of 300,000 cm²/Vsec and an electron density of 3×10^{11} cm^{-2}. On top of the heterojunction we holographically define a photoresist grating which then is covered with a semitransparent layer of NiCr serving as gate. Application of a negative gate bias V_g causes depletion of electrons at the heterojunction interface primarily in those regions where the gate is closest to that interface. Increasing negative gate bias thus results in an increasing periodic modulation of the inversion electron density N_s until at $V_g = V_d$ the electron system is transformed into a series of narrow parallel inversion channels. With further negative bias $V_g < V_d$ the linear electron density N_ℓ and width W of these inversion channels are further reduced until the connection to source and drain contacts are pinched and the electron number in a given channel becomes fixed. In the following I want to discuss at first the transition from 2D to

1D electronic behavior as manifested by magnetotransport experiments (3,4) and its effect on the infrared electronic excitations of the system (1,5). In the last section I want to discuss recently observed novel magnetoresistance oscillations (6,8,9) that occur in the regime where the 2D electron density is only moderately modulated and that reflect lateral superlattice effects.

II. FROM TWO-DIMENSIONAL TO ONE-DIMENSIONAL BEHAVIOR

The gated heterojunction device sketched in Fig. 1 is nearly ideally suited to study the transition from 2D to 1D electronic behavior since such a transition can be induced by solely changing the gate bias. At $V_g = V_d$ the differential capacitance of the device decreases substantially reflecting the transition from a density-modulated 2DES to an array of parallel inversion wires. One way to establish lateral quantization in such inversion wires as their electronic width W becomes comparable to the Fermi wavelength of the inversion electrons is to study Shubnikov-de Haas (SdH) oscillations of the magnetoresistance in a magnetic field perpendicular to the interface plane. For a 2DES these are known to be periodic in 1/B and the periodicity is a measure of electron density N_s. For a 2DES with modulated electron density they still remain periodic in 1/B but now reflect an average electron density $\overline{N}_s$. When lateral quantization sets in there are only a finite number of 1D subbands occupied already at magnetic field B=0. With increasing magnetic field these 1D subbands increase their energetical spacing and become depopulated (10). This gives rise to SdH-like oscillations which are no longer periodic in 1/B. This is illustrated in Fig. 2a where for a device as in Fig.1 the quantum index of the magnetoresistance maxima is plotted versus their position in reciprocal magnetic field at gate voltages just above and below $V_d = -0.5$ V. The deviation from 1/B periodicity is clearly visible at low magnetic fields and can be well described by a parabolic confinement model as entered in Fig. 2a.

Though the confining potential is only expected to be parabolic at very low densities N_ℓ and should become more square-well like with increasing N_ℓ (11), we believe the parabolic model to yield resonably accurate values for the density N_ℓ, the 1D subband spacing $\hbar\Omega$, and the electronic width at the Fermi energy. In Fig. 2b the 1D subband spacing extracted from the analysis of the magnetoresistance data in Fig. 2a as-

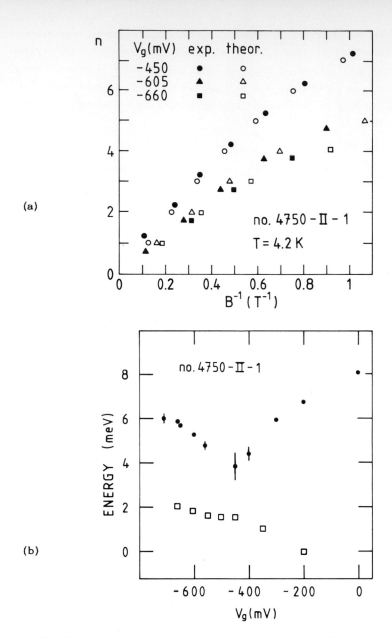

Fig. 2. Evaluation of the magnetoresistance oscillations in a GaAs–AlGaAs heterojunction with a laterally periodic gate of period a =400 nm. (a) quantum index n of the maxima in dR/dV_g vs. reciprocal magnetic field as observed (filled symbols) and fitted (open symbols) with a parabolic confining potential at various gate voltages V_g. (b) Energy of the 1D subband spacing as derived from (a) (open squares) in comparison to energy of infrared resonances at B=0 (filled circles) vs. V_g (from Ref.4).

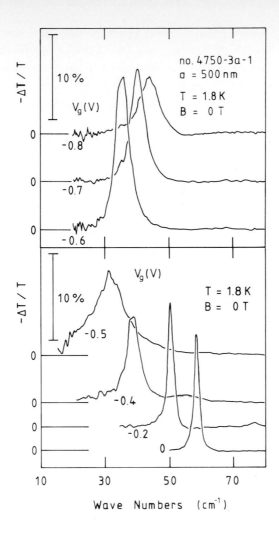

Fig. 3. Infrared transmission spectra of a GaAs–AlGaAs heterojunction with laterally periodic gate of period a=500 nm. The lower part displays spectra in the 2D regime ($V_g \geqslant V_d = -0,5V$), the upper part in the 1D regime (from Ref. 2).

suming parabolic confinement is displayed as squares versus gate voltage. Lateral quantization becomes already visible at gate voltages just above V_d and for $V_g \leqslant V_d$ yields subband spacings of about 2 meV for this particular device.

Another way to study the transition from 2D to 1D electronic behavior is to investigate the infrared electronic excitations in a transmission experiment (1,2,7). For laterally periodic structures with periods a which are much smaller than the infrared wavelengths one can observe both q=0 modes driven by the spatially uniform component of the infrared field as well as modes with in-plane wave vector q=(2π/a)n (n=1,2,...) excited by the spatially modulated components of the infrared field. For a 2DES at magnetic field B=0 the q=0 component gives rise to Drude absorption whereas the q≠0 components excite 2D plasmons with well defined wave vector. Such 2D plasmons excited at q=2π/a are clearly visible in the V_g=0V spectrum of Fig. 3. There the relative change in transmission $-\Delta T/T = -(T(V_g)-T(V_t))/T(V_t)$ with V_t=-1V is displayed vs. infrared frequency. The Drude contribution (at q=0) to $-\Delta T/T$ is not visible at frequencies above 10 cm^{-1} for high mobility heterojunctions such as the one here. With decreasing V_g, i.e. decreasing $\overline{N}_s$, the 2D plasmon resonance with frequency $\omega_p \propto (\overline{N}_s)^{1/2}$ initially shifts as expected to lower frequency. Around V_g=V_d the resonance changes character and suddenly starts to increase in frequency with decreasing V_g. Also, the signal strength at $V_g \leq V_d$ is much stronger than at V_g=0V. The latter reflects that the resonance now becomes excited by the uniform (q=0) component of the infrared field. In fact, the distinction between q=0 and q=2π/a modes becomes unnecessary once the electron system develops into a set of parallel wires of period a. The changed character of the resonance and the observation that it increases in frequency with decreasing $\overline{N}_s$ makes it natural to associate this resonance with a transition between 1D subbands. However, as also known from 2D intersubband resonance, such an infrared induced intersubband transition is usually shifted to energies above the subband spacing by collective effects.

In the 1D regime the infrared resonance energy may be written as $\hbar\omega_0 = \hbar(\Omega^2+\omega_d^2)^{1/2}$, where $\hbar\Omega$ is the 1D subband spacing and $\hbar\omega_d$ denotes collective contributions. Such collective effects which in simplest approximation are associated with classical depolarization (1,4) are naively expected to become less important with decreasing carrier density and to vanish as $\omega_d \propto (\overline{N}_s)^{1/2}$. Insofar it is surprising that the comparison of 1D subband spacings and the infrared resonance energies as displayed in Fig. 2(b) for a particular sample suggests that collective effects here are most important at the lowest carrier density. This comparison also shows that the size of the collective effects in such GaAs-AlGaAs heterojunctions is so large that a reliable determination of 1D subband spacings from infrared spectra alone becomes virtually impossible. Though the quantitative description of such collective effects is improving (12) it still is not possible to correctly predict the collective shifts in a particular

72

experimental situation.

Spectroscopic studies on inversion wires created in periodic MOS-structures on InSb (1,3) show that in such structures collective effects can be significantly smaller. In these devices a Schottky grating used to define the 1D channels presumably screens most of the collective effects. Also here the collective effects decrease with decreasing electron density N_ℓ and vanish, as naively expected, for vanishing N_ℓ. Nevertheless, a quantitative understanding of such collective phenomena is still lacking but necessary to allow a reliable interpretation of infrared spectra without the need of additional transport experiments.

III. MAGNETORESISTANCE IN A LATERAL SUPERLATTICE

In high mobility heterojunctions with a moderate periodic modulation of the 2D electron density Weiss et al. recently observed a new type of magnetoresistance oscillations that directly reflect the effect of the periodic density modulation (8). In their experiment they induce the density modulation via the persistent photoeffect by illuminating the sample with a light grating. With devices as in Fig. 1 tunable electrostatic modulation of the carrier density is achieved by just applying gate voltages $V_g > V_d$. In suitably shaped Hall bar samples, that allow four point probing of the magnetoresistance component ρ_{xx} perpendicular to the density grating (see coordinates in Fig. 1) these novel magnetoresistance oscillations are directly observable as illustrated in Fig. 4. These oscillations have the following characteristics. They usually are observed in ρ_{xx} at magnetic fields below the SdH-oscillations and are much weaker as well as phase shifted in ρ_{yy}. They are periodic in 1/B and yield maxima in ρ_{xx}, whenever the classical cyclotron diameter $2R_c = 2m^* \cdot v_F/(eB) = (m+0.25)a$ with m=1,2,..., where a is the period of the potential modulation. They occur at magnetic fields with $\mu \cdot B > 1$, where μ is the electron mobility. In this field regime on has approximately $\rho_{xx} \approx \sigma_{yy} \cdot B^2/(N_s^2 e^2)$, i.e., resistance perpendicular to the density grating reflects conductance parallel to the grating.

The origin of these novel oscillations was independently associated to the effect of the density superlattice on the magnetic band structure by Winkler et al. (6) and Gerhardts et al. (9). In a simple approach where the effect of a weak periodic potential modulation on the magnetic band structure is considered in lowest order perturbation theory (13) one can

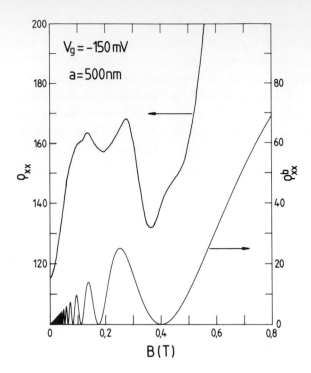

Fig. 4. Measured magnetoresistance ρ_{xx} in a GaAs–AlGaAs heterojunction with periodically microstructured gate (left scale) in comparison to the calculated oscillatory band contribution ρ_{xx}^b (right scale) (from Ref.6).

see that Landau levels transform into Landau bands since the periodic potential lifts the energetical degeneracy of Landau orbits with different center coordinate x_0. As illustrated in Fig. 5 the bandwidth of these Landau levels varies with energy and has a maximum at the Fermi energy E_F when $2R_c=(m+0.25)a$. In addition to the scattering induced conductivity that usually controls ρ_{xx} for $\mu B \gg 1$ one obtains a band conductivity σ_{yy}^b reflecting the bandwidth of the Landau bands. With reasonable approximations one can calculate σ_{yy}^b analytically for the regime $\mu B > 1$. In Fig. 4 the associated band resistivity ρ_{xx}^b thus calculated with experimentally determined parameters is compared to the experimental observation and is found to yield the correct period, phase, and amplitude (6). Since the dispersion of the Landau bands $\delta E/\delta k_y = \hbar v_y$ gives rise to an $\vec{E} \times \vec{B}$ drift velocity v_y it can be at least qualitatively understood that a semiclassical calculation by Beenakker (14) yields the identical expression for the oscillatory conductivity σ_{xx}^b that has been calculated by Winkler et al. as sketched above. Essential for the appearance of the novel magnetoresist-

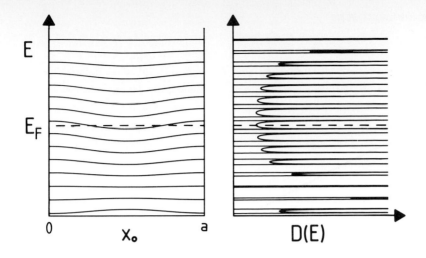

Fig. 5. Sketch of the dispersion of Landau bands $E(x_o)=E(\hbar k_y/eB)$ (left) and the corresponding density of states D(E) (right) in the absence of broadening (from Ref. 6).

ance oscillations is the condition $\mu B > 1$ at $2R_c/a=1$, meaning that the orbiting electron must test at least one superlattice period a without being scattered. Insofar these magnetoresistance oscillations directly reflect the effect of a lateral superlattice potential on the magnetoresistance.

IV. CONCLUSION

The above discussion has served to demonstrate that laterally periodic nanostructures in which the confining potential is modulated via field effect are excellent devices to investigate effects of lateral quantum confinement on both electronic transport properties as well as on infrared excitations. With field effect lateral confining dimensions can be well controlled in the 100 nm regime and give rise to lateral quantization energies of several meV. Since electrostatic confinement preserves good device mobilities these quantization energies can be much larger than level broadening. The novel electronic properties of quantum wires, quantum dots, as well as lateral superlattices thus become accessible for detailed experimental studies.

ACKNOWLEDGEMENT

The work summarized here was carried out in close collaboration with J. Alsmeier, F. Brinkop, W. Hansen, U. Merkt, K. Ploog, Ch. Sikorski, and R. Winkler. I also wish to thank C. W. J. Beenakker, A. V. Chaplik, R. R. Gerhardts, D. Heitmann, and D. Weiss for stimulating discussions. The financial support of the Stiftung Volkswagenwerk is gratefully acknowledged.

REFERENCES

1. Hansen, W., Horst, M., Kotthaus, J. P., Merkt, U., Sikorski, Ch., and Ploog, K. (1987). Phys. Rev. Lett. 58, 2586.
2. Kotthaus, J. P., Hansen, W., Pohlmann, H., Wassermeier, M., and Ploog, K. (1988). Surf. Sci. 196, 600.
3. Alsmeier, J., Sikorski, Ch., and Merkt, U. (1988). Phys. Rev. B 37, 4314.
4. Brinkop, F., Hansen, W., Kotthaus, J. P., and Ploog, K. (1988). Phys. Rev. B 37, 6547.
5. Kotthaus, J. P. (1988). In "The Physics of Semiconductors", (W. Zawadzki, ed.), Vol. 1, p. 47. Polish Academy of Sciences, Warsaw.
6. Winkler, R. W., Kotthaus, J. P., and Ploog, K. (1989). Phys. Rev. Lett. 62, 1177.
7. Kotthaus, J. P. (1987). Physica Scripta T19, 120.
8. Weiss, D., v. Klitzing, K., Ploog, K., and Weimann, G. (1989). Europhys. Lett. 8, 179.
9. Gerhardts, R. R., Weiss, D., v. Klitzing, K. (1989). Phys. Rev. Lett. 62, 1173.
10. Berggren, K. F., Thornton, T. J., Newson, D. J., and Pepper, M. (1986). Phys. Rev. Lett. 57, 1769.
11. Laux, S. E., Frank, D. J., and Stern, F., (1988). Surf. Sci. 196, 101.
12. Que, W. and Kirczenow, G. (1988). Phys. Rev. B 37, 7253.
13. Chaplik, A. V. (1985). Solid State Commun. 53, 539.
14. Beenakker, C. W. J. . (1989). preprint.

FABRICATION AND OVERGROWTH OF QUANTUM WIRES AND DOTS FOR OPTOELECTRONIC APPLICATIONS

Steven P. Beaumont

Nanoelectronics Research Centre
Department of Electronics and Electrical Engineering
University of Glasgow
Glasgow, Scotland UK

Introduction

By altering the dimensionality of a semiconductor, amongst other things we can expect to modify its optical properties. Thus in quantum wells, carriers are confined by double heterojunctions to a two-dimensional sheet: the confinement changes the density of states distribution and moves the lowest energy level available in the conduction and valence bands to higher values. These and other effects can be exploited in semiconductor lasers and optical modulators. Reducing the dimensionality still further to 1-D (quantum wires) and 0-D (quantum dots or boxes) might confer additional benefits. For example, it has been suggested that quantum dot lasers might have narrower linewidths and greater temperature stability[1], and that nonlinear effects might be enhanced[2].

Quantum boxes are made by patterning quantum well layers. They have a δ-function density of states distribution where the separation between each discrete level depends on the degree of lateral confinement. If this is weak, the density of states becomes a quasi-continuum; if strong, discrete lines should appear in the luminescence spectra of the dots. There will also be a blue shift in the luminescence relative to the quantum well from which the dots are made. Of these two effects, fine structure in the luminescence is a less ambiguous test of lateral confinement than a blue shift, which can arise from other agencies such as quantum well interdiffusion or strain. For visibility of fine structure, one needs to fabricate quantum dots less than about 30nm in diameter for GaAs/AlGaAs quantum well starting material and less than 40-50nm for InGaAs/InP[3]. These size ranges challenge the capabilities of nanolithography using electron beam techniques, but nonethless can be attained by it.

Two fabrication schemes have been employed for making quantum dots. One uses quantum well intermixing induced by masked ion implantation followed by annealing[4].

Fine structure attributed to lateral confinement was observed in the cathodoluminescence of quantum wires made in this way. The second method employs dry ion etching to form free-standing quantum dots. It has the supposed disadvantage of creating exposed and possibly damaged semiconductor surfaces: but as we shall show, the results obtained are of interest and at odds with expectations. This paper reports on the luminescence of quantum dots fabricated by this route, and the first successful attempts to embed these structures in a high bandgap semicconductor matrix - the starting point for useful device fabrication - by overgrowth.

Electron beam lithography

Direct write electron lithography has been shown to have the highest resolution of all currently available lithographic techniques. Most e-beam nanolithography is carried out in modified scanning microscopes which have the correct combination of accelerating voltage and beam diameter. Typically 50-100kV electrons are used to minimise resolution loss by forward electron scattering in the resist and interaction effects between adjacent elements due to backscattering from the substrate. High voltage also increases the brightness of the electron source resulting in a larger beam current for a given spot diameter, making it easier to image small registration marks and other predefined features on the substrate to be patterned.

To write a mask for ion etching, two alternative schemes can be employed. The first involves exposing the pattern in positive resist followed by liftoff of an inorganic film such as a metal or an oxide which has a high resistance to the etching process. The second simply uses a negative e-beam resist as a direct dry-etch mask. In principle the first route offers the highest resolution and also the greatest resistance to the etch process as one can optimise the combination of a radiation sensitive high resolution electron resist with an insensitive, low-etch rate mask. However, it can be difficult to remove the etch mask after processing, and its presence can interfere with optical experiments and subsequent processes such as overgrowth. Furthermore, stress in the mask can contribute undesirable artifacts, such as blue shifts, in the luminescence spectrum. On the other hand negative organic resist is easily removed by oxygen plasma etching after dot fabrication and very high resolution can be obtained[5], but care has to be taken to ensure that the resist is sufficiently thick to avoid damage to the underlying semiconductor by penetrating ions, and one may have to compromise resolution as a result.

Fig (1) 30nm diameter quantum dots fabricated by SiCl4 reactive ion etching from a negative resist mask

Dry etching

Dry etching techniques are too numerous to survey in this paper. The basic principle involves bombardment by energetic ions which either erode by sputtering or react chemically with the substrate to form volatile compounds that easily desorb from the surface. If the ions are formed from an inert gas such as argon, the etching is wholly mechanical: but if a reactive gas is employed, the etching mechanism is usually a combination of mechanical erosion and volatilisation, with some enhancement of the chemical reaction rate by the ion bombardment. Etch gases used to fabricate quantum dots include BCl_3[6], $SiCl_4$ and CH_4/H_2 (this work), Ar/Cl_2[7], Ar and CCl_2F_2/Ar for AlGaAs/GaAs and Ar/O_2[8] (InGaAs/InP). The dots are either etched completely through the quantum well layers, or in one example, partially through the cladding layer[9].

Fig (1) shows quantum dots written in a converted TEM (JEOL 100CXII, 100kV, 2nm spot diameter) using PMMA followed by liftoff of NiCr and $SiCl_4$ etching. The dot diameter is ~20nm and we believe these to be some of the smallest quantum dots ever fabricated using this route.

There is one important side-effect of dry etching which needs to be considered for quantum structure fabrication, that of the damage done to the sidewalls of the etched structure. It has been known for some time that all ion etching processes cause damage to the directly etched surface of a semiconductor. The extent of the damage varies from process to process and usually scales with the energy of the bombarding ions, but is normally much deeper than might be expected from the range of low energy ions in solids. For example, the depth to which GaAs is damaged by 1keV Ar^+ ion bombardment is ~100nm, yet the range of these ions is only a few nanometres at most. Damage is poorly understood but easily measured by Schottky diode characterisation, CV profiling and optical techniques including luminescence[10] and Raman spectroscopy[11].

Less attention has been paid to the issue of sidewall damage, yet it is of considerable importance for optical devices such as lasers where damage to dry etched mirrors increases threshold currents and decreases lifetime, in integrated optical waveguides, and in quantum dots where the damaged surface may result in excessive nonradiative recombination and a reduction in the effective size of the dot.

A few studies of sidewall damage have emerged in recent years. Laser threshold currents have been used to compare the damage caused by different dry etch techniques. Our own work has concentrated on the damage done to quantum wires fabricated for electron transport measurements[12]. As the width of a quantum wire is reduced one would expect its conductance to fall to zero when its width is approximately twice the surface depletion depth if all its surfaces are perfect. Wet etched wires, however, cut off at larger widths and there is a correlation between the excess width and the energy of the etching ions. For $SiCl_4$ etching we measure damage thicknesses in the range 15-30nm in n^+ GaAs wires were measured. We might therefore expect to have to take damage into effect when interpreting data from sub-micron quantum dot structures.

Dry etched quantum dots

Our own work on dry etched quantum dots has concentrated on a careful study of the luminescence (PL and PLE) of structures fabricated by e-beam lithography and dry etching, using different masks and etch gases in an attempt to understand the influence of processing on the optical behaviour. In our studies[13], MBE-grown quantum well material, consisting of 5, 10, 20 and 80nm thick GaAs wells separated by 34nm thick $Al_{0.3}Ga_{0.7}As$ barriers was employed. Patterning into quantum dots was performed by electron beam direct write lithography using our Philips PSEM 500 SEM at 50kV and a spot diameter of 8-10nm. Positive resist/liftoff and negative organic (HRN) etch masks were laid down, as described above. In the former case titanium or nichrome films were lifted off from 125nm thick PMMA bilayers. In the latter, 120nm thick HRN was tried first, but it was found that although it was possible to fabricate well-formed dots from these films, there was severe broadening of the optical transitions suggesting that low energy ions or electrons had penetrated the mask to damage the quantum well. Increasing the resist thickness to 280nm eliminated this effect and confirmed the penetration hypothesis. 100μm x 100μm arrays of dots having nominal diameters of 40-400nm were defined: the separation between dots was 3 to 5 times the dot diameter. Several 100μm x 100μm unpatterned mesas were included as controls.

Dots were etched by reactive ion etching using $SiCl_4$ in a Plasma Technology RIE-80 machine at 13.56MHz, 10mT pressure: the bias voltage was 300V at a power density of $0.65W/cm^2$, to give an etch rate of 200nm/min. Other samples were etched by 1:5 CH_4/H_2 in an Electrotech SRS Plasmafab 340 machine at 15mT; the dc bias in this case was 1000V, the power density $0.75W/cm^2$ and the etch rate 25nm/min. Nichrome masks were used for $SiCl_4$ and Ti with CH_4/H_2 to optimise etch contrast. The etch depth was 0.2-0.6μm/min for silicon tetrachloride and 0.1-0.2μm for CH_4/H_2: in all cases both the 5 and 10nm wells were etched through.

PL and PLE spectra were excited at 4K using a tunable dye laser and spectrally dispersed using a double grating spectrometer with GaAs photomultiplier and photon counting electronics. A microscope and TV camera were employed to aid alignment of the 30μm diameter focal spot of the laser with each dot array.

Figs (2) and (3) show the integrated PL intensity of the 5 and 10nm wells excited at 1.65eV (below the AlGaAs bandgap) from arrays of dots fabricated using $SiCl_4$ and CH_4/H_2 respectively. The intensities are scaled by the fractional surface coverage of the quantum dots and normalised to the emission from the mesas. Thus a relative intensity of unity corresponds to a luminescence efficiency per unit area equal to the unpatterned quantum well material.

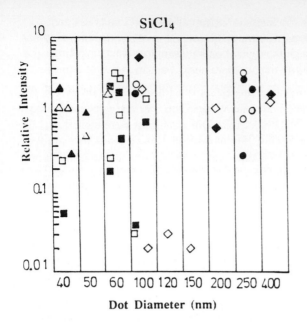

Fig (2) Integrated intensity of luminescence excited at 1.65eV from arrays of dots of various diameters etched with SiCl$_4$ relative to control mesas. Luminescence intensities are scaled to the volume of well material excited. Open symbols represent the 5nm well data: filled symbols the 10nm well data.

△ Etch depth 0.6μm, 0.12μm HRN mask.
□ Etch depth 0.2μm, 0.12μm HRN mask.
○ Etch depth 0.2μm, 250Å NiCr mask.
◇ Etch depth 0.3μm, 0.28μm HRN mask.

The majority of the data points are scattered about a relative intensity of 1, and each point is reproducible to a factor of 2 on successive measurements. It therefore appears that the luminescence intensities scale appoximately with the area of well excited. Over many samples with different preparation techniques, there was no evidence either of enhanced luminescence as reported by Kash et al.[14], or of a rapid and systematic attenuation of luminescence efficiency as reported by Forchel[15]. Moreover, any dry etched damage around the periphery of the dots appears not to affect the optical output of the dots. Initially, our intepretation of these results rested on the passivation of surface states by the etch process as it appeared that these could not be operative in our structures. There are many reports in the literature of the passivation of defects, including dopants, by hydrogen[16]: indeed, hydrogen passivation is an important factor in the production of high quality amorphous silicon films for device applications, where it passivates dangling bonds. It is also believed that chlorine is a passivating agent too, although this view seems to be supported more by folklore than by hard scientific evidence. Therefore it is possible that the gas mixtures we are using cause surface passivation and a reduction in nonradiative

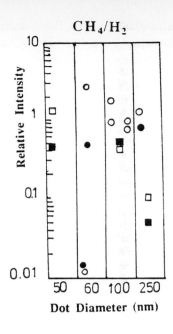

Fig (3) As Fig (2), but dots etched by CH_4/H_2.

O Etch depth 0.1μm, 0.28μm HRN mask.
☐ Etch depth 0.1μm, 250Å Ti mask.

recombination. Fig (4) shows that the measured luminescence is directly proportional to optical excitation over six orders of magnitude, suggesting that saturable traps are not important in determining luminescence behaviour. Alternatively, the surface states are not active because the diffusion of photoexcited carriers is inhibited. Recent measurements[17] report such an effect, which is particularly important in thin quantum wells at low temperatures.

Overgrowth of dry-etched quantum dots

If quantum dots of good luminescence efficiency can be fabricated by dry etching it is tempting to try to bury them in a matrix of overgrown AlGaAs as the first step towards the fabrication of a quantum dot device. We have carried out preliminary experiments on quantum dot overgrowth by metallorganic chemical vapour deposition (MOCVD) on structures fabricated in both MBE and MOCVD-grown quantum well layers[18]. The MBE material in these experiments consisted of 11.5, 5.7, 2.8 and 1.4nm thick GaAs wells separated by 20nm $Al_{0.3}Ga_{0.7}As$ barriers grown in that order on an n^+ GaAs substrate with an intervening undoped GaAs buffer layer and a 1.4nm thick GaAs cap. The results presented here apply to the 1.4nm well only. The MOCVD sample comprised a single 10nm thick quantum well surrounded by 50nm thick $Al_{0.3}Ga_{0.7}As$ barriers, grown at $750^{\circ}C$.

These layers were patterned using HRN resist and $SiCl_4$ reactive ion etching in the same manner as the previous experiments. Again, 100μm x 100μm arrays of dots were fabricated and beside each array was a 100μm square control mesa. Dot diameters were 70nm, 110nm and 350nm (300nm in the MBE sample). After luminescence assessment of the as-etched dots and removal of

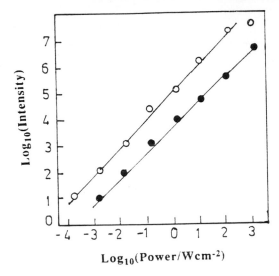

Fig (4) Integrated PL intensity from 10nm well in 60nm diameter dots (filled circles) and the 100μm^2 mesa (filled circles) etched using CH$_4$/H$_2$ as a function of excitation at 1.65eV

the residual resist by oxygen plasma ashing in a Plasmafab 505 oxygen barrel etcher, Al$_{0.4}$Ga$_{0.6}$As was overgrown by MOCVD at 750°C. Fig (5) shows an array of dots partially overgrown with AlGaAs: the dots are beginning to planarise as the surrounding gaps fill in. The overgrowth layer on the MOCVD quantum dots was 0.2μm thick and the MBE dots were overgrown with 0.6μm of AlGaAs, both followed by a 40nm thick GaAs cap. Low temperature photoluminescence (PL) spectra were obtained by excitation with the 633nm line of a HeNe laser. MBE samples were assessed at 4K, MOCVD samples at 60K. The spectral resolution of the monochromator employed was 1nm.

There were significant differences between the spectra obtained from MBE and MOCVD as-etched dots. Whilst as before, the MBE dots luminesced with approximately the same efficiency as the unpatterned material, the MOCVD dots *did not luminesce detectably*. Only the control mesas on the MOCVD samples luminesced.

After overgrowth there was an overall reduction in luminescence from both samples by a factor of 7-10. On the MBE sample, luminescence was detected only from the control mesas and the 300nm dots. On the MOCVD sample, the mesas luminesced as before, *but luminescence was recovered from the 350nm dots.* In both MBE and MOCVD samples, after overgrowth the luminescence scaled with the volume of material.

Figs (6) and (7) show the QW emission from the largest QDs and corresponding mesas in the MBE (spectra A and B) and MOCVD (spectra C and D) starting material before (solid line) and after (dashed line) overgrowth. In the MBE case the control and QD emission are at identical wavelengths after etching with a slight broadening of the QD emission. After overgrowth the dot and mesa emission shift by 10nm and 7.5nm respectively to higher energy. For the MOCVD

Fig (5) Array of partially overgrown quantum dots showing the development of a planar surface as the space between dots is filled in.

sample there are similar shifts in the control and dot emission after overgrowth. In neither sample could emission be detected from the smaller dots.

The significant results here are: a) the absence of emission from the MOCVD QDs b) the recovery of emission from large dots on that sample after overgrowth and c) the blue shift observed on mesas and dots. c), being common to both mesas and dots, is either due to aluminium migration during overgrowth or strain, or a combination of both. It seems unlikely that the former mechanism could be responsible for the blue shift on the mesas, as the MOCVD starting material was grown at the same temperature as the overgrown layer and the MBE layers are known to be of high quality: as was pointed out above, such quantum wells are very stable against high temperature annealing. So strain seems most likely to be the cause of the blue shift though it is possible that lateral migration of aluminium from the overgrown layer is responsible for the additional blue shift observed in the dots, and it is conceivable that this diffusion, enhanced perhaps by disorders at the etched surfaces, is the cause of the complete absence of luminescence from the smaller overgrown MBE dots.

Why should the MOCVD QDs not luminesce after etching? The QW in this material is much larger than the QW investigated in the MBE dots (10nm vs 1.4nm) and the luminescence was

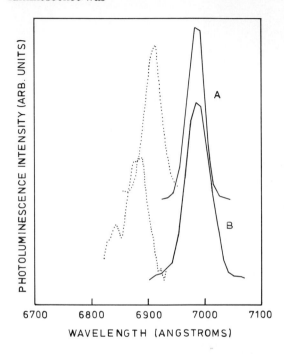

Fig (6) QW emission from the mesa (A) and 300nm diameter QDs (B) patterned in MBE material before regrowth (solid line) and after regrowth (dashed line)

recorded at higher temperature (60K vs 4K). Therefore we expect photoexcited carriers to have greater thermal energy and may be able to diffuse to the surface to recombine nonradiatively there. If this model is correct, then the effect of overgrowth must be at least to passivate the surface states on the MOCVD grown quantum dots and possibly to reconstruct the lattice between the dots and the overgrown layer: in which case we believe this to be the first evidence that free quantum dot surfaces created by ion etching can be successfully overgrown to leave a high-quality interface.

Conclusions

Lateral quantisation should be detectable in the optical spectra of quantum dots fabricated from quantum well layers over a range of sizes accessible to electron beam lithography. Whilst fine structure has been attributed to lateral quantisation in structures confined by impurity induced disorder, in dots fabricated by ion etching the only detectable effect is an ambiguous blue shift. The free surfaces created by the etching process do not necessarily cause a loss of luminescence because, at low temperatures, photoexcited carriers can be localised on fluctuations in narrow quantum wells; and it may be possible to passivate etched surfaces by overgrowth of a high bandgap matrix, allowing QD luminescence to be observed at higher temperatures from thick wells.

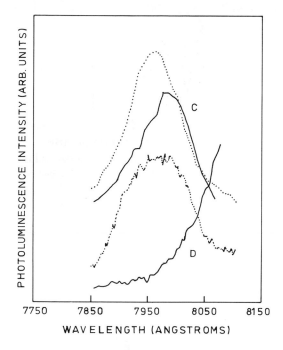

Fig (7) QW emission from the mesa (C) and 350nm diameter QDs (D) patterned in MOCVD starting material before regrowth (solid line) and after regrowth (dashed line)

The technology of quantum dot fabrication is a severe and exciting challenge to nanostructure engineering. The future of this topic may not only result in the successful fabrication of quantum dot lasers or modulators by the techniques discussed in this paper, but also to the stimulation of new resist processes or controlled epitaxial growth the obtain dots consisting of just a few unit cells, where the oscillator strength will be concentrated in just a few discrete optical transitions.

Acknowledgements

The work reported in this paper is largely that of my colleagues, Ms. Hazel Arnott, Drs. Clivia Sotomayor-Torres and Stephen Thoms at Glasgow University; Dr. Steve Andrews of GEC Hirst Research Laboratory who performed all the PL and PLE measurements on free-standing etched quantum dots, and his colleague Dr. Tom Kerr who grew the MBE samples for patterning; and Dr. Rick Glew of STC Technology Ltd. who supplied the MOCVD layers and undertook all the overgrowth. The work is largely supported by the UK Science and Engineering Research Council, and Ms. Arnott holds a SERC CASE studentship with GEC.

References

[1]Arakawa Y. and Sakaki H. Appl. Phys. Lett. **24**, 195-7, 1982

[2]Chemla D.S. and Miller D.A.B. Optics Lett. **11**, 522-4, 1986

[3]Vahala K.J. IEEE J. Quantum Electron. **24**, 523-530, 1988

[4]Cibert J>, Petroff P.M., Dolan G.J., Pearton S.J., Gossard A.C. and English J.H. Appl. Phys. Lett. **49**, 1275-1277, 1986

[5]Thoms S., McIntyre I., Beaumont S.P, Cheung R., Al-Mudares M. and Wilkinson C.D.W. J. Vac. Sci. Technol **B6**, 127-30, 1988

[6]Reed M.A., Bate R.T., Bradshaw K., Duncan W.M., Frensley W.R., Lee J.W. and Shih H.D. J. Vac. Sci. Technol **B4**, 358, 1986

[7]Maile B.E., Forchel A., Germann R., Menschig A., Streubel K., Scholz F., Weimann G. and Schlapp W. Microelectronic Engineering **6**, 163, 1987

[8]Temkin H., Dolan G.J., Panish M.B. and Chu S.N.G. Appl. Phys. Lett. **50**, 413-415, 1987

[9]Kash K., Worlock J.M., Sturge M.D., Grabee P., Harbison J.P., Scherer A. and Lin P.S.D. Appl. Phys. Lett. **53**, 782-4,1988

[10]Wong H.F., Green D.L., Lin T.Y., Lishan D.G., Bellis M., Hu E.L., Petroff P.M., Holtz P.O. and Merz J.L. J. Vac. Sci. Technol **B6**, 1906-1910, 1988

[11]Watt M., Sotomayor-Torres C.M., Cheung RT., Wilkinson C.D.W. and Beaumont S.P. J. Modern Optics **35**, 365-370, 1988

[12]Thoms S., Beaumont S.P., Wilkinson C.D.W., Frost J. and Stanley C.R. Microelectronic Engineering **4**, 249-256, 1986

[13]Arnot H., Andrews S.R. and Beaumont S.P. Microelectronic Engineering **7** (1988)

[14]Kash K., Scherer A., Worlock J.M., Craighead H.G. and Tamargo M.C. Appl. Phys. Lett. **49**, 1043, 1986

[15]Forchel A., Leier H., Maile B.E. and Germann R. Festkorperprobleme **28**, 99-119, 1988

[16]Pearton S., Corbett J.W. and Shi T.S. Appl. Phys. A **43**, 153 (1987)

[17]Hillmer H., Hansmann S., Forchel A., Morohashi M., Lopez E., Meier H.P. and Ploog K. Appl. Phys. Lett. **53**, 1937-1939, 1988

[18]Arnot H.E.G., Watt M., Sotomayor-Torres C.M., Glew R., Cusco R., Bates J. and Beaumont S.P. to appear in Superlattices and Microstructures

SURFACE PHONON STUDIES OF NANOSTRUCTURES [1]

M Watt, H E G Arnot, C M Sotomayor Torres and S P Beaumont

Nanoelectronics Research Centre,
Department of Electronics and Electrical Engineering,
University of Glasgow, Glasgow, G12 8QQ, Scotland

1. INTRODUCTION

Recent advances in lithographic techniques and processes have enabled the physical realisation of fabricated nanostructures. In this work we report on Raman scattering investigations of the phonons in quantum cylinders of GaAs. The study of such structures in GaAs is a prerequisite for the development of similar structures in AlGaAs-GaAs quantum well material. This latter material is a potential candidate for the fabrication of a quantum dot laser with improved properties over conventional quantum well lasers.

---

[1] This work has been supported by the U.K. Science and Engineering Research Council.

2. THEORY

The phonons of non-infinite crystals can be described using the dielectric continuum model developed by Englman and Ruppin (1968). In this model, general equations for the frequencies of the lattice vibrations are obtained from first principles and yield bulk TO, bulk LO and surface phonon solutions. Sample geometry is incorporated by imposing appropriate boundary conditions and this model has already been applied successfully to microcrystalline spheres of GaP (Hayashi and Kanamori, 1982).

We used the solution obtained for an infinite cylinder and applied it to cylinders of GaAs in air. Retardation effects were neglected. The calculated surface phonon frequencies lie between the bulk TO and LO frequencies and are shown in figure 1 as a function of cylinder radius. The frequency of the lowest order mode depends strongly on cylinder radius in the region $r \leq 30$nm while for cylinders with $r \geq 60$nm, its frequency is almost independent of cylinder radius. The frequencies of the higher order modes are almost constant irrespective of cylinder radius.

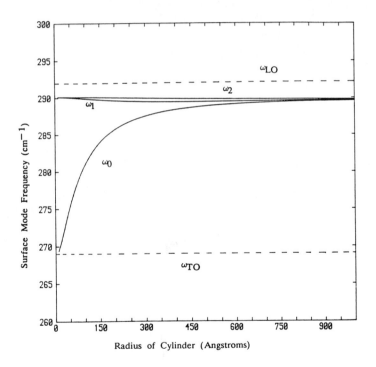

Figure 1. Surface phonon frequencies as a function of cylinder radius for the first three order modes (after Englman and Ruppin, 1968). The dashed lines show the bulk LO and TO frequencies for comparison.

Figure 2. SEM micrograph of a typical quantum cylinder sample. The cylinder diameter is about 70nm.

3. EXPERIMENTAL

The quantum cylinders were fabricated in arrays of 200 microns square on undoped GaAs using conventional electron beam lithography techniques and subsequent reactive-ion-etching (RIE) with $SiCl_4$ (see S P Beaumont, this conference, for details of the fabrication process). An SEM micrograph of a typical sample is shown in figure 2.

The Raman scattering experiments were performed at room temperature using the 488nm line of an Argon ion laser. Spectral resolutions and polarisation configurations are given in the figures. Spectra were obtained with the laser spot both on and off the patterned area; the off-pattern spectra provide independent assessment of the damage caused by the fabrication processes, in particular the RIE process, which is known to cause significant damage to the crystal if the parameters are not properly optimised (see, e.g., Kirillov et al, 1986 and Watt et al, 1988).

In the backscattering geometry from a (100) face of GaAs, first-order Raman scattering selection rules permit only the observation of the LO phonon. Surface roughness and damage to the crystal relax this selection rule thus allowing weak scattering from the TO phonon. In addition, the structured surface of the samples alters the direction of the incident light as shown in figure 3 so that a nearly-backscattering geometry does not hold good for the area patterned with cylinders. In

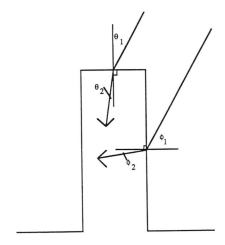

Figure 3. This shows the effect of the structured surface on the coupling of the incident light. The part of the beam hitting the top of the cylinder can be regarded as in a nearly-backscattering geometry, however, that part of the beam hitting the side of the cylinder results in an almost right-angle scattering geometry.

the new experimental geometry of the patterned area, the TO phonon becomes an allowed scattering mechanism and a strong peak is observed in the spectra.

Many samples were studied in the course of this work and showed the enhanced TO phonon intensity from the patterned area. Some of the samples also exhibited an additional feature in the Raman spectrum between the TO and LO phonon frequencies. These samples fulfilled three criteria which the other samples did not meet: the cylinders (a) were 80nm in diameter or smaller, (b) were 200nm long or longer and (c) had vertical sidewalls. Samples which did not meet these three criteria showed only a weak tail to the low energy side of the LO phonon rather than a well-defined feature. Spectra from one of the samples, QD12, is shown in figure 4 for various scattering geometries. As the incident light moves perpendicular to the cylinders, the intensity of the additional feature in the Raman spectrum increases in intensity with respect to the LO and TO phonon intensities. In the off-pattern spectra, there is no additional feature and no TO phonon.

4. DISCUSSION

Two possible assignments for the additional Raman feature will be considered here: damage effects and surface phonons.

It is well known that crystalline damage relaxes the first-order Raman scattering selection rules and results in a broadening and shifting of the LO phonon (see, e.g., Burns et al, 1987 and Holtz et al, 1988). In addition, contributions from acoustic phonons at the Brillouin zone edges become evident in the spectra. If the damage is increased sufficiently to bring the sample into an amorphous state the spectrum obtained will reflect only the GaAs phonon density of states which has three broad peaks at 80, 175 and 250 cm^{-1}.

The displacement of the additional feature in the Raman spectra is between 3.5 and 7cm^{-1} from the LO phonon and is comparable with the shifts in the LO phonon frequency observed after ion implantation of GaAs (Tiong et al, 1984 and Holtz et al, 1988). The damage due to such implantation, however, is expected to be much greater than for RIE: not only is the accelerating voltage three orders of magnitude smaller for RIE, but the ions involved are smaller and carry less momentum. Three experimental observations support our interpretation of low damage. We have studied the Raman spectra of the patterned areas down to 25cm^{-1} and we observe no evidence of broad damage-induced peaks. We also observe low or zero intensity of the TO phonon in our off-pattern spectra thus suggesting that the RIE has left a smooth surface. We do not observe the additional Raman feature in samples with comparatively large cylinder diameters (i.e. $\geq$ 100nm) although if the additional feature resulted from surface damage due to the fabrication processes, we would expect these to be equally affected.

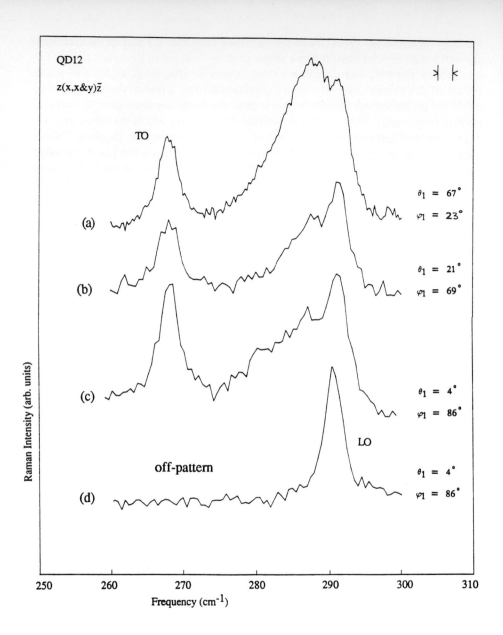

Figure 4. Raman spectra of one of the samples (QD12) for various scattering geometries (as defined in figure 3). Spectra (a) - (c) are from the patterned area while spectrum (d) is from the unpatterned part of the sample.
$\lambda = 488nm$, *resolution* $= 2cm^{-1}$.

In view of these three observations, we reject the assignment of the additional Raman feature to a damage-induced effect.

If the experimental points are plotted on the same scale as the theoretical curves of surface phonon frequency, good agreement is observed with the lowest order surface phonon (see figure 5). The spectra of figure 4 also show consistency with the identification of this feature as a surface phonon. The theory developed by Englman and Ruppin (1968) predicts a triply-degenerate surface phonon with vibrations in the r, θ and z directions. Light incident parallel to the cylinders excites only the z vibrations whereas light incident perpendicular to the cylinders excites both the r and the θ phonons thus showing a stronger scattering intensity. The effect on the surface phonon frequency of changing the dielectric constant of the surrounding material is also shown in figure 5. Here it can be seen that the phonon moves to lower energies as the surrounding dielectric constant increases. This suggests that coating the cylinders with a material which has a dielectric

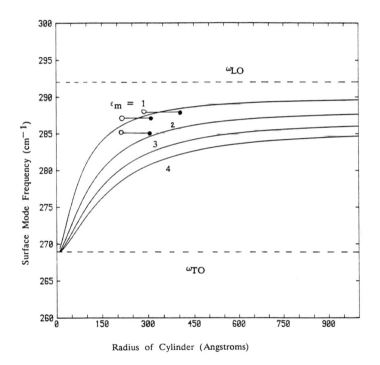

Radius of Cylinder (Angstroms)

Figure 5. This is a plot of the experimental data points of the additional Raman feature and the theoretical calculations for the lowest-order surface phonon frequency. The calculations are based on a value of $\cos\theta = 1/\sqrt{2}$ where θ is the angle of incidence. In order to account for experimental deviations from this, the solid symbols correspond to the measured radii while the open symbols are plotted using cylinder radii which have been scaled to compensate for a $\cos\theta$ value of 1.

constant greater than one would result in an observed shift of the additional Raman feature to lower energies. We have investigated this effect and preliminary results on patterns overgrown with silicon nitride indicate that the additional feature does indeed shift to lower frequencies.

5. CONCLUSIONS

We have fabricated cylinders of 60nm diameter and 200nm height in GaAs. The Raman spectra from arrays of these cylinders has an additional spectral feature between the TO and LO phonons and a greatly-enhanced TO phonon scattering intensity. Neither of these effects is observed from unpatterned areas of the sample. The enhanced TO phonon intensity can be explained in terms of a new scattering geometry introduced by the cylinders. Two possible assignments for the additional spectral feature were discussed: damage effects and a surface phonon of the cylinders. We identify the peak as a surface phonon since this is consistant with our experimental results.

REFERENCES

Burns, G., Dacol, F.H., Wie, C.R., Burstein, E., and Cardona, M. (1987).
    Sol. St. Comm. 62, 449
Englman, R., and Ruppin, R. (1968). J.Phys. C1, 614
Hayashi, S., and Kanamori, H. (1982). Phys. Rev. B26, 7079
Holtz, M., Zallen, R., and Brafman, O. (1988). Phys. Rev. B37, 4609
Kirillov, D., Cooper III, C.B., and Powell, R.A. (1986).
    J. Vac. Sci. Technol. B4, 1316
Tiong, K.K., Amirtharaj, P.M., and Pollak, F.H. (1984).
    Appl. Phys. Lett. 44, 122
Watt, M., Sotomayor Torres, C.M., Cheung, R., Wilkinson, C.D.W.,
    Arnot, H.E.G., and Beaumont, S.P. (1988). J. Mod. Opt. 35, 365

MAGNETIC EFFECTS IN QUANTUM DOTS[1]

W. Hansen
T. P. Smith III
J. A. Brum[2]
J. M. Hong
K. Y. Lee
C. M. Knoedler
D. P. Kern
L. L. Chang

IBM, T. J. Watson Research Center,
Yorktown Heights,
N.Y. 10598

INTRODUCTION

Investigations of few-electron systems confined in semiconductors are not only of basic physical interest [1-9] but may also turn out to be important for laser applications[10] and new device concepts in ultra large scale integration of electronic devices.[11, 12] So far optical spectroscopy, transport measurements, and capacitance measurements have focussed on the detection of the energy quantization caused by the spatial confinement. Here we discuss spatial energy quantization in quantum dots in the presence of a magnetic field. This problem has been considered theoretically for

---

[1]   *This work is partly supported by the ONR.*

[2]   *Permanent address: Universidade de Campinas, Departemento de Física Do Estado Sólido e Ciência dos Materiais, and Laboratório Nacional de Luz Síncrotron, 13081, Campinas (SP) Brazil*

97

more than 50 years, mainly to clarify the contribution of surface states to the magnetization.[7, 13-15] The length scale introduced by the magnetic field is the cyclotron orbit $l_c = \sqrt{(\hbar/eB)}$ and its ratio to the extent of the confinement potential determines the character of the electronic states in the quantum dots. In high fields, where the magnetic length is much smaller than the diameter of the quantum dot, the energy states condense into highly degenerate Landau levels and the system exhibits bulk like, i.e. two-dimensional, behavior. On the other hand, in the absence of a magnetic field the degeneracy of the discrete energy levels is low and depends only on the symmetry of the confinement potential. A very small magnetic field lifts this degeneracy in analogy to the Zeeman effect of electrons in atomic orbitals. Consequently, in intermediate magnetic fields, where the magnetic length is comparable to the extent of the confinement potential, the energy levels rearrange and transform into Landau levels.

Figure 1. . SEM photograph of a part of the quantum dot matrix on a heterojunction surface. The diameter of the dots is 300nm.

SAMPLE FABRICATION

Sample preparation starts with a modulation-doped heterojunction grown by molecular beam epitaxy. The buffer layer contains a heavily doped GaAs layer separated by 50-100 nm undoped GaAs from the AlGaAs-GaAs interface where the electron system resides. This doped GaAs layer provides the back contact for charging the electron system at the heterojunction interface at low temperatures without diminishing its mobility. The epitaxial growth is terminated with a comparatively thick (30nm) GaAs cap-layer. This cap layer is microstructured in a subsequent reactive ion etching process through a gold mask defined by electron beam lithography. Details of the fabrication process are described elsewhere.[16] The SEM photograph in Fig.1 shows a portion of the dot matrix of a sample with 300nm dot size. From the micrograph we infer, that the shape of the dots is a square with rounded corners. Variations of the dot size are beyond the resolution of the SEM. The sample preparation is finished with the evaporation of a Schottky gate on top of the microstructured sample surface. Lateral confinement of the electrons into 0D states at the heterojunction interface is provided by the band bending of the undulating Schottky barrier. The number of carriers can be controlled via a gate voltage applied between the front gate and the back contact.

EXPERIMENTAL RESULTS

The differential capacitance of the samples is measured with lock-in technique at modulation frequencies of about 10kHz and modulation voltages below 10 mV . In order to enhance the capacitance oscillations, the gate voltage derivative is recorded. Fig. 2 shows traces of the capacitance derivative taken on samples with 300 nm and 400 nm dots size with magnetic fields applied in the direction of strong confinement, i. e. the direction perpendicular to the heterojunction interface. The gate voltage derivative of the capacitance in the 300nm dot samples is zero at gate voltages below $V_g = -270mV$ and rises sharply at $V_g = -270mV$. indicating that electrons start to occupy the 0D electron states at the heterojunction interface. Corresponding behavior is observed at $V_g = -370mV$ in the 400 nm dot samples and at $V_g = -50mV$ in the 200nm samples. An increase of the threshold voltage with decreasing dot size is expected, since fringing fields lift the potential minima underneath the GaAs dots. The oscillatory behavior of the capacitance signal above the threshold voltage reflects the quantization of the system into discrete energy states. The contribution of the density of states to the capacitance oscillates

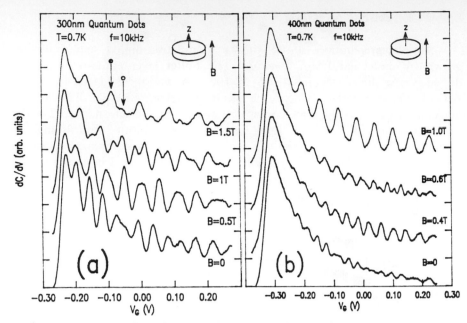

Figure 2. . Gate voltage derivative of sample capacitance in heterostructure quantum dot samples with (a) 300nm and (b) 400nm dot diameter and different magnetic fields applied perpendicular to the heterjunction interface.

with maxima whenever the Fermi-energy crosses a 0D energy level. The oscillation period is larger in samples with smaller dot size, since the level spacing increases with decreasing dot size.

When a small magnetic field is applied, at high quantum numbers the positions of the oscillation maxima shift rapidly. At magnetic fields of about 0.6 T and 1.0 T in samples with 400 nm and 300 nm dot size respectively a pronounced splitting of the oscillation maxima is observed. This splitting occurs in samples with 200 nm dot size at an even higher field of 1.5 T. The corresponding magnetic lengths of these field values are approximately half the radius of the confinement potential. At higher fields the spacing between the oscillation maxima increases linearly with magnetic field, indicating that a Landau level like behavior is regained.

Fig. 3 shows the magnetic field behavior of the oscillation maxima in detail for a sample with 300 nm dot size, where the gate voltage position of the maxima is plotted vs. the magnetic field. Whereas at low quantum numbers, i. e. gate voltages below $V_g = -0.1V$, the levels shift only slightly in a low magnetic field, a complex behavior is observed for higher quantum numbers. As can be seen in Fig. 3., the splitting of the oscillation maxima in the B = 1 T trace of Fig. 2 results from an anti-crossing like behavior of

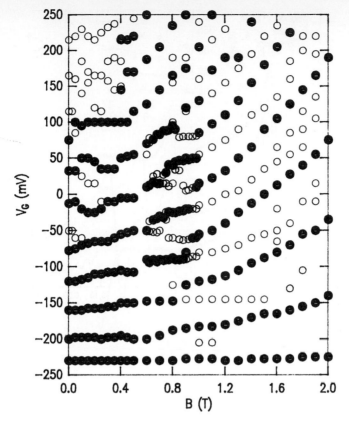

Figure 3. . Gate voltage positions of the oscillation maxima in a 300nm dot sample plotted vs. the magnetic field.

---

the levels. At fields above $B = 1.6$ T the spacing between dominant maxima increases linearly with the magnetic field, indicating the onset of Landau-level-like behavior.

Spin splitting is observed at magnetic fields above $B = 4$ T. Since, in the high field limit, we expect the electron systems to exhibit two-dimensional behavior, fractional quantization should be observable in the magnetic quantum limit as they are in two-dimensional systems[18, 19] In Fig. 4 we show experimental traces taken on samples with 300 nm dot size at magnetic fields up to $B = 30$ T. So far fractional states have been observed in samples with 300 nm and 400 nm dot size at filling factors 2/3 and 1/3. We estimate the number of electrons in the 300 nm dot at $B = 30$ T and filling factor 1/3 to be about 40. Observation of fractional quantization demonstrates the outstanding quality of our samples.

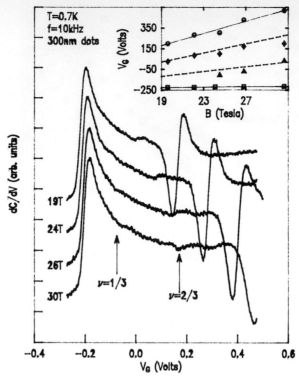

Figure 4. . Gate voltage derivative of the capacitance oscillations at very high magnetic fields. The insert shows the gate voltage positions of the capacitance structures as a function of the magnetic field. Dashed lines are calculated for filling factors $v = 1/3$ and 2/3 basing on the gate voltage positions of the $v = 0$ and 1 structures.[17]

MODEL CALCULATIONS

A first order explanation of the experimental data can be obtained by calculating the magnetic field dependence of single-particle energy spectra in idealized potentials. For these calculations we assume that motion in the direction perpendicular to the heterojunction interface (z-direction) is decoupled from the motion parallel to the interface and that the electrons are in the lowest quantum level in z-direction. Two different geometries are considered, with the real potential in our structures somewhere in between these two. In Fig. 5a we plot the energy spectrum of a single electron in a harmonic potential with rotational symmetry with respect to the z-axis $(V(x,y) = m^*\Omega^2(x^2 + y^2)/2$, with $\hbar\Omega = 2meV)$. In Fig. 5b we present model calculations for a square-shaped, square-well potential with finite walls (600meV) and a width of 120 nm. The energy spectra are obtained by

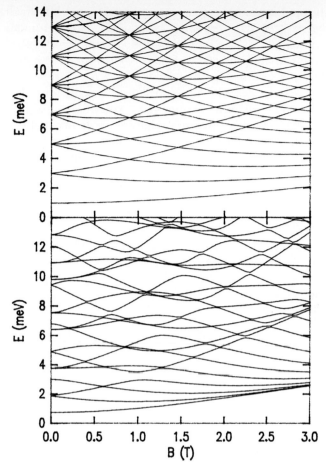

Figure 5. . Magnetic field dependence of the calculated single electron energy spectrum in a harmonic potential with rotational symmetry (top) and square shaped square-well potential with finite walls (bottom).

projecting the Hamiltonian in a magnetic field into a finite basis set consisting of solutions of the zero-field Hamiltonian and numerical diagonalization. The main difference between the spectra in Fig.5a and 5b is that the levels in Fig. 5a cross each other without interfering, whereas in the square shaped potential the levels anti-cross. Level crossing as seen in Fig. 5a occurs only if the magnetic field does not destroy the symmetry of the solutions at B = 0 T. Although both spectra look quite different in detail they are qualitatively similar to one another and the experimental data as well. In all three cases, we observe a complex behavior at low magnetic fields and high quantum numbers, a cross-over regime at intermediate fields, where the magnetic length becomes comparable to the radius of the confinement potential, and a condensation of the electron states into Landau like states at high magnetic fields. The onset of the condensation into Landau levels is seen in Fig 5b at fields above $B = 2.0T$.

To quantitatively describe the capacitance oscillations these models must be refined. The effective confinement potential must be calculated self-consistently (the effective potential changes with occupation and thus with the gate voltage), electron-electron correlation should be taken into account[6], and a large number of slightly different quantum dots (differences arising from potential fluctuations resulting either from scattering centers or from lithographic imperfections) should be averaged. Details of the magnetic-field behavior of the experimental spectra might provide valuable insight for improving existing models.

REFERENCES

[1] J. Cibert, P. M. Petroff, G. J. Dolan, S. J. Pearton, A. C. Gossard, and J. H. English, Appl. Phys. Lett. **49**, 1275 (1986).

[2] Y. Miyamoto, M. Cao, Y. Shingai, K. Furuya, Y. Suematsu, K. G. Ravikumar, and S. Arai,
Jpn. Journ. Appl. Phys. **26**, L225 (1987).

[3] T. Inoshita, S. Ohnishi, and A. Oshiyama,
Phys. Rev. Lett. **57**, 2560 (1986).

[4] M. A. Reed, J. N. Randall, R. J. Aggarwal, R. J. Matyi, T. M. Moore, and A. E. Wetsel, Phys. Rev. Lett. **60**, 535 (1988).

[5] T. P. Smith, III, K. Y. Lee, C. M. Knoedler, J. M. Hong, and D. P. Kern, Phys. Rev. B **38**, 2172 (1988).

[6] G. W. Bryant, Phys. Rev. Lett. **59**, 1140 (1987).

[7] U. Sivan and Y. Imry, Phys. Rev. Lett. **61**, 1001 (1988).

[8] W.-M. Que and G. Kirczenow,
Phys. Rev. **B38**, 3614 (1988).

[9] C. Sikorski and U. Merkt, (preprint).

[10] K. J. Vahala,
IEEE Journ. of Quantum Electronics **24**, 523 (1988).

[11] K. Obermayer, W. G. Teich, and G. Mahler,
Phys. Rev. **B37**, 8096 (1988).

12 J. N. Randall, M. A. Reed, R. J. Matyi, T. M. Moore, R. J. Aggarwal, and A. E. Wetsel,
SPIE Advanced Processing of Semiconductor Devices II **945**, 137 (1988).

13 C. G. Darwin, Proc. Cambridge Phil. Soc. **27**, 86 (1930).

14 R.B. Dingle,
Proc. Roy. Soc. London **A216**, 463 (1953).

15 M. Robnik, J. Phys. A: Math. Gen. **19**, 3619 (1984).

16 K. Y. Lee, T. P. Smith, III, H. Arnot, C. M. Knoedler, J. M. Hong, and D. Kern, J. Vac. Sci. Tech. **B6**, 1856 (1988).

17 W. Hansen and et al., (to be published).

18 T. W. Hickmott,
Phys. Rev. Lett. **57**, 751 (1986).

19 T. P. Smith, III, W. I. Wang, and P. J. Stiles, Phys. Rev. B **34**, 2995 (1986).

ANALYTIC SELF-CONSISTENT CALCULATIONS FOR INHOMOGENEOUS TWO-DIMENSIONAL ELECTRON GASES

John H. Davies

Department of Electronics and Electrical Engineering
Glasgow University, Glasgow, G12 8QQ, U.K.

1. INTRODUCTION

Structures where the motion of electrons is confined to less than three dimensions are playing an ever-increasing rôle in semiconductor physics. This is particularly true of nanostructures, which are typically made by patterning electrons in a two-dimensional electron gas (2DEG). Electrons can be removed from unwanted areas by etching or by the electrostatic potential from a patterned gate. Other methods of confining the electrons, such as the use of low-energy ions, will not be treated in this paper.

A wide range of approximate formulas is available for treating three-dimensional systems with simple geometries. One example is the depletion region in a p-n diode or a Schottky barrier, whose thickness is calculated in any undergraduate course on semiconductor physics. A corresponding problem in a two-dimensional system is to calculate the width of the depletion region at the edge of a semi-infinite wafer of modulation-doped material. This proves to be much more difficult, and the simple 'abrupt' approximation used successfully in three-dimensional systems fails badly.

The more complicated geometry of low-dimensional systems means that numerical techniques must be used if precise numbers are needed. A complementary approach is taken in this paper: the aim is to develop simple, approximate analytical methods in which the physics is more transparent. A key ingredient of these calculations is the treatment of the surface states on GaAs and (Al,Ga)As. It is assumed that these states provide perfect pinning of the chemical potential at a fixed energy below the conduction band at the surface, and that the surface states remain in thermal equilibrium with electrons in the bulk of the semiconductor. The accuracy of these assumptions is unclear at the low temperatures where most measurements on these structures are made. This is not merely a defect of the theory described here: correct treatment of the surface states is vital in any modelling technique, and more experimental information on the behaviour of the surface states at low temperature is urgently needed.

2. NARROW WIRES AT THRESHOLD

There are many techniques for fabricating narrow 'wires' in modulation-doped semiconductor layers. The aim is to restrict the 2DEG to a narrow channel so that motion in only one dimension is free, and few (ideally only one) transverse modes are permitted perpendicular to this. The ideas behind the calculation are best illustrated by considering band diagrams along a direction normal to the surface of uniform modulation-doped layers, shown in figure 1.

If we integrate Poisson's equation in from large z for the case where electrons are present, the bands begin to curve because of the negative charge density in the 2DEG. There is a uniform electric field in the spacer, and opposite curvature caused by the positive donors in the n-AlGaAs. A self-consistent calculation is clearly needed for the 2DEG. This requirement vanishes if conditions at threshold are treated, because very few electrons are present. Almost no charge density is encountered until the n-AlGaAs region is reached and the band must therefore be flat to the right of this point. Electrons are about to enter the band if the system is at threshold, so E_c in the GaAs just brushes the chemical potential μ. The two discontinuities in E_c cancel, which means that the electrostatic potential generated by the donors exactly balances the height ϕ_s of the Schottky barrier on the surface. This idea can be applied to wires to determine the conditions for threshold. The main assumptions are as follows.

a) The doped regions are fully ionized and contain no electrons, free or trapped.
b) The structure is in equilibrium, so μ is flat throughout; in particular, the surface states are assumed to be in equilibrium with electrons in the bulk.
c) There is perfect pinning of the chemical potential at all points on the surface; in other words, E_c is always $e\phi_s$ above μ on the surface.
d) The structure is at low temperature, so that the 2DEG is always degenerate.

Combining (b) and (c) shows that E_c must be at the same energy at all points on the surface, which means that the surface can be treated as an equipotential — a vital simplification when the calculations are extended from uniform layers to wires, because a variety of tricks is available to treat electrostatic problems with equipotential boundaries. Details of the calculations have already been published[1].

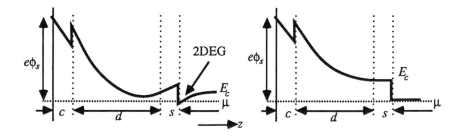

Figure 1. Bands for uniform modulation-doped layers. Electrons are present in the 2DEG in the first case, while the second shows the bands at threshold.

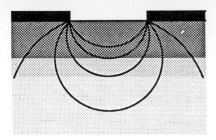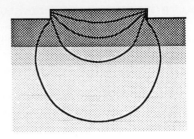

Figure 2. Confining equipotentials for gated and shallow-etched wires.

The confining part of the total electrostatic potential is shown in figure 2 for two cases, gated and shallow-etched wires. The fields are seen to be very similar and this is also true of the energy levels and spatial extent of the wavefunctions.

The Schrödinger equation separates approximately into terms for motion normal and parallel to the surface. Energy levels normal to the surface are separated by about 40 meV, while those for motion parallel to the surface are much closer — typically 10 meV or less apart. These energies change considerably as more electrons enter the wire, as shown by the self-consistent numerical calculations of Laux, Frank and Stern[2]. An important practical point is that the width of a shallow-etched rib at threshold is a rapidly-varying function of the depth of etching, indicating that precise control of fabrication is needed if a very narrow wire is to be produced.

3. DEPLETION REGION IN A 2DEG

Figure 3 shows a physical system with a depleted 2DEG at the edge of a semi-infinite modulation-doped structure. Two simplifications are introduced to model the 2DEG.

a) The thickness of the 2DEG is taken to be infinitesimal.
b) The density $N(x)$ of the 2DEG is given by a Thomas-Fermi approximation.

The electrons contribute a term ϕ_e to the total potential, given by the integral form of Poisson's equation:

$$\phi_e(x) = \frac{-e}{4\pi\varepsilon\varepsilon_0} \int_{-\infty}^{\infty} dx' \, N(x') \ln\left\{\frac{(x-x')^2 + 4p^2}{(x-x')^2}\right\}, \tag{1}$$

where p is the depth of the electrons below the surface, and the term containing p is an image to satisfy the equipotential boundary condition due to the surface states.

Combining (1) with the external potential and the Thomas-Fermi result gives an integral equation for $N(x)$. It should be possible to solve this analytically or numerically for a simple external potential, and is in progress.

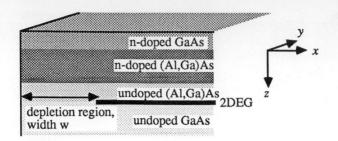

Figure 3. Depletion region in a 2DEG at the edge of a semi-infinite wafer.

A first guess at a solution for the depletion region is the 'abrupt' approximation, universally used in three-dimensional systems: $N(x) = N_\infty \Theta(x-w)$, where w is the width of the depletion layer and N_∞ is the density of electrons in the 2DEG far from the surface. Unfortunately this proves to a poor approximation to the correct solution and gives rise to a total potential that is not monotonic. The rational function

$$N(x) = N_\infty \frac{x-w}{x-w+x_\infty} \Theta(x-w)$$ (2)

is much better. The approach of $N(x)$ to N_∞ for large x is controlled by x_∞ which can be calculated analytically, while w is chosen so that $E_c(w) = \mu$. Typically w lies in the range 100–200 nm, much bigger than bulk depletion widths which are about 30 nm for the highly doped materials used in modulation-doped layers. An important feature of (2), which also holds for the exact solution, is that the deviation of $N(x)$ from N_∞ only decays as x_∞/x for large x. The effect of the surface is felt a long way into the material, unlike the exponential decay in three dimensions.

4. ELECTRONS UNDER A PATTERNED GATE

Many recent experiments, notably those that demonstrated quantization of the ballistic resistance, used patterned gates to restrict the 2DEG to narrow channels and 'point contacts'. It appears that a self-consistent calculation is needed to describe these devices, but we shall see that a good results can be obtained without this.

The logarithm in (1) is peaked at $x' = x$, and decays like a Lorentzian of width p for large $|x - x'|$. It can be replaced with a δ-function provided that $N(x)$ varies slowly on the scale of p. Equation (1) becomes

$$\phi_e(x) \approx -\frac{ep}{\varepsilon\varepsilon_0} N(x).$$ (3)

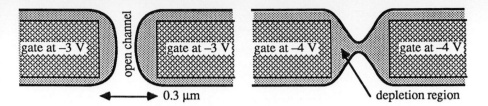

Figure 4. Sketch of the channel under a short split gate (ref. 3). Electrons travel vertically from source to drain. The gates are 0.3 μm long and 0.3 μm apart. A gate bias of −3 V leaves a narrow channel while −4 V gives complete pinch-off.

This is a remarkable simplification because it is a *local* relation between $\phi_e(x)$ and $N(x)$. It can be applied at the edge of the occupied region, where $N(x) = 0$ so $\phi_e(x) = 0$ and the potential is given entirely by the external potential [unfortunately this is the situation where (3) is least accurate]. Thus $E_c(x)$ can be calculated from the external potential alone. Results for the depletion region in §3 show that the error is around 10–20 nm, which is acceptable for a wide channel. Equation (3) can easily be extended to systems patterned in both x and y, and figure 4 shows the results for the two-dimensional 'point contact' used by Wharam *et al*[3]. The most important feature is the smooth narrowing of the channel between the gates, with gently rounded corners. It is difficult to see how 'over-the-barrier' resonances could be set up within such a channel because the reflection coefficients at the ends must be very small.

5. CONCLUSIONS

The theory described in this paper provides a first step towards developing simple pictures for understanding the behaviour of electrons in strongly inhomogeneous two-dimensional systems. The method for finding the edge of the occupied region under a patterned gate is particularly useful, as a full three-dimensional calculation is a major numerical undertaking. It may be possible to improve the accuracy of the local approximation [equation (3)] by including a gradient term in the expansion, which would enable it to be applied to a wider range of devices.

REFERENCES

1. J. H. Davies, *Semicond. Sci. Technol.* **3** 995–1009 (1988)
2. S. E. Laux, D. J. Frank and F. Stern, *Surf. Sci.* **196** 101–6 (1988)
3. D. A. Wharam, M. Pepper, H. Ahmed, J. E. F. Frost, D. G. Hasko, D. C. Peacock, D. A. Ritchie and G. A. C. Jones, *J. Phys. C* **21** L887–891 (1988)

OPTICAL INVESTIGATION OF A ONE-DIMENSIONAL ELECTRON GAS

G. Danan
J. S. Weiner
A. Pinczuk
J. Valladares
L. N. Pfeiffer
K. West

AT&T Bell Laboratories
Murray Hill, NJ 07974

ABSTRACT

We report the observation of one-dimensional confinement in optical emission and inelastic light scattering of the electron gas in laterally patterned modulation-doped heterostructures. The optical emission lineshapes arise from confinement of the electrons and holes in a type II lateral superlattice. The inelastic light scattering spectra exhibit peaks at energies corresponding to the spacings between the one-dimensional subbands.

I. INTRODUCTION

Recently, there has been great interest in the properties of a one-dimensional electron gas (1DEG). 1DEG structures have previously been studied by means of transport (1) and infrared absorption (2). These techniques only give information about the majority carriers. We have studied a 1DEG by means of band-to-band optical emission, which gives information about the valence band as well, and inelastic light scattering. The spectral lineshapes show that the electrons and holes are confined in a type II quantum wire superlattice.

II. QUANTUM WIRE FABRICATION

The structures were fabricated from a high mobility modulation doped GaAs/AlGaAs single quantum well heterostructure by the procedure shown in figure 1. The as-grown two-dimensional electron gas structure consisted of a 250 Å GaAs quantum well, separated from the surface by a 1000 Å n-doped AlGaAs layer and a 40 Å GaAs cap layer. The electron density and mobility of the as-grown structure were $n = 3 \times 10^{11}/cm^{11}$ and $1.4 \times 10^{6} cm^{2}/sec$ respectively. PMMA was spun onto the samples. 1000 Å lines with 2000 Å period were patterned on the sample using a JEOL JBX-5D electron beam writer. 200 Å of chromium was angle evaporated onto the PMMA to provide resistance to dry etching. The sample was then ion milled to selectively deplete the free charge in the quantum well. Quantum wires were also made by reactive ion etching a similar pattern into a silicon nitride coating. The electron density was found to be substantially reduced under the silicon nitride and restored where it had been removed. In both samples the lateral modulation of the charge density induces a periodic lateral potential which confines the electron gas to one dimension. Qualitatively similar spectra are observed in both samples.

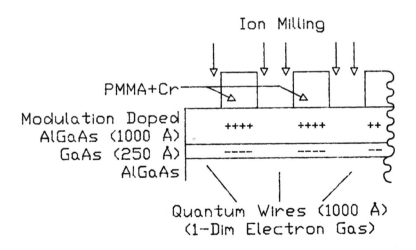

Fig. 1. Fabrication of modulation doped quantum wires by electron beam lithography, followed by ion milling to selectively deplete the electron gas.

III. OPTICAL SPECTRA OF QUANTUM WIRES FABRICATED BY ION MILLING

The inelastic light scattering spectra of the quantum wires show peaks that are attributed to one-dimensional intersubband transitions. They indicate subband spacings of ~2 meV, consistent with the spacings estimated for the lateral confinement potential. The photoluminescence spectrum (PL) of the as-grown modulation doped quantum well structure is shown in figure 2a. It exhibits a sharp onset at the bandgap, decreasing in intensity with increasing energy up to a cutoff at the Fermi level. The PL spectrum of the quantum wires, shown in figure 2b, is qualitatively different from that of the as-grown sample. The spectrum exhibits an onset at the bandgap, but, the PL intensity increases with increasing energy, reaching maximum intensity just before cut-off at the Fermi level. From the position of the Fermi level and the subband spacings we estimate that five conduction band subbands are occupied. The spectral lineshape of the quantum wires arises from the spatially separate confinement of the electrons and holes in a type II lateral superlattice. The lowest transitions are weak because of the small spatial overlap of the lowest electron and hole subbands. At higher energies the PL intensity increases due to the increasing spatial overlap of the electron and hole wavefunctions, which are less well-confined.

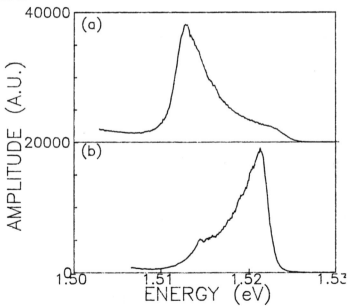

Fig. 2. (a) Photoluminescence spectrum of as-grown modulation doped quantum well. (b) Photoluminescence spectrum of quantum wires.

IV. CONCLUSIONS

In conclusion, we have observed, for the first time, confinement of electrons and holes in a type II quantum wire superlattice. This is a result which cannot be obtained from transport or infrared absorption measurements. We observe new structure in the inelastic light scattering spectra which we attribute to transitions between one-dimensional conduction subbands.

REFERENCES

1. G. Timp, this conference; A. M. Chang and T. Y. Chang, this conference.
2. W. Hansen, M. Horst, J. P. Kotthaus, U. Merkt, Ch. Sikorski and K. Ploog, Phys. Rev. Lett. 58, 2586 (1987); D. Heitmann, T. Demei, P. Grambow and K. Ploog, this conference.

QUANTUM PHASE COHERENT EFFECTS IN THE PHOTOLUMINESCENCE SPECTRA OF DISORDERED MESOSCOPIC STRUCTURES [1]

S. Bandyopadhyay

Department of Electrical and Computer Engineering

University of Notre Dame

Notre Dame, Indiana 46556

At low enough temperatures, when the *inelastic* scattering time in a quantum dot exceeds the radiative recombination lifetime of photoexcited electrons and holes, the photoluminescence spectrum of the dot becomes sensitive to the exact *locations* of the elastic scatterers within the dot. This is a result of quantum interference whose nature is determined by the precise configuration of the elastic scatterers inside the dot. Several features of the photoluminescence spectra are influenced by the configuration, the most remarkable of which is the fact that the usual red-shift of the peak frequency, associated with bandgap renormalization, can change into a blue-shift depending on the configuration. The dependence of the optical spectra on the internal configuration is basically the same effect that makes (universal) conductance fluctuations sample-specific. An important consequence of this effect is that different quantum dots in a lateral surface superlattice will exhibit slightly different spectra if they merely have different impurity configurations, but are otherwise identical. The resulting inhomogeneous broadening can be comparable to the energy spacing between the subbands, so that it can sometimes mask the discreteness of the optical spectra expected of quasi-zero dimensional structures.

I. INTRODUCTION

It is well-known that elastic scattering does not destroy an electron's phase-memory so that quantum interference effects are not inhibited by impurity scattering at low enough temperatures. In a disordered semiconductor nanostructure,

[1]Supported by the Air Force Office of Scientific Research under grant no. AFOSR-88-0096 and by an IBM Faculty Development Award.

if the inelastic scattering times of both electrons and holes exceed the radiative recombination lifetime of a photoexcited electron-hole pair, then the optical dipole constituted by the pair never loses its phase-memory during its entire lifetime. In that case, the time-dependent decay of the optical dipole, due to impurity scattering, is apt to be influenced by quantum interferences between the electron and hole states in the system. The decay of the optical dipole moment determines the lineshape of the optical spectrum - the lineshape is in fact obtained by Fourier transforming the decay characteristic - and consequently the lineshape will also be influenced by quantum interference effects and depend on the precise configuration of the elastic scatterers that determines the nature of this interference [1]. The important implication here is that in the presence of phase-coherence, not only are macroscopic *transport* properties (such as the conductance of a sample) a function of such microscopic details as the impurity configuration [2], but so are *optical* properties like the photoluminescence spectra. In other words, *"mesoscopic physics" includes not only transport phenomena, but also optical phenomena.*

A practical consequence of the latter is that different quantum dots in a lateral surface superlattice will exhibit slightly different spectra if they merely have different impurity configurations but are otherwise identical. This phenomenon therefore induces a unique kind of inhomogeneous broadening in the photoluminescence linewidth of a superlattice structure whose origin is purely quantum-mechanical and specifically arises from phase-coherence. In some instances, this inhomogeneous broadening can be so large that it can even mask the discreteness of the optical spectra expected of quasi-zero dimensional structures.

II. THEORY

The time-dependent decay of the optical dipole moment $\mathbf{P}(t)$ associated with transitions between a conduction band state and a valence band state in a disordered quantum dot is given by [1]

$$P(t) = \frac{1}{2}P(0) \sum_p c_{ep}^*(t)c_{hp}(t) + c.c., \tag{1}$$

where $c_{ep}(t)$ and $c_{hp}(t)$ are the time-dependent complex amplitudes of the pth electronic state and the pth hole state that the photoexcited electron and hole couple to at time t via the impurity interaction, and the summation over p is carried out to include all such states in the system. The lineshape of the photoluminescence spectra $\mathcal{F}(\omega)$ is obtained by Fourier transforming the time-dependent decay of $\mathbf{P}(t)$ into the frequency domain ω of the incident photons.

The task here is to evaluate the amplitudes $c_{ep}(t)$ and $c_{hp}(t)$. For both electrons and holes, these amplitudes are found from [1]

$$[c(t)] = exp\left[-\frac{i\mathcal{H}t}{\hbar}\right][c(0)], \tag{2}$$

where $\mathcal{H}$ is the Hamiltonian matrix for the disordered system and $[c(t)]$ is a column vector whose elements are the complex amplitudes $c_p(t)$ of the various electronic or hole states that the photoexcited electron or hole couples to. The Hamiltonian $\mathcal{H}$ can be expressed as

$$\mathcal{H} = \mathcal{H}_0 + \mathcal{H}' , \tag{3}$$

where $\mathcal{H}_0$ is the unperturbed Hamiltonian (a diagonal matrix) whose elements are the kinetic energies of the various subband states that the photoexcited electron and hole couple to, and $\mathcal{H}'$ is the impurity interaction Hamiltonian whose elements are given by

$$H'_{pq} = -\frac{\Gamma q^2}{4\pi\epsilon} \sum_i < \phi_p(r)|\delta(r-r_i)|\phi_q(r) >= -\frac{\Gamma q^2}{4\pi\epsilon} \sum_i \phi_p^*(r_i)\phi_q(r_i) \tag{4}$$

In the above equation, $\phi_p(r)$ is the wavefunction of the pth subband state that the electron or hole couples to as a result of the impurity interaction, i.e. it is the pth eigenfunction of $\mathcal{H}_0$. The impurity potentials were assumed to be δ potentials located at coordinates at r_i. The summation is carried out over the coordinates of all the impurities in the system and the sum obviously *depends on the exact locations of the impurities*. The parameter Γ is a parameter representing the strength of the interaction. The choice of δ-scatterers (instead of screened Coulomb scatterers) in our model is merely a matter of convenience; it does not alter the essential physics.

Since the Hamitonian for the system is now clearly dependent on the coordinates of the impurities, it is obvious that the amplitudes $c_p(t)$ (see Equation (2)) and hence the time-dependent optical dipole moment $\mathbf{P}(t)$ (see Equation (1)) will also depend on the exact *locations* of the impurities within the system. Consequently, the optical spectrum of a sample will be a "fingerprint" of the internal configuration of the scatterers.

III. EXAMPLE

By way of an example, we have calculated the photoluminescence lineshape (corresponding to an electron-light hole transition) for a two-dimensional quantum dot with a parabolic confining potential. The material was assumed to be GaAs. The impurity coordinates were generated by random number generators and the concentration was 5×10^{11} cm^{-2}. In our calculation, we included 36 electronic states and 36 hole states[2]. The calculated lineshape is plotted in Fig. 1.

---

[2]These states need not be degenerate in energy, since strong impurity scattering, even though elastic, can couple an electron or hole between states that are non-degenerate in energy over short periods of time.

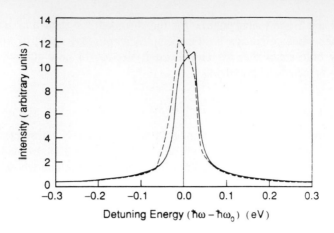

Fig.1 The photoluminescence spectrum (corresponding to an electron-light hole transition) for a quantum dot with parabolic confining potential. The incident photon energy $\hbar\omega_0$ is 1.4623 ev (the bandgap of GaAs is assumed to be 1.42 ev). The solid and dashed lines are for two different impurity configurations. Note that for one configuration, the peak frequency is red-shifted and for the other, it is blue-shifted. The difference corresponds to an energy of $\sim$ 4 mev which is comparable to the energy separation between the subband states ($\sim$ 6 mev). Consequently, the inhomogeneous broadening caused by varying impurity configuration, in different quantum dots in a lateral surface superlattice, may mask the discreteness of the optical spectra expected of quantum dots.

IV. DISCUSSION

To understand the nature of the photoluminescence lineshape in Fig. 1., we have to first recast Equation (1) in the form

$$P(t) = \frac{1}{2}P(0)exp(i\omega_0 t)\sum_p b_{ep}^*(t)b_{hp}(t) + c.c., \qquad (5)$$

where

$$b_{ep}(t) = exp\left[i\frac{E_e t}{\hbar}\right]c_{ep}(t) \quad ; \quad b_{hp}(t) = exp\left[i\frac{E_h t}{\hbar}\right]c_{hp}(t) \qquad (6)$$

In Equation (6), E_e and E_h are the energies of the states to which the electron and hole are photoexcited by the incident radiation and ω_0 is the resonant photon frequency corresponding to this transition, i.e.

$$\hbar\omega_0 = E_e - E_h \qquad (7)$$

120

We can now recast Equation (5) as

$$P(t) = \frac{1}{2} P(0) exp(i\omega_0 t) A(t) e^{-i\theta(t)} + c.c.$$
$$= P(0) A(t) cos [\omega_0 t - \theta(t)] , \qquad (8)$$

where $A(t)$ is the magnitude and $\theta(t)$ the phase of the complex product $\sum_p b_{ep}^*(t) b_{hp}(t)$ representing the decay of the dipole moment. Obviously, both $A(t)$ and $\theta(t)$ depend sensitively on the impurity configuration.

The photoluminescence lineshape $\mathcal{F}(\omega)$ is obtained by Fourier transforming $P(t)$

$$\mathcal{F}(\omega) = \frac{1}{2\pi} \int e^{i\omega t} P(t) dt \qquad (9)$$

Hence we see that the phase $\theta(t)$ in Equation (8) has two effects. Firstly, it makes the lineshape asymmetric about the peak frequency, and secondly, it shifts the peak frequency away from the resonant frequency ω_0. This shift is associated with the real part of the self-energy correction (for both electrons and holes) due to impurity interaction which renormalizes the effective bandgap. Ordinarily, one would expect a shift to lower frequencies, i.e. a red-shift. However, we find from Equation (8) and (9) that depending on $\theta(t)$, or the precise details of the impurity configuration, the shift can be either a red-shift or a blue shift! That means that in the phase-coherent regime, quantum interference effects influence even the bandgap renormalization! This is truly a surprising result and is verifiable experimentally. A change in the sign of the shift is a remarkable effect of quantum interference and an intriguing case of microscopic features affecting macroscopic observables in a non-trivial way.

Finally, the only issue that remains to be discussed is the temperature at which such an effect could be observed. Inelastic scattering times of $\sim$ 10 ps have been measured at 4.2 K in GaAs samples [3] with a carrier concentration exceeding 7 x 10^{11} cm^{-2}, whereas a radiative recombination lifetime of $\sim$ 20 ps has been calculated for quantum dots [4]. Since the only requirement to observe the above effect is to ensure that the inelastic scattering time exceeds the radiative recombination lifetime, it is conceivable that this effect can be observed at temperatures not too far below liquid helium temperature. This makes it practical to verify this effect in semiconductor quantum dots.

REFERENCES

1. Bandyopadhyay, S. (1988). Phys. Rev. B15 38, 7466.
2. Al'tshuler, B. L. (1985). Pis'ma Zh. Eksp. Teor. Fiz. 41, 530 [JETP Lett. 41, 363].
3. Lin, B. J. F., Paalanen, M. A., Gossard, A. C., and Tsui, D.C. (1984). Phys. Rev. B. 29, 927.
4. Takagahara, T., (1987). Phys. Rev. B15. 36, 9293.

FLUCTUATIONS IN SUB-MICRON SEMICONDUCTING DEVICES CAUSED BY THE RANDOM POSITIONS OF DOPANTS

John A. Nixon
John H. Davies
John R. Barker

Department of Electronics and Electrical Engineering
Glasgow University, Glasgow, G12 8QQ, U.K.

1. INTRODUCTION

Improvements in fabrication techniques are continually driving down the size of semiconducting devices. Effects that were 'averaged away' in large structures become observable, and even dominant, in small structures. In this paper we consider fluctuations in the classical electrostatic field that arise from the random positions of the impurities in a doped semiconductor. The traditional approximation of treating a doped region as a uniform charge density is inadequate in a sub-micron structure. We show that the characteristics of low-dimensional devices near threshold can be dominated by the fluctuations from the discrete, random nature of the impurities. The ideas apply to any sub-micron structure, but low-dimensional devices are particularly susceptible because their electronic screening is weak, and the high mobility of modulation-doped layers renders them susceptible to disorder that would otherwise be masked. The ideas are illustrated by specific calculations for two devices. The first is a short gate MODFET, a device which has been studied in recent experiments. The same ideas are then extended in the second model to a self-consistent calculation for a heterojunction including the electronic potential.

2. SHORT-GATE MODFET

A MODFET with a narrow channel and a short gate is shown in figure 1a. The threshold characteristic, conductance of the channel as a function of bias on the gate, was found to be broad, 'noisy', and different for nominally identical devices[1]. We shall show that potential fluctuations can explain the main results of the experiments.

The active electrons in the device are in a quasi-two dimensional electron gas (2DEG) in the x-y plane which is about 1 μm wide along y. The source-drain bias (along x) is kept small to ensure that the linear conductance of the channel is measured, and the electrons are degenerate at the temperatures used (a few K). The length of

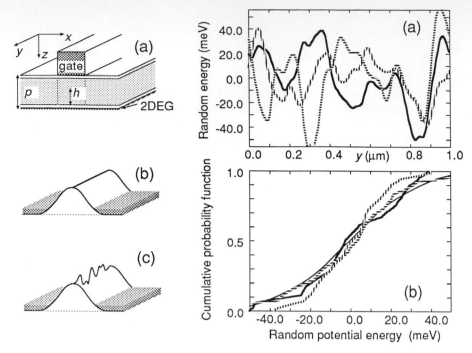

Figure 1. (a) Schematic diagram of the short gate MODFET. (b) Idealized potential barrier under the gate. (c) Fluctuations in the height of the barrier caused by the randomly positioned donors.

Figure 2. (a) Random potential energy as a function of y for 3 different configurations of donors. (b) Cumulative probability density, proportional to conductance G for the same configurations. The smooth curve would result if self- averaging occured.

the gate in x is much smaller than the mean free path of electrons in a uniform 2DEG. Thus the potential under the gate acts like a simple barrier *perpendicular* to the direction of flow, such that a typical electron will scatter from it only once as it travels from source to drain. This is the extreme short-gate limit of a FET.

A negative bias on the gate creates a potential barrier to the electrons in the wire. It is usually assumed that the channel turns on uniformly across its width (in y) as the gate voltage becomes less negative and the barrier is lowered. This is the case of the idealised barrier shown in figure 1b. However, superimposed upon the potential due to the gate is the potential from the donors which are randomly distributed throughout the doped layer of the structure, and the potential from these depends on y. Thus the peak of the potential barrier will look something like figure 1c. As the barrier is lowered, some parts of the channel turn on before others and the threshold is smeared. The detailed characteristics depend on the distribution of impurities.

We have made a simple model to estimate the size of the fluctuations in potential under the gate. It is applicable near threshold, when there are few electrons in the part of the channel underneath the gate. The model has the following features.

i) We only consider donors under the gate, assuming that the effect of donors away from the gate is screened by the high density of electrons under these regions.

ii) Surface states on GaAs, and the metal gate, pin the chemical potential at the surface to an energy in the middle of the gap. Thus an image charge of each donor is created.

iii) The potential is calculated only under the mid-point of the gate along x, at the summit of the barrier produced by the gate. It is therefore only a function of y.

We expect the form of the fluctuations that we calculate to be accurate, but the scale will be less precise because of the approximate treatment of the screening given in (i). The surface that generates the image charges is a distance p above the 2DEG and the random charges are a distance h above the 2DEG. A donor i at y_i, together with its image charge, then contributes

$$\varphi(y; y_i) = \frac{e}{4\pi\varepsilon\varepsilon_0} \left[\frac{1}{\sqrt{(y-y_i)^2 + h^2}} - \frac{1}{\sqrt{(y-y_i)^2 + (2p-h)^2}} \right] \quad (1)$$

to the potential at y. Summing over all donors gives the total potential $\phi(y)$.

The probability density function for ϕ in a system with infinite length along y can be calculated analytically. For the experimental devices[1], $n_1 \approx 2\times10^9$ m^{-1}, $h \approx 30$ nm and $p \approx 60$ nm which gives a standard deviation of 27 mV.

Figure 2a shows numerical calculations of the potential energy $-e\phi(y)$ for three configurations of a finite system whose parameters were chosen to match those of the experimental device. Equation (1) was summed over 2000 charges randomly positioned under the gate, and the mean potential energy was subtracted out. The fluctuations are long ranged, and the standard deviations fell around the analytic result.

The conductance G was calculated from the random potential energy by assuming that electrons can pass under the gate at a given point if the potential energy there is less than the chemical potential μ. Thus G is proportional to the fraction of the width of the gate where the chemical potential is greater than the potential energy plus the energy barrier due to the gate. This in turn is proportional to the cumulative probability function for the random potential energy, shown in figure 2b for the systems simulated numerically. The channel turns on as the chemical potential is raised through about 0.1 eV, and the conductance has the general shape seen in the first experimental sample[1]. The overall smooth curve is interrupted by jumps in the slope that occur when the chemical potential reaches one of the valleys in the random energy, or submerges one of the peaks. The fluctuations in potential energy shown in figure 2a typically occur over about 0.1 μm in y, which is twice the length of the gate in x. This lends support to the 'short-gate' approximation made earlier. To take account of fringing of the field under the gate, the scale of energies in figure 2b needs to be multiplied by about 4.4 to be compared with the gate voltage in the experiments. This makes the theoretical curve rather too broad.

These calculations show that fluctuations in the electrostatic potential caused by the random positions of the donors have a major effect on the characteristics of a MODFET with a narrow channel and short gate. Our simple model reproduces the main features seen in experiments[1]. The fine detail arises from other causes, probably from quantum-mechanical interference between different paths under the gate which could be verified by measuring the magnetoresistance. A fuller description of this model is given in Davies and Nixon[2].

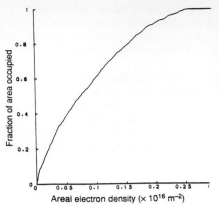

Figure 3. Fraction of area of plane of 2DEG submerged under electrons as a function of areal electron density.

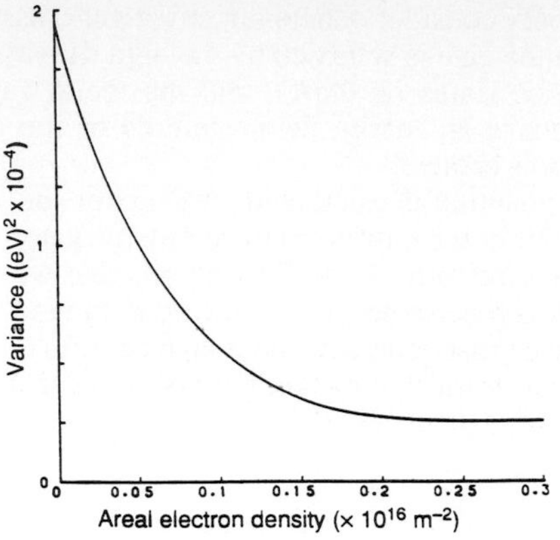

Figure 4. Variance of the random potential energy for the same example as a function of areal electron density.

3. SELF-CONSISTENT CALCULATION OF AlGaAs/GaAs HETEROJUNCTION

The ideas on potential fluctuations can be extended to other devices based on AlGaAs/GaAs heterostuctures. In patterned devices, where electrons in the 2DEG are further confined, screening will be greatly reduced. It is then necessary to consider *all* donors in the doped regions, not just those under a gate, and the random potential will be enhanced. We have developed a semiclassical model to treat such structures, with the electron density given by a (local) Thomas-Fermi approximation. The model can be used to study   gated devices and quantum wires.

Consider the case of a uniform gate covering the heterojunction surface. The above method for summing the potential due to individual donors as in equation (1) is extended for this case where the sum is over a 3-dimensional distribution. The potential is calculated on a finite rectangular lattice in the plane of the 2DEG. It will vary as a function of x and y. The potential energy in the heterojunction well can be calculated by including the surface Schottky barrier height and the conduction band discontinuity between AlGaAs and GaAs.

The potential profile can be shifted up or down relative to the chemical potential by altering the gate voltage. When parts of the potential energy are lowered through the chemical potential electrons appear in the 2DEG. These electrons are included classically . The presence of electrons changes the energy profile — this is calculated and the procedure is iterated until a self-consistent solution is obtained. This method of including electrons implicitly includes their screening effect on the random potential and specifically takes account of the highly inhomogeneous nature of the 2DEG at low electron densities.

By varying the gate voltage the variation of the random potential and electron density can be studied as the number of electrons increases. The first electrons which appear in the 2DEG will lie in the deep troughs of the random potential energy and all states will be localised. Band tailing due to this is clearly seen. As the potential energy is lowered relative to the chemical potential the area covered by electrons increases. This inhomogeneous nature of the 2DEG is shown in figure 3 which shows the fraction of area covered by electrons against areal density. As the gate voltage increases more electrons appear in the 2DEG and the peaks of the potential energy gradually become submerged, and eventually extended states exist.

The screening effect of the electrons is illustrated in figure 4 which shows how the variance of the potential energy is reduced as more electrons enter the 2DEG. The potential energy is smoothed out. However there remains a residual constant fluctuation which is independent of electron density at high electron densities but does depend on the original unscreened potential. This implies there will always be some localised states. They are unlikely to affect transport at high electron densities. Rorison *et al*[3] have considered fluctuations from random donors in the limit of a uniform 2DEG.

A simple percolative model was used to study transport, assuming that only regions with electrons present conducted and that these had uniform conductivity. A small bias was applied across two sides of a rectangle and the relative conductances as a function of gate voltage was calculated. The results again show a broadened noisy threshold which differs for different donor distributions. Further work is under way to study devices with finite gates.

4. CONCLUSIONS

Our calculations have shown that fluctuations in the electrostatic potential caused by the random positions of donors have a major effect on the characteristics of a MODFET with a narrow channel and short gate. The same ideas applied to an AlGaAs/GaAs heterojunction show important effects although specific calculations for patterned devices have yet to be performed. Our simple model in section 2 reproduces the main features in an experiment although there is fine detail which arises from other causes, probably quantum-mechanical interference between different electronic paths, as in 'universal' conductance fluctuations.

The analysis emphasises that devices can now be built so small that long-standing approximations such as treating a doped region as a uniform charge density are no longer accurate, and the effects of fluctuations caused by the random, discrete nature of dopants are directly measurable. The implications for practical devices are serious, because the trend is towards lower operating voltages where the broader threshold caused by fluctuations under shorter gates cannot be tolerated.

REFERENCES

1. S. Washburn, A.B. Fowler, H. Schmid and D. Kern, *Phys. Rev. B* **38** 1554 (1988).
2. J.H. Davies and J.A. Nixon, *Phys. Rev. B* **39** 3423 (1989)
3. J.M. Rorison, M.J. Kane, M.S. Skolnick, L.L. Taylor and S.J. Bass, *Semicond. Sci. Technol.* **3** 12 (1988).

LATERALLY CONFINED RESONANT TUNNELING DIODE WITH ADJUSTABLE QUANTUM-DOT CROSS-SECTION*

W. B. Kinard[1], M. H. Weichold[1],
G. F. Spencer[2], and W. P. Kirk[1,2]

[1] Department of Electrical Engineering,
and [2] Department of Physics,
Texas A&M University
College Station, Texas

A three terminal device has been fabricated in double barrier $GaAs/Al_{0.3}Ga_{0.7}As$ heterostructure material. Self-aligned photolithographic techniques were used to define a gate potential region around the periphery of a resonant tunneling diode. When sufficient gate voltage was applied, the resonant tunneling carriers became depleted in the vertical well. Consequently, the cross-section of the tunneling current could be laterally constricted, demonstrating that three-dimensional confinement of carriers with a gate potential is feasible. The structure of this device is unique in that it allows both measurements and direct comparisons of in situ adjustment of the tunneling cross-section of a two-dimensional resonant tunneling diode.

I. INTRODUCTION

With advances in molecular beam epitaxy (MBE) over the last decade, it is now possible to grow atomically thin layers on a semiconductor substrate. These layers can be grown thinner than the characteristic wavelength of a conduction electron, thus allowing the fabrication of finite quantum wells with discrete energy levels and the investigation of unique tunneling processes.

Exploitations of these wells and the tunneling processes produce very interesting I-V characteristics in devices defined with double barrier het-

* This work was supported in part by the Texas Advanced Technology Program under grant No.3611 and by the National Science Foundation under grant DMR 8800359

erostructures. One result of the tunneling phenomenon is the appearance of negative resistance at a certain external bias. As early as 1974, Chang, Esaki and Tsu [1] observed resonant tunneling through a double barrier heterostructure. Because of the negative differential resistance properties of these structures, and because these structures are capable of operation in the terahertz range, they have many applications,[2] among which are high frequency oscillators, mixers, and detectors, as well as, very high density memory elements.

In addition to device applications, one now has the capability of effectively confining charged carriers to two dimensions, one dimension, or even zero dimension. Until recently, investigations simply addressed two-dimensional structures in which the momentum of the charged carrier was quantized normal to the heterostructure interface. In the past, fabrication limitations have restricted researchers to these two-dimensional electron gas structures. With recent advances in lithography and other microfabrication techniques, researchers are now able to further confine electrons to one dimension, thus creating quantum wires. These structures allow investigation of very interesting phenomena such as electron-electron interactions and universal conductance fluctuations.[3] Reed *et al.* have hypothesized that "confinement to zero dimensions will yield equally intriguing phenomena".[4]

A gated GaAs/AlGaAs resonant tunneling diode has been fabricated and characterized. It allows *in situ* adjustment of the tunneling cross-section. The results reported here indicate that the carrier confinement is due to constriction of the current channel via application of a gate potential. Additionally, this work points the way to more intriguing transport phenomena in which the gate potential confines the tunneling carriers to zero dimension, producing quantum dots. Experimental results as well as projected phenomena are reported here.

II. DEVICE DESCRIPTION

A cutaway view of the gated resonant tunneling diode (RTD) is shown in Fig. 1. A standard pre-metallization cleaning was used to remove organic and metallic contaminants from the GaAs MBE material before depositing AuGe/Ni on the wafer's backside. An image reversal photolithography liftoff step was used to define AuGe/Ni device arrays on the wafer's topside. Arrays of circles, with diameters 2 μm, 4 μm, and 6 μm were fabricated to increase the magnitude of the current while retaining small geometries. The AuGe/Ni layers were annealed to produce ohmic contact arrays of circles on the MBE material and an ohmic backside.

The arrays of circular ohmic contacts on the wafer's topside served a three-fold purpose. First the ohmic metal layer on top of the devices was used to efficiently inject carriers into the device for tunneling investigations. Second, the ohmic metal circles served as an etch mask for an anisotropic, reactive-ion etch (RIE) step as indicated in Fig. 1. The RIE etch step was followed by an isotropic liquid etch in a citric acid

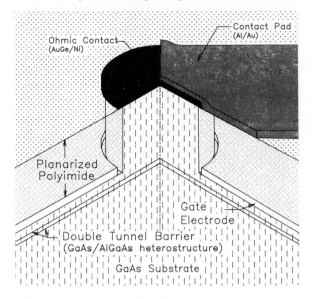

Figure 1: Cutaway view of gated resonant tunneling diode.

solution. The isotropic etch was highly selective with respect to the ohmic metal circles and produced circular arrays of mesas in which the diameters of the ohmic metal circles on the top of the mesas were greater than the diameters of the GaAs posts underneath. This enabled the ohmic metal circles to fulfill their third role as a self-aligned shadow mask to non-conformal deposition of chrome on the wafer's surface. The chrome deposited on the ohmic metal did not affect the ohmic contact, but the chrome deposited on the GaAs at the base of the mesas created a rectifying contact around the periphery of the mesas (Fig. 1). Note that because of the unique undercut feature created by the dual etch step, the metal at the base of the device mesas was not in electrical contact with the ohmic metal on the mesa tops. This allowed efficient injection of carriers from the mesa tops into the device for tunneling yet provided a potential region around the periphery of the device to regulate the tunneling current. Separation of the two metal contact layers was achieved via a polyimide planarization step. The device operates similar to a vertical GaAs FET structure with the addition of resonant enhancement of current due to tunneling through the discrete energy level(s) in GaAs/AlGaAs vertical potential well.

III. EXPERIMENTAL RESULTS

Figure 2 shows the room temperature I-V characteristics versus gate potential. Note that an increase in gate potential decreased the current through the device without an appreciable shift in the resonant

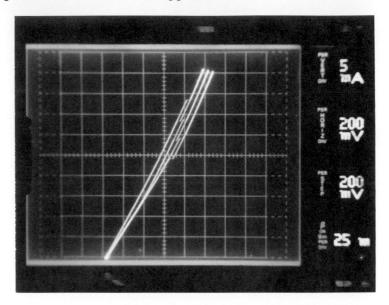

Figure 2: Room temperature I-V characteristics versus gate voltage of the gated RTD

tunneling bias. This demonstrates that the tunneling cross-section of the resonant tunneling diode can be adjusted via application of a gate potential. From these curves, the transconductance was computed to be 20 mA/V which indicated unwanted leakage in the gate electrode. However, work is underway on a new design to minimize this problem.

Assuming homogeneous doping in the well, axial symmetry in the device with respect to depletion width, and no potential drop between the gate and the well, Poisson's equation was solved to yield the lateral confining potential in the well. The results are plotted in Fig. 3. In this plot the confining potentials for the three gate voltages used in the I-V traces of Fig. 2 are shown. The consequent decrease in the tunneling cross-section versus gate potential is shown on the abscissa. These results indicate that the continuous constriction of the RTD conducting channel will lead to zero-dimensional confinement of electrons.

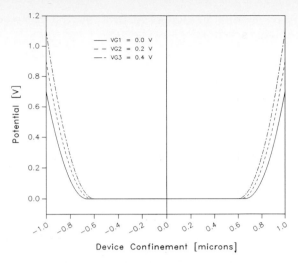

Figure 3: Lateral confining potential in the gated RTD well for three gate voltages.

IV. CONCLUSION

A gated GaAs heterostructure resonant tunneling device has been fabricated and used to demonstrate constriction of the tunneling current through the vertical structure. Whereas transconductance values indicated problems with gate leakage currents, work is underway to improve the gating function. These results show that a continuous *in situ* transition from a two-dimensional structure to a zero-dimensional (or quantum-dot) structure can now be realized.

ACKNOWLEDGEMENT

The authors wish to thank J. Gardner, B. Gibler, R. Atkins, and V. Swenson for their technical support and would especially like to thank D. Yost for his assistance with the reactive-ion etching process. We also thank Mark Reed and John Randall at Texas Instruments for providing helpful discussions and the MBE material.

REFERENCES

1. L. L. Chang, L. Esaki, and R. Tsu, *Appl. Phys. Lett.* **24**, 593 (1974).
2. S. Luryi, *Appl. Phys. Lett.* **47**, 490 (1985).
3. W. J. Skocpol, P. M. Mankiewich, R. E. Howard, L. D. Jackel, D. M. Tennant, and A. D. Stone, *Phys. Rev. Lett.* **56**, 2865 (1986).
4. M. A. Reed, J. N. Randall, R. J. Aggarwal, R. J. Matyi, T. M. Moore, and W. E. Wetsel, *Phys. Rev. Lett.* **60**, 535 (1988).

TIME-DEPENDENT MODELING OF RESONANT TUNNELING STRUCTURES USING THE 3-DIMENSIONAL SCHRODINGER EQUATION: INVESTIGATION OF THE INTRINSIC TIME CHARACTERISTICS OF A ZERO-DIMENSIONAL SEMICONDUCTOR NANOSTRUCTURE.

G. Neofotistos and K. Diff

Physics Department and Center for Advanced Computational Science,
Temple University, Philadelphia, PA 19122

and

J. D. Gunton

Physics Department, Lehigh University, Bethlehem, PA 18105.

Time-dependent modeling of resonant tunneling has so far been performed in devices with confining potentials in only one direction. However, recently fabricated nanostructures exhibit quantum confinement in all three dimensions and require three-dimensional methods to adequately model these new devices. Here we present, for the first time, such a simulation method based on the numerical solution of the three-dimensional time-dependent Schrödinger equation. Computationally effective Fourier transform methods have been employed for the integration of the Schrödinger equation. Results of simulations performed on one-dimensional double barrier structures (in order to test the Fourier transform methods with respect to standard numerical schemes), and three-dimensional double barrier structures and quantum dots (in order to obtain the intrinsic time characteristics of these devices), are presented.

I. INTRODUCTION

Recent advances in microfabrication technology have made possible devices that confine carriers quantum-mechanically in all three dimensions [1]. Computationally effective three-dimensional quantum mechanical simulations are already considered to be necessary to model the effects of lateral confinement and to describe carrier transport in these devices [2].

In this paper, we describe such a simulation, which is based on the numerical solution of the time-dependent Schrödinger equation. Two methods have been employed for the time-evolution of the Schrödinger equation: a simple Fourier transform method [3], and a split-operator Fourier transform method [4], both utilizing the fast Fourier transform algorithm. These methods offer the advantage of using larger time steps and mesh spacings than conventional numerical methods and are, therefore, more effective to deal with computationally intensive three-dimensional problems. In order to test the accuracy of the Fourier transform (FT) methods we have performed simulations on one-dimensional double barrier structures. We have used both FT methods and a standard Crank-Nicholson (CN) scheme. The results have been found to be in excellent agreement and the FT methods faster than CN by a factor of 30. We have then proceeded to investigate the time dependent characteristics of a three-dimensional double barrier structure without lateral potentials, and a nanostructure with a harmonic oscillator type laterally confining potential in the quantum well.

We should note that the main aim of this work at this stage is to demonstrate the feasibility of the numerical solution of the three-dimensional time-dependent Schrödinger equation for quantum transport in electronic devices rather than to attempt a detailed and complete simulation of an experimentally realized particular device.

135

II. NUMERICAL METHODS USED

We solve the time-dependent Schrödinger equation in the effective mass approximation

$$i \, \hbar \, \frac{\partial \psi(x,y,z,t)}{\partial t} = H \, \psi(x,y,z,t) \tag{1}$$

where H is the Hamiltonian

$$H = - \frac{\hbar^2}{2m_i^*} \left(\frac{\partial^2}{\partial x^2} + \frac{\partial^2}{\partial y^2} + \frac{\partial^2}{\partial z^2} \right) + V(x,y,z) = T + V \tag{2}$$

and all symbols retain their standard meaning. The solution of Eq. (1) is

$$\psi(x,y,z,t) = U(t,t_o) \, \psi(x,y,z,t_o) \tag{3}$$

where $U(t,t_o) = e^{\frac{-iH\delta_t}{\hbar}}$ ($\delta_t = t - t_o$) is the evolution operator. Since the Hamiltonian consists of a differential operator (kinetic energy operator T) as well as of a scalar (potential energy operator V) it is in general impossible to compute the action of the evolution operator *as it is* on an arbitrary wave function and we must, therefore, find approximate schemes of expressing $U(t,t_o)$. Following earlier works on decomposition formulas of general exponential operators [5], we approximate the evolution operator by either the simple decomposition scheme

$$e^{\frac{-iH\delta_t}{\hbar}} \approx e^{\frac{-iT\delta_t}{\hbar}} \, e^{\frac{-iV\delta_t}{\hbar}} \tag{4}$$

(introducing an error of order δ_t^2 [3]), or by the symmetrically split-operator scheme [7]

$$e^{\frac{-iH\delta_t}{\hbar}} \approx e^{\frac{-iT\delta_t}{2\hbar}} \, e^{\frac{-iV\delta_t}{\hbar}} \, e^{\frac{-iT\delta_t}{2\hbar}} \tag{5}$$

(introducing an error of order δ_t^3 [6]). In both schemes the kinetic energy operator T is discretized by using 3-point differencing and Fourier transformed to momentum space. At each time step the wavefunction is also Fourier transformed to momentum space where the kinetic energy operation takes place and then transformed back to real space where the potential energy operation takes place. The norm and energy of the wavefunction are preserved as well as the time-reversal symmetry of the Schrödinger equation.

III. RESULTS

The temporal behavior of the resonant tunneling process is studied by following the time evolution of a Gaussian wavepacket tunneling through the structure (the use of an incoming wavepacket instead of a plane wave is considered [7] to be consistent with the finite spatial dimensions of the device). We study GaAs/AlGaAs double barrier structures under no external bias, focusing, quantitatively, on the electron's probability amplitude inside the quantum well $P(t,E_o) = \int_{q.well} |\psi(x,y,z,t)|^2 \, dxdydz$, where E_o is the initial average energy of the wavepacket in the direction normal to the heterojunction interfaces. $P(t,E_o)$ contains the essential information about the intrinsic temporal behavior of the resonant tunneling process [8]. As can be seen in Fig. 1, after an initial "build-up" time during which the electron density in the well builds up via multiple reflections [9], $P(t,E_o)$ reaches its maximum and then decays (following the familiar exponential decay law [10]). We have considered parabolic conduction bands, Γ-Γ tunneling channel, no effective mass variations due to different heterojunction materials, no scattering, and no space-charge effects. In order to choose the mesh spacings and the time step for the FT methods, we have to consider minimization of the discretization error, accomodation of all k-components of the wavepacket by the mesh chosen, accomodation of the entire energy spectrum of the problem at hand, sampling requirements, and stability and convergence criteria. We have found that a time step of 0.3 fs and a mesh spacing of 5 A in the direction normal to the heterojunction interfaces, satisfy the above conditions.

The computational efficiency and accuracy of the FT methods are tested by performing resonant tunneling simulations of one-dimensional double barrier structures. The simple FT method gives results that are almost identical with those obtained by the split-operator FT method (their difference is of the order of 10^{-4}) and in excellent agreement with those obtained by using a standard CN scheme to integrate the Schrödinger equation [8]. The FT methods are faster than CN by a factor of 30. We proceed to study resonant tunneling in a three-dimensional double barrier without lateral potentials and a model quantum dot consisting of a double barrier with a parabolic potential (due to Fermi-level pinning [1]) in the lateral dimensions centered on the post axis. A Gaussian wavepacket

$$\psi(x,y,z,t=0) = \frac{1}{(\sigma_x\sigma_y\sigma_z)^{1/2}} \frac{1}{(2\pi)^{3/4}} e^{ik_oy} e^{\frac{-(x-x_o)^2}{4\sigma_x^2}} e^{\frac{-(y-y_o)^2}{4\sigma_y^2}} e^{\frac{-(z-z_o)^2}{4\sigma_z^2}} \qquad (6)$$

is injected with initial average momentum in the direction normal to the heterojunction interfaces. Details of the simulations are given in Table I. We have considered on-resonance tunneling with $E_o = E_r$, and off-resonance tunneling with $E_o = E_r + \hbar\omega$, where $\hbar\omega$ is the quantum dot subband splitting. As it can be seen in Figs. 1,2, the temporal characteristics of the quantum dot are the same as those of the simple double barrier structure (of course the peak of the probability amplitude inside the quantum well of the dot will be smaller due to sidewall depletion caused by the lateral parabolic potential). It must be mentioned that these results are representative of a particular choice of the initial wavefunction (no transverse momentum) and a complete time dependent study must consider, among others, the effects of the transverse momentum on the tunneling probability.

IV. SUMMARY

We have presented a quantum mechanical simulation method capable of modeling carrier transport in three dimensional structures. This method is based on the numerical solution of the three dimensional time-dependent Schrödinger equation. We have also presented initial results on the intrinsic time characteristics of a model quantum dot. A complete time dependent study of resonant tunneling in a quantum dot is currently under way.

This work was supported by the U.S. Office of Naval Research under grant N00014-83-K-0382 and by an allocation of computer time by the Pittsburgh Supercomputing Center.

REFERENCES

[1] M. A. Reed, J. N. Randall, R. J. Aggarwal, R. J. Matyi, T. M. Moore, and A. E. Wetsel: Phys. Rev. Lett. 60, 535, 1988.
[2] G. W. Bryant: Phys. Rev. B 39, 3145, 1989.
[3] C. J. Sweeney and P. L. DeVries: Comp. in Phys. Jan/Feb 1989.
[4] M. D. Feit, J. A. Fleck, and A. Steiger: J. Comp. Phys. 47, 412, 1982.
[5] M. Suzuki: J. Math. Phys. 26, 601, 1984.
[6] H. De Raedt: Comp. Phys. Rep. 7, 1, 1987.
[7] A. P. Jauho: Proceedings of Advanced Summer School on Microelectronics, Kivenlahti, Espoo, Finland, June 1987, Eds. T. Stubb and R. Paanaren.
[8] H. Guo, K. Diff, G. Neofotistos, and J. D. Gunton: Appl. Phys. Lett. 53, 131, 1988.
[9] R. Ricco and M. Ya. Azbel: Phys. Rev. B 29, 1970, 1984.
[10] A. Peres: Annals of Physics 129, 33, 1980.

TABLE I

NUMERICAL AND DEVICE PARAMETERS FOR A 3-DIMENSIONAL

TIME-DEPENDENT SIMULATION OF AN AlGaAs-GaAs-AlGaAs QUANTUM DOT

Device dimensions:		
1) along the lateral directions	224	nm
2) along the longitudinal (-y) direction	2048	nm
Barrier widths	4	nm
Well width	5	nm
Barrier material	$Al_{0.25}Ga_{0.75}As$	
Well material	GaAs	
Barrier potential height	0.187	eV
Resonant energy level	0.0725	eV
Fermi-level pinning	0.7	eV
Subband energy splitting	11	meV
Energy spread of the incoming wavepacket	6	meV
Mesh size	$8\times4096\times8$	
Mesh spacings:		
1) lateral dimensions	28	nm
2) longitudinal direction	0.5	nm
Time step	0.3×10^{-15}	s

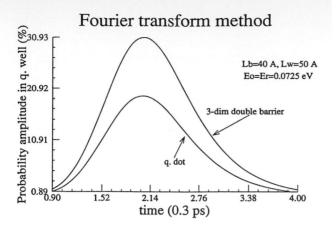

Fig. 1: Probability of finding an electron inside the quantum wells of a three dimensional double barrier structure and a quantum dot when the initial average wavepacket energy E_o is equal to the resonant energy level E_r (on-resonance tunneling). The time needed for the probability amplitude to reach its peak ("build-up" time) is the same in both structures as well as the decay constant ($\tau \approx 155$fs) of the exponential decays.

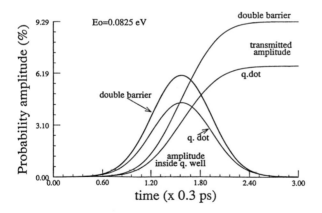

Fig. 2: Probability of finding an electron inside the quantum wells of a three dimensional double barrier structure and a quantum dot when the initial average wavepacket energy E_o is equal to $E_r + \hbar\omega$ where $\hbar\omega$ is the quantum dot subband splitting.

139

INTERLEVEL PLASMONS IN
QUASI-ONE-DIMENSIONAL STRUCTURES

Sergio E. Ulloa[1]

Department of Physics and Astronomy and
Condensed Matter and Surface Sciences Program
Ohio University
Athens, Ohio 45701, USA

Y. C. Lee and B. S. Mendoza

Physics Department, SUNY at Buffalo
Amherst, New York 14260, USA

ABSTRACT

We present calculations of the collective electronic modes of a
quasi-one-dimensional wire structure. Intersubband-like plasma
modes are shown to exist in addition to the well known plasmon. These *interlevel* modes are associated with groups of electrons populating the different transverse-motion levels introduced
by the extreme quantum-confinement conditions and are thus
unique to these systems. Dispersion relations are approximately
linear. Oscillator strengths typically peak for wavenumbers $q \approx$
10^5 cm^{-1}, readily accessible via optical gratings.

Plasma oscillations are a prevalent and well known phenomenon in electronic systems, providing one of the most visible manifestations of particle-
particle interactions. As such, the plasma modes have been studied both
theoretically and experimentally in a variety of metals and semiconductor
systems in three- and two-dimensional structures [1,2], in layered charged
systems [3,4], in the surfaces of metals and overlayers [5], and even in one-
dimensional systems [6,7]. The plasma modes typically studied in systems of

---

[1]Supported in part by U.S. Department of Energy, Grant No. DE–FG02–87ER45334

141

reduced dimensionality are the analogue of the optical branch of plasma oscillations present in a three-dimensional case, in the sense that the whole electronic system oscillates in phase. On the other hand, the extreme quantization of the level structure introduced by the nanoscale confinement of presently available one-dimensional systems opens up the possibility of plasma modes associated with the different transverse-motion levels. In that case, one could picture the resulting modes as electrons in one level oscillating out of phase with those in others, a completely different situation to the typical 'bulk' modes. Here we present calculations of the modes in a quasi-one-dimensional system with only a few occupied transverse-motion-levels [7]. We show that these novel plasma modes are well defined over a wide range of wavenumbers, their dispersion curves are approximately linear, and have a non-negligible spectral weight, making them likely to be detected in experiments which couple to density oscillations.

The normal modes are calculated using the longitudinal dielectric response function derived via the self-consistent field approximation [7,8]. This gives a dielectric function $\epsilon(q, \omega) = \epsilon_1 + i\epsilon_2$, which at zero temperature, and neglecting retardation effects, can be written as [9]

$$\epsilon_1 = 1 + \frac{v(qa)}{\pi q a_o^*} \sum_{\mathcal{E}_n \leq \mathcal{E}_F} \log \left| \frac{\omega^2 - q^2 u_{n+}^2}{\omega^2 - q^2 u_{n-}^2} \right|, \tag{1}$$

and

$$\epsilon_2 = \frac{v(qa)}{\pi q a_o^*} A_\omega. \tag{2}$$

Here, q is a wavevector along the channel, $a_o^* = \hbar^2 \kappa / m e^2$ is the effective Bohr radius, κ is the background dielectric constant, $\mathcal{E}_n$ denotes the transverse-motion energy levels, and $u_{n\pm} = u_n \pm \hbar q / 2m$, with $u_n = [2(\mathcal{E}_F - \mathcal{E}_n)/m]^{1/2}$. $\mathcal{E}_F$ is the Fermi energy, and $A_\omega = 1$ if $q u_{n+} > \omega > q|u_{n-}|$, $A_\omega = 0$ otherwise. Also, $v(x) = K_o(x^2/8\pi) \exp(x^2/8\pi)$ is the effective Coulomb potential in the quasi-one-dimensional structure, and K_o is the modified Bessel function. This form of the potential results from assuming a gaussian envelope function for the electronic distribution across the channel [10]. It can be shown [9] that this expression of $v(qa)$ reduces to the expected limiting two-dimensional (one-dimensional) form of $2\pi/qa$ $(-\log qa)$ for $qa \gg 1$ $(qa \ll 1)$ [2,7].

The poles of ϵ_1 are remnants of the single-particle excitations at $\omega = q|u_{n\pm}|$. The collective excitations, on the other hand, are given by the zeros of the dielectric function [1]. From the form of ϵ_1, it is easy to show that there is a collection of zeros, each of them lying between consecutive poles of ϵ_1, *in addition* to the regular mode appearing at large frequencies. As a consequence, *the total number of collective modes is equal to the number of populated transverse-motion levels in the system.* Furthermore, the interlevel modes will be free of Landau damping as long as the poles of ϵ_1 are well separated, which is the case for wavenumbers smaller than a critical value, $q < q_c = m(u_n - u_{n+1})/\hbar$. This provides an argument as to why the analogue interlevel plasmons in two-dimensional systems do not occur, since there the number of levels is too large for the modes to be free of damping [11]. Similarly, although metallic nanoscale wires are attainable [12], the typically large electronic densities prevent these modes from being well resolved. Consequently, the ability to control the overall charge density and channel width while maintaining very high mobilities in narrow heterojunction layers [6,13], makes these quasi-one-dimensional structures the prime experimental system for the observation of the interlevel plasma modes.

Dispersion relations for the collective modes are obtained numerically. The zeros of the dielectric function are calculated for a given set of structural parameters such as the effective mass, dielectric constant, channel width and electron density (or number of occupied transverse-motion levels), and for different wavenumbers. Also of importance are the so-called oscillator strengths for the different modes, F_j, which are a measure of the spectral weight and therefore an estimate of the observable intensities in a typical experiment. The oscillator strengths can be calculated from the residues of the inverse dielectric function, or equivalently from [14],

$$F_j(q) = \left(\omega^2 \frac{\partial \epsilon_1(q,\omega)}{\partial \omega^2} \right)^{-1}_{\omega^2 = \omega_j^2(q)}, \tag{3}$$

where $\omega_j(q)$ are the different mode frequencies.

Figure 1 shows dispersion curves for a narrow heterojunction channel with parameters appropriate to InSb ($m = 0.014, \kappa = 14.7$), width $a = 1000$ Å and surface charge density $n_s = 6 \times 10^{11}$ cm^{-2}. In this case there are five transverse-motion levels filled (assuming square-well level spacings [10]) and correspondingly, four interlevel plasma modes in addition to the usual plasmon at higher frequencies. The linear dispersion is evident for the interlevel plasmons, with phase velocities only a fraction of the Fermi velocity. The corresponding oscillator strengths are also shown. Similar calculations for a 1000 Å GaAs channel ($m = 0.067$), for example, give four interlevel modes as well, however the energy scale is reduced by about a factor of four, the value of the mass ratio. This is understandable, considering that the smaller mass makes the quantum confinement effects more apparent, shifting upwards all the mode frequencies as well as the F_j functions [9]. Notice also in Fig. 1 that the interlevel modes with higher frequencies (and lower strengths) cease to exist for smaller q values since they have a lower q_c. However, the low energy modes are well defined and strong for easily accessible wavenumbers. Indeed, for a wide range of charge densities and other parameters, the oscillator strengths in the low frequency modes peak for $q \approx 1$–3×10^5 cm^{-1}. We have also studied the effect of variations in the background dielectric constant which could arise from varying screening effects from the metal gates in

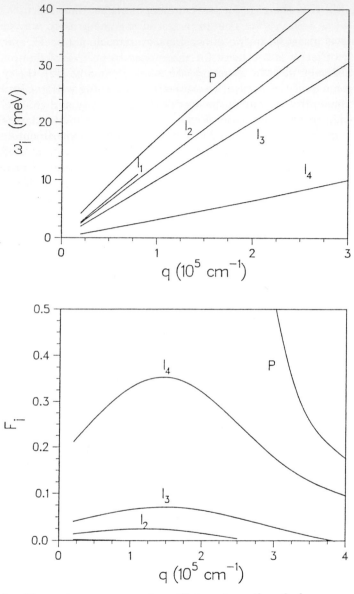

Figure 1: *Dispersion curves and oscillator strengths of plasma modes for a 100 nm InSb channel, $n_s = 6 \times 10^{11}$ cm^{-2}. P denotes the usual 'bulk' mode and I_n the interlevel plasmons. Notice I_1 and I_2 are damped beyond $q_c \approx 0.8$ and 2.5 (10^5 cm^{-1}), respectively.*

the system [6]. The 'bulk' plasmon frequency is shifted linearly with changing κ, while the interlevel modes are not, constrained as they are to have a phase velocity always in between two adjacent values of u_n. The oscillator strengths are affected only slightly. On the other hand, varying the channel width produces significant changes. For example, reducing the channel width decreases the number of occupied transverse levels, producing in turn fewer modes although stronger and easier to detect.

In conclusion, interlevel plasma modes introduced by the extreme quantization of the single-particle levels exist in quasi-one-dimensional systems. Characteristic energies and the large values of the oscillator strengths for typically accessible grating parameters suggest a strong coupling to infrared radiation probes. Experiments to explore the characteristics of these unique modes would be of great interest. The study of a possible novel mechanism for superconductivity in slender structures, where the electrons would interact via the exchange of interlevel plasmons, will be presented elsewhere.

REFERENCES

[1] P. M. Platzmann and P. A. Wolff, Solid State Phys. **13** (1973).
[2] T. Ando, A. B. Fowler and F. Stern, Rev. Mod. Phys. **54**, 437 (1982).
[3] J. J. Ritsko and M. J. Mele, Phys. Rev. Lett. **42**, 666 (1979).
[4] G. Fasol, N. Mestres, H. P. Hughes, A. Fischer, and K. Ploog, Phys. Rev. Lett. **56**, 2517 (1986); X. Zhu, X. Xia, J. J. Quinn, and P. Hawrylak, Phys. Rev. B **38**, 5617 (1988), and references therein.
[5] W. L. Schaich and K. Kempa, Physica Scripta **35**, 204 (1987).
[6] W. Hansen, M. Horst, J. P. Kotthaus, U. Merkt, and Ch. Sikorski, Phys. Rev. Lett. **58**, 2586 (1987); F. Brinkop, W. Hansen, J. P. Kotthaus, and K. Ploog, Phys. Rev. B **37**, 6547 (1988).
[7] W. I. Friesen and B. Bergersen, J. Phys. C **13**, 6627 (1980); Y. C. Lee, S. E. Ulloa, and P. S. Lee J. Phys. C **16**, L995 (1983); S. Das Sarma and W. Lai, Phys. Rev. B **32**, 1401 (1985); W. M. Que and G. Kirczenow, Phys. Rev. B **37**, 7153 (1988).
[8] H. Ehrenreich and M. H. Cohen Phys. Rev. **115**, 786 (1959).
[9] Y. C. Lee, B. S. Mendoza, and S. E. Ulloa, unpublished.
[10] S. E. Laux and F. Stern, Appl. Phys. Lett. **49**, 91 (1986).
[11] Modes in coupled two-dimensional systems do exist, however. See S. Das Sarma and A. Madhukar, Phys. Rev. B **23**, 805 (1981), as well as reference [3].
[12] J. T. Masden and N. Giordano, Phys. Rev. Lett. **49**, 819 (1982).
[13] T. P. Smith, III, H. Arnot, J. M. Arnot, C. M. Knoedler, S. E. Laux, and H. Schmid, Phys. Rev. Lett. **59**, 2802 (1987); M. L. Roukes, A. Scherer, S. J. Allen, Jr., H. G. Craighead, R. M. Ruthen, E. D. Beebe, and J. P. Harbison, Phys. Rev. Lett. **59**, 3011 (1987); G. Timp, H. U. Baranger, P. de Vegvar, J. E. Cunningham, R. E. Howard, R. Behringer, and P. M. Mankiewich, Phys. Rev. Lett. **60**, 2081 (1988).
[14] D. A. Kirzhnits, E. G. Maksimov, and D. I. Khomskii, J. Low Temp. Phys. **10**, 79 (1973). The longitudinal f-sum rule gives $\sum_j F_j \omega_j^2 = \omega_P^2$.

Chapter 4

Quantum Devices and Transistors

THEORY OF THE QUANTUM BALLISTIC
TRANSPORT IN CONSTRICTIONS
AND QUANTUM RESONANCE DEVICES

Dick van der Marel

Faculty of Applied Physics

Delft University of Technology

Lorentzweg 1

2628 CJ Delft, the Netherlands

I. INTRODUCTION

Recent experimental developments on quantized conductance in 2D point contacts (1) have prompted a flurry of theoretical activity (2) in the field of quantum ballistic transport. Although the theoretical approaches differ in various respects, they all have in common that they are based on, or can be traced back to, the Landauer formula

$$G = e^2/h \sum_{ij} T_{ij} \qquad (1)$$

which applies to an idealized two-terminal geometry. A detailed discussion about the consequences of applying this formula to non-ideal geometries, e.g. having a finite width of the wide parts of the sample, was given by Landauer (3). Also in the present paper I will stick to idealized geometries, where the microstructure is flanked by infinite 2D halfplanes. The electrical contacts to these planes are assumed to lie far from the constriction(s) compared to the inelastic mean free path.

The formalism underlying the numerical results that I will discuss below is given in Ref. 4. In the present paper I will only briefly point out the basic physics. The linear conductance through a constricted region in an otherwise unperturbed 2DEG can be calculated from the eigenfunctions in the following way: Due to the presence of a constriction the eigenfunctions having their principle weight in the left half plane (left-side lobes) contribute an infinitesimal leakage current through the constriction. Similarly the right-side lobes contribute a particle current pointing to the left. In the absence of an external voltage these currents cancel. Due to a small external voltage between left and right the left-side lobes will acquire a surplus occupation given by the density of states multiplied by the change in potential. The resulting current is now given by this extra density multiplied with the flux of each eigenfunction (i.e. the rate at which a unit area passes from left to right) integrated over all left-side lobes. For the left-side and right-side lobes the angle of incidence is still a good quantum number, which is due to the infinite extent of the continuum solutions compared to the finite scattering cross section of the constriction. Therefore the summation over eigenstates in terms of an angular integral can be expressed as follows:

$$G = e^2 \int_{-\pi/2}^{\pi/2} \Phi(E_F, \alpha) \frac{\partial^2 n}{\partial E \partial \alpha} \, d\alpha \tag{2}$$

A set of tight binding wave functions is used to represent the electronic states, which is convenient for computational reasons and has the possibility of working in the limit of free electron bands ($\lambda_F \gg a$, a is the lattice parameter) and nonparabolic bands ($\lambda_F \approx a$). The Hilbert space is a 2D square lattice with nearest neighbor hopping parameter t, which corresponds to $\hbar^2 / (2 \, m^* \, a^2)$ where m^* is the effective mass near the center of the Brillouin zone. For such a tight binding lattice the flux can be expressed in the value of the eigenfunctions at two adjacent rows of lattice points, which can be chosen at x=0 and x=a respectively (The x and y directions are chosen perpendicular and parrallel to the barrier respectively) :

$$\Phi(\psi) = \frac{4\pi t a^2}{h} \, \text{Im} \left(\sum_n \langle \psi | 0, n \rangle \langle 1, n | \psi \rangle \right) \tag{3}$$

In Ref. 4 it was shown, that in a geometry with a set of apertures in an infinetely high barrier at x = 0 the conductance is exactly given by:

$$G = \frac{2e^2}{h} \, \text{Tr} \{ (\text{Im}\Gamma^*) \, (\text{Im}\Gamma^{-1}) \} \tag{4}$$

where

$$\Gamma_{n,n'} \equiv \pi^{-1} \int_0^\pi \cos([n-n']\phi) \, \sqrt{(E/2t - 2 + \cos(\phi))^2 - 1} \, d\phi$$

In the more general case, where a microstructure (a tube with or without delta function impurities, a box, a horn etc.) is added to the aperture(s), the expression is slightly more complicated. A detailed description is given in Ref. 4. Part of the results discussed below were calculated with the latter expression.

II. CONDUCTANCE QUANTIZATION IN POINTCONTACTS

In Fig. 1 the result is given for a few configurations where the conductance as a function of constriction width is calculated. The following observations can be made from this figure:

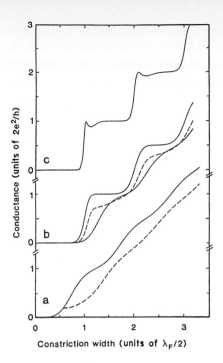

Fig.1. Conductance versus constriction width. (a)L=0. Solid/dashed curve: constriction without/with impurity. (b)L=0.48λ_F. Upper solid curve: channel without impurity. Lower solid curve: impurity inside the constriction (x=0.32λ_F, y=0.32λ_F). Dashed curve: impurity outside the constriction (x=0.57λ_F, y=0.32λ_F). (c)L=0.99λ_F.

In the first place there are oscillations with inflection points at precise integer multiples of $2e^2/h$ even for a constriction of zero length. On increasing the length of the constriction the plateaus start to flatten, until at a critical length L_c the plateaus become horizontal. On further increasing the length oscillations below the plateaus set in due to interference between waves reflected from the back and front end of the narrow region. The critical length where the n'th plateau becomes horizontal obeys the following scaling behaviour :

$$L_c = 0.45 \sqrt{W\lambda_F} = 0.32 \lambda_F \sqrt{n} \qquad (5)$$

The square root dependance of L_c on W implies that in order to have quantization the narrow region can be much shorter than wide. Note that this is somewhat counter intuitive: In order to have 1D subbands one tends to assume that a channel must be much longer than wide. This turns out to be an unnecessary condition. More important is, that the channel length is of the order of (or smaller than) the decay length $1/\kappa$ of the electrons tunnelling below the lowest unoccupied 1D subband in the narrow region, i.e. $L_c^{-2} \sim \kappa^2 \sim 2m\,\hbar^2\,(E_{n+1} - E_n)$. In a square well confinement potential the latter energy difference is proportional to the channel index: $\kappa^2 \sim n\,W^{-2}$. Using the fact, that the channel index n is approximately given by $2W/\lambda_F$ we can now qualitatively understand, that L_c is proportional to the square root of the width. Also shown in Fig.1 is the effect of impurities, which clearly is to destroy quantization.

III. QUANTUM RESONANCE DEVICES

I now turn to the problem of a small box connected to 2D half planes on both sides through point contacts. Technically such a device can be realized by means of a set of gates that defines the shape, and a second gate covering the box and its surroundings. The latter gate can be used to define the position of the Fermi level. In Fig. 2 the result is given for a round box. The peaks coincide with the energy positions of the localized levels inside the box and are caused by resonant tunneling. In the situation depicted in Fig. 2 the apertures on both sides act as mode selectors which allow only for partial transmission of the lowest subband in the narrow region. This effectively makes the problem 1D, even though the continuum states to which the localized states in the box couple are in the 2D half planes. The only relevant quantum number is now parity with respect to the y mirror plane. The states can be characterized by 'atomic' quantum numbers: 1s, $1p_x$, $1d_{x^2-y^2}$, 2s etc. . For a box of 200 nm diameter and an effective mass of $0.07m_e$ these levels are at 3.6, 9.3, 16.7 and 18.5 K respectively. The $1p_x$ and $1d_{x^2-y^2}$ levels are degenerate with the $1p_y$ and $1d_{xy}$ levels respectively, but the latter do not couple to the modes selected by the point contacts, as these levels have odd symmetry around the x-axis, whereas the selected modes are even. With regards to mirror symmetry around the y axis we observe, that 1s, $1d_{x^2-y^2}$ and 2s are even, whereas $1p_x$ is odd.

As parity is conserved by the coupling of the localized states to the continuum (The slight asymmetry in the geometry of Fig. 2 has neglegible physical influence) one can write $|k_p> = (|-k_R> \pm |k_L>)/ \sqrt{2}$, where p indicates the parity quantum number and the +/- sign refers to even and odd parity. In(out) coming waves have posive (negative) values for k. Due to unitarity the outcoming waves acquire a parity- and energy-dependent phase shift $\eta_p(E)$. A wave entering from the left is given by a linear combination of even and odd-parity waves, so that the scattered wave is: $| k_{scatt} > = 2^{-1/2} (e^{2i\eta_e} | k_e > + e^{2i\eta_o} | k_o >)$. The transmission is now:

$$T = | <- k_R | k_{scatt} >|^2 = \sin^2 (\eta_e - \eta_o) \tag{6}$$

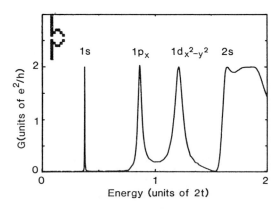

Fig.2. Conductance versus Fermi energy for the geometry indicated in the top left corner. In the calculation the barrier extends from y = -∞ to ∞. Indicated above each resonance are the orbital quantum numbers of the corresponding virtual bound states.

In the more general case of an asymmetrical box other linear combinations of L and R waves have to be taken, which results in a reduced transmitted amplitude compared to Eq. (6). The energy dependance of the phase shifts follows directly from the Friedel Sum Rule (5):

$$d\eta_p / dE = \pi \Delta\rho_p \tag{7}$$

where $\Delta\rho_p$ is the impurity induced change in the density of states. Generally speaking this implies, that the phase shift increases with π each time the Fermi level crosses a quasi localized level, giving rise to a resonance peak with a maximum transmission of exactly one. If an even and an odd level are energetically close to each other, complete or partial cancellation can occur. On the other hand, if there is a succession of two or more states of the same parity, there must be a point of exactly zero transmission between two subsequent peaks. Especially if the energy positions are close to each other this gives rise to a very strong energy dependency of the conductance. In its simplest form the change in density of states connected with each localized state has a Lorentzian line shape:

$$\Delta\rho_p = \pi \frac{d}{dE} Arg(E - E_{ip} - i \Gamma_{ip}) \tag{8}$$

This immediately leads to the following form of the phase shifts:

$$\eta_p = \sum_i cotg^{-1}(\Gamma_{ip} / (E - E_{ip})) \tag{9}$$

A plot of η_e, η_o, η_e-η_o and T is displayed in Fig. 3 , taking 3, 14, and 18 K for the energy position of the even peaks and 0.2, 1.5 and 2.0 K for the corresponding Γ. The odd peak was positioned at 7K with Γ=1.0 K. Clearly most of the physics of Fig.2 is contained in the simple scattering phase shift considerations explained above.

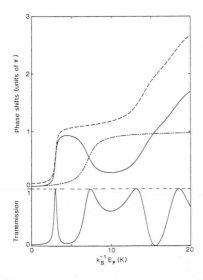

Fig. 3 . The phase shifts and the transmission versus Fermi energy of a QRD containing 3 even levels and 1 odd level. Dashed, chained and solid curves are η_e, η_o and η_e- η_o .

IV. ELECTRON CORRELATION IN QRD'S

So far I have neglected electron correlation effects. Although this is reasonable in the 2D wide parts of the geometry, as interactions are strongly screened there, this is no longer correct in a small box, or even in a 1D constriction. Especially in a small box the situation is reminiscent to atomic physics. Let us first take a look at the relevant energy scales. We take the energy difference between the two lowest levels (1s and 1p) as representative of the level spacings without electron-electron interactions:

$$\Delta E = E_p - E_s = \frac{\hbar^2}{2\,m^*}\,\frac{0.71}{\pi\,R^2} \tag{10}$$

the width of the s level scales as

$$\Gamma_s \approx E_s\,\frac{W}{R}\,\exp(-2\,\pi\,\frac{L}{W}\,\sqrt{1-\frac{2\,W^2}{\pi R^2}}\,) \tag{11}$$

i.e., keeping the shape and the relative sizes of the QRD fixed, the linewidths and the energies scale with R in the same way. The Coulomb integral of two electrons in a 2D circular box scales as:

$$U = f\,\frac{e^2}{\varepsilon R} \tag{12}$$

where ε is the dielectric constant and f is a dimensionless factor of the order one which depends on the details of the wavefunctions involved. If two electrons occupy states with different orbital quantum numbers, we also have to take into account the exchange integral J, favouring parallel spin alignment. The exchange integral follows the scaling behaviour of the unscreened Coulomb interaction U given in Eq. (12). Both an increase of the number of occupied levels and a reduction of ΔE leads to a reduced value of U due to screening. From the analogy with transition metal impurities I expect a much weaker effect of screening on the exchange interaction J (6). The scaling behaviour of the bare (unscreened) parameters is displayed in Fig. 4 in a temperature versus length "phase diagram". We recognize from this plot, that for boxes larger than 110 (nm) $\Delta E < U$, whereas for smaller boxes $\Gamma_s \ll U < \Delta E$. *The latter condition leads to the formation of a magnetic state localized inside the box on partial filling of the s level.* The Hamiltonian describing the interacting system is the well known Anderson impurity Hamiltonian:

$$H = \sum_k \varepsilon_k\,\psi_{ke}^\dagger\,\psi_{ke} + \sum_\sigma n_{\sigma s} + U\,n_{\downarrow s}n_{\uparrow s} + \sum_k (V_k\,\psi_{ke}^\dagger\,\psi_s + V_k^*\,\psi_s^\dagger\,\psi_{ke}) \tag{13}$$

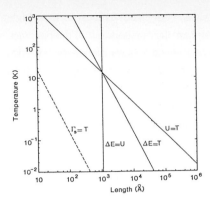

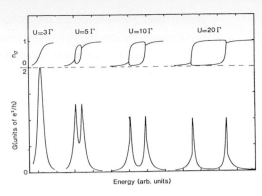

Fig. 4. Phase diagram of temperature versus diameter of a circular QRD (using typical GaAs parameters for the effective mass and the dielectric constant), indicating the various regions in parameter space .

Fig. 5. Plot of the conductance (lower panel) and the occupation number for both spin directions in the localized state (upper panel) versus Fermi energy for various values of U/Γ.

The easiest approximation is to solve the Schrodinger equation using the self consistent Hartree-Fock approximation (7). As a result the s-virtual bound state will divide up in two spin split peaks as soon as the density of states at the Fermi level exceeds a certain critical value. This can only occur if $U > \pi\Gamma$. In other words, if the latter so called Anderson criterion is fulfilled, the Hartree-Fock approximation gives a sharp transition between the magnetic and the non magnetic regime as a function of s-level occupation. It is important to point out that this sharp transition is an artifact of the Hartree-Fock approximation. More advanced ground state calculations lead to a gradual crossover between the two regimes (8). I have performed the selfconsistent Hartree-Fock calculation as a function of Fermi energy, thus extracting a conductance per spin channel from the s-level occupation numbers (i.e. by employing the Friedel sum rule). The resulting conductances and occupation numbers are displayed in Fig. 5. The leftmost example is at the border of the above mentioned Anderson criterion and there is only a single peak with its maximum at $2e^2/h$. For larger U the Anderson criterion is satisfied and there are now two peaks, with their maxima between e^2/h and $2e^2/h$.

The rightmost example is representative of the box of about 200 nm diameter discussed in section 3. If in this example the Fermi level is inbetween the two peaks, the box contains approximately a single spin, and we have to face the fact, that our previous assumption of independent channels for the two different spin quantum numbers no longer holds. The reason is as follows: A current flowing through the box has to be envisaged as a sequence of virtual hops of electrons and/or holes into and out of the box. An applied voltage favours hopping in one direction. An electron hopping into the box must have its spin antiparallel to the spin that is already there (A more correct way of stating this is: both electrons must form a singlet state). In the second step of the virtual process, where an electron hops out of the box, there are now two energetically equivalent possiblities: Either an up-spin remains in the box and a downspin hops out , or vice versa. In other words: Each electron contributing to the current has a 50% chance of undergoing a spin flip, which is the strongest form of spin flip dephasing. In fact this behaviour will lead to the formation of a Kondo state at low temperatures in which the electron in the box builds up a collective singlet state with electron-hole pairs in both halfplanes (8). This leads to the characteristic logarithmic behaviour of the transport properties with the Kondo scaling temperature: $k_B T_K = E_F \exp (- (|E_s| |E_s+U|) / (U \Gamma_s))$, where E_s and E_s+U are the energies of the occupied and the unoccupied spin levels relative to the Fermi energy. With the gated QRD structure one has the unique possibility of tuning these energies and hence of having an experimental handle on T_K. In the case where the Fermi level is precisely midway E_s and E_s+U, one can easily estimate that the Kondo temperature is far below experimentally accessible temperatures. On moving the Fermi level closer to one of the peaks T_K rises sharply and the anomalous temperature dependency should enter the observable temperature range. Therefore I believe, that Quantum Resonance Devices are a unique tool for studying collective behaviour induced by spin flip scattering.

I gratefully acknowledge the Max-Planck-Institut fuer Festkoerperforschung in Stuttgart for hospitality and financial support during preparation of part of this work. I have benefitted from stimulating discussions with O. Gunnarsson and J. Zaanen on the subject of magnetic impurity scattering.

REFERENCES

1. B.J. van Wees, H. van Houten, C.W.J. Beenakker, J.G. Williamson, L.P. Kouwenhoven, D. van der Marel and C.T. Foxon, Phys. Rev. Lett.60, 848 (1988); D.A. Wharam, T.J. Thornton, R. Newbury, M. Pepper, H. Ajmed, J.E.F. Frost, D.G. Hasko, D.C. Peacock, D.A. Ritchie and G.A.C. Jones, J. Phys. C21, L209 (1988).

2. Y.Avishai and Y.B.Band,preprint ; L.I.Glazman, G.B. Lesovick, D.E. Kmelnitskii and R.I. Shekter, P'isma Zh. Eksp.Teor.Fiz. 48, 218(1988); E.G. Haanappel and D. van der Marel, Phys. Rev. B, Rapid Commun. 39,xxxx(1989); Y.Isawa, J.Phys.Soc.Jpn 57,3457 (1988); R.Johnston and L. Schweizer,J. Phys.C21,L861(1988); A. Kawabata, preprint ;J. Masek and B. Kramer,preprint ; G.Kirczenow, Solid State Commun. 68,715 (1988); A. Szafer and A.D. Stone, Phys. Rev. Lett. 62,300(1989);

3. R.Landauer, Phys. Lett. 85A,91(1981); R. Landauer, "Which version of the formula for conductance as a function of transmission probabilities is correct?", unpublished;

4. D. van der Marel and E.G. Haanappel, Phys Rev. B 39,yyyy(1989);

5. J.Friedel, Nuovo Cimento 7,287(1958);

6. D. van der Marel and G.A.Sawatzky, Phys. Rev. B,37,10674(1988);

7. P.W. Anderson, Phys. Rev. 124, 41(1961);

8. O. Gunnarsson and K.Schonhammer,Phys.Rev.Lett.50,604(1983); P.Schlottmann, Phys.Rev.Lett.50,1697(1983); K.G.Wilson,Rev.Mod.Phys.47,773 (1975);

CRITERIA FOR TRANSISTOR ACTION BASED ON QUANTUM INTERFERENCE PHENOMENA

Fernando Sols [1]

Department of Physics and Coordinated Science Laboratory
University of Illinois at Urbana-Champaign
Urbana, Illinois

M. Macucci, U. Ravaioli, and Karl Hess

Department of Electrical and Computer Engineering
and Coordinated Science Laboratory
University of Illinois at Urbana-Champaign
Urbana, Illinois

1. INTRODUCTION

The study of electronic transport in structures based on very small one-dimensional conductors is currently receiving considerable attention [1]. Nanostructure technology has progressed enormously in recent years, and experiments have shown that it is possible to reduce the size of devices and lower the temperature to the point where the electron wavefunctions maintain phase coherence and quantum interference phenomena are observed. These "mesoscopic" systems [2] cannot be described by standard classical transport theory, where self-averaging over many microscopic configurations

---

[1]Present address: Departamento de Fisica de la Materia Condensada, Universidad Autonoma de Madrid, E-28049 Spain.

is assumed. Due to their similarities with guided electromagnetic propagation, semiconductor quantum wires whose width is comparable to the electron wavelength are usually called "electron waveguides".

There is growing interest in the possibility of realizing electronic devices which are based on quantum interference effects. We have recently proposed a new transistor principle in which the electron transmission in a quantum wire is controlled by a remote gate voltage, which modifies the penetration of the electron wavefunction in a lateral stub, affecting in this way the interference pattern [3]. Devices based on a similar principle have been independently proposed by Fowler [4] and Datta [5]. The schematic structure of the Quantum Modulated Transistor (QMT) that we propose is shown in Fig. 1(a). For convenience, we use the terminology of the Field Effect Transistors (FET's) and refer to the electron reservoirs as source (S) and drain (D), while the electrode on the stub termination is the gate (G). We want to emphasize, however, that the QMT is not based on the field effect principle but purely on quantum interference. The voltage applied to the gate varies the depletion length associated to the metal-semiconductor junction, and this modifies the effective length of the stub. The interference pattern of the electron wavefunctions can be varied considerably and calculations have shown that the electron transmission probability can be varied between 0 and 1 if propagation takes place in the fundamental transverse mode (single-channel regime) [3]. A similar principle is used for frequency measurements in electromagnetic waveguides, where a piston is mechanically moved to vary the resonating frequency of a lateral cavity. The idea of using a lateral stub to influence the electron transmission through a quantum wire was hinted by Gefen et al. [6] and more strongly noted by Landauer [7]. A study of scattering due to a finite lateral chain inserted in an infinite tight-binding chain has been given in [8].

In this paper we present a theoretical investigation of QMT structures. Our main goal is to identify configurations that may lead to a clear experimental observation of the proposed transistor effect, and to formulate design criteria for practical device applications. In particular, we study the effect of the stub length and width on the transmission probability, and we show that in devices with multiple stubs the control over the transmission probability may be improved.

2. THEORETICAL MODEL

In this section we formulate the scattering problem posed by geometrical modifications in quantum wires and describe the calculation method of the

S-matrix elements (transmission and reflection coefficients). The system is idealized by assuming that the geometry is delimited by hard wall potentials and that the potential is otherwise flat. We assume that inelastic scattering is negligible, since this is the fundamental condition for quantum interference to occur, and focus on the one-electron problem of scattering by boundaries. We neglect in this work the effects of electron-electron interaction except for the use of an effective Fermi liquid picture. We study only the electron energy range for single channel propagation, where quantum modulation of the transmission can be more clearly observed [3].

In order to calculate the transmission probability T, we take advantage of the fact that plane wave motion can be emulated by a nearest neighbors tight-binding (TB) Hamiltonian in a lattice with periodicity a, much smaller than the electron wavelength and the device dimensions. The geometry of Fig. 1(a) is filled by a sufficiently dense square lattice, and a hopping energy between nearest neighbors $V = \hbar^2/2Ma^2$ is assumed, so that in the continuum limit we describe the motion within the wire boundaries of an otherwise free particle with positive effective mass M. To calculate the transmission coefficient we use a Green's function method. Formally, the problem is similar to that of a TB chain where each "site" represents a transverse chain (slice) of the TB stripe, which contains several real lattice sites. This Green's function approach has been a standard procedure in the literature to calculate the transmission through disordered samples and an iterative method [9] has often been used. In our problem, the geometry (in particular, the change of width in a portion of the electron wave guide) rather than the presence of impurities is the cause of scattering. We can then take advantage of the relatively simple geometrical structure of Fig. 1(a) and decompose the lattice into left and right semi-infinite stripes and a central finite rectangular lattice formed by the sidearm plus the adjacent portion of the main wire. The eigenmodes and propagators within each of these three isolated regions can be calculated exactly and this fact simplifies the calculations considerably.

In our generalized recursive Green's function method, the exact knowledge of the propagation within a finite stripe is exploited and a portion of wire of uniform width is integrated at every step. This technique can be reduced to the standard recursive Green's function method [9] by simply replacing the finite perfect stripe by a single slice. The method is easily extended to more complicated geometries including multiple stubs. The relation between the Green's functions and the S-matrix has been given by Fisher and Lee [10] and has been generalized to an arbitrary scattering structure by Stone and Szafer [11]. The Green's function method described has been applied to calculate the transmission probability T for an electron impinging on structures with one or more stubs. In all the calculations we used a nominal width $W = 100\text{Å}$ for the main wire and a typical electron effective mass $M = 0.05 \, m_o$, but we want to stress that the results are general and can be applied to structures with a different width, provided that the energies and all the other lengths in the problem are changed according to the following scaling transformations:

i) $\lambda' = \eta \, \lambda$ and $L' = \eta \, L$, where λ stands for the wavelengths involved in the problem while L stands for the lengths defining the geometrical structure. Note that, for a fixed electron effective mass M, the scaling of the wavelengths requires also the scaling of all the energies in the problem, according to the rule $E' = E/\eta^2$.

ii) $M' = \eta \, M$ and $E' = E/\eta$. This transformation leaves the wavelengths unchanged.

3. NUMERICAL RESULTS AND DEVICE CONSIDERATIONS

We present here numerical results for the transmission probability for an electron impinging on a single stub structure, as a function of the electron energy E and the effective length of the stub L^* for various stub widths L_z. In Fig. 1(c) ($L_z = W = 100\text{Å}$), it can be seen that, for a given energy, the transmission exhibits a periodic pattern as a function of the effective length L^*, the period being one half of the wavelength which the electron assumes along the stub (in the case of $L_z = W$ it coincides with the incident wavelength). One can see that a relatively small change in the effective length L^* can induce dramatic changes in the transmission. This remarkable fact constitutes the basis for our conjecture that transistor action by quantum interference should be possible. The transmission probability $T(E, L^*)$ for $L_z = 120\text{Å} > W$ (wide stub) is shown in Fig. 1(d). At high energies, $T(E, L^*)$ exhibits a quasiperiodic pattern due to the fact that the electron can propagate along the stub with two different wavelengths, corresponding to the two available transverse modes (an increase of the stub width lowers the energy of the transverse modes, making more of them available for propagation at a given energy). For narrow stubs (see Fig. 1(b), where $L_z = 80\text{Å}$), electrons with a relatively low energy cannot penetrate the stub due to the higher zero-point energy marking the threshold for propagation in the stub. The result is that the low energy maximum in $T(E, L^*)$ becomes broad and independent of the effective length L^*, as Fig. 1(b) clearly shows, since low energy electrons penetrate very little in the sidearm, and are thus insensitive to changes in the effective stub length. This geometry is therefore incapable of transistor action at low electron energies. The major conclusion of our investigation about the interplay between the length and the width of the stub is that operation

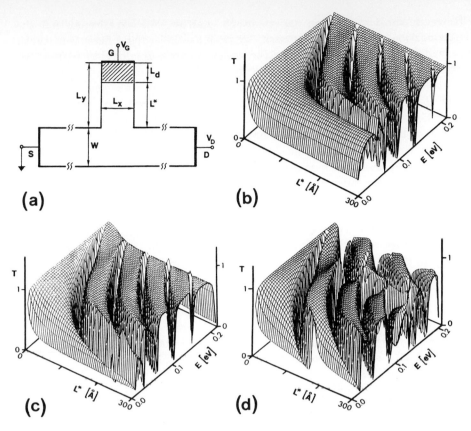

Fig. 1. (a) Schematic view of the Quantum Modulated Transistor structure; (b) Transmission probability $T(E, L^*)$ for a single stub structure with $W = 100$ Å and $L_x = 80$ Å, (c) $L_x = 100$ Å, and (d) $L_x = 120$ Å.

in the single channel regime is desirable and that the width of the sidearm should be about the same as the one of the main wire.

For purposes of modulation by quantum interference, it is desirable to have $T(L^*)$ with broad regions of high and low transmission at the Fermi energy $(E = E_F)$. These regions should be separated by short intervals in which $T(L^*)$ varies rapidly. In addition, this structure should be as insensitive to energy variations as possible. We focus here on the first requirement. Some important features of $T(E, L^*)$ can be better appreciated in a contour plot form. In Fig. 2(a) we show $T(E, L^*)$ for a stub of width $L_x = W = 100$Å. One can quickly recognize that the most promising range of energies lies roughly in the middle of the single channel regime.

However, the transmission valleys are too narrow. We have found that the introduction of a slight mismatch, by making the stubs different in length or width, gives a broadening of the valley, in total analogy with standard microstrip filter design. We consider in the following only stubs with same width and different length. We show in Figs. 2(b) the contour plot of $T(E,L^*)$ for a structure with two stubs of width $L_z = 100\text{Å}$, separated by $S = 95\text{Å}$. The stub length difference is $\Delta = 10\text{Å}$ (L^* is the effective length of the shortest stub). The broadening of the transmission valleys can be seen clearly. We have finally studied the case of three stubs in series separated by 95Å, with relative length difference of 10Å between consecutive ones. The results shown in Fig. 2(c) display a triple splitting of the narrow valleys in the high energy region, as expected. The transmission valleys in the middle energy region acquire a considerable width, which is comparable to the width of the high transmission plateaus. This makes the structure very attractive for possible switching applications and adds flexibility to the design of configurations for more complicated logic functions.

Operation in the single channel regime is essential for practical applications. For a wire of width $W = 1000$ Å this implies a carrier density of $n_s \cong 10^{10} \ cm^{-2}$, which is very low, and in addition one would have to operate below liquid helium temperature in order to have a sharp Fermi surface. On the other hand, for a width $W \cong 100\text{Å}$, n_s could be on the order of $10^{12} \ cm^{-2}$ and operation at liquid nitrogen temperature or above would be feasible.

Theoretical estimates of the QMT performance show that the electron transit time through the device may be be well below one picosecond and that cutoff frequencies up to the Terahertz range may be achieved. A detailed analysis will be presented elsewhere [12]. An important area to be investigated involves also the possibility of creating logic networks based on the devices discussed here. The design of connections which do not destroy the interference effects will be of primary importance. Due to the low current levels, which is an intrinsic limitation for this type of devices, it is also important to determine configurations which do not require large currents for logic operation. The best example of low power devices in conventional integrated circuits is probably given by the CMOS family. The basic CMOS inverter is realized by connecting two complementary p- and n-channel MOS transistors, so that one of the devices is turned off for either logic input (0 or 1). No drain current flows, since the devices are connected in series, except for a small charging current during the switching transients. It is conceivable to realize a similar elementary inverter using the proposed QMT. The three stub structure is probably the most attractive one, since, as already discussed, it is possible to obtain plateaus and valleys of transmission with comparable width (Fig. 2(c)). The idea is to use two devices with slightly different stub length, so that, for the same voltage applied to the stub termination, one device is ON and the other one is OFF, and a basic inverter is obtained. This could be envisioned as the basic building block to obtain an entire logic family.

ACKNOWLEDGEMENTS

This work was supported by the Office of Naval Research and the Army Research Office. F. Sols acknowledges the support from the Joint Program between the Fulbright Commission and Spain's Ministerio de Educación y Ciencia and U. Ravaioli the support of the National Science Foundation. Computations were performed on the CRAY X-MP/48 of the National Center for Supercomputing Applications of the University of Illinois.

REFERENCES

1 For a general review see the May 1988 issue of IBM J. Res. Develop.
2 Y. Imry, *Directions in Condensed Matter*, G. Grinstein and E. Mazenko, eds., 101 (World Publishing Co., Singapore, 1986).
3 F. Sols, M. Macucci, U. Ravaioli, and K. Hess, Appl. Phys. Lett. **54**, 350 (1989).
4 A.B. Fowler, Workshop on Quantum Interference, Atlanta (1988).
5 S. Datta, to appear on Superlattices and Microstructures.
6 Y. Gefen, Y. Imry, and M. Ya. Azbel, Phys Rev. Lett. **52**, 129 (1984).
7 R. Landauer, IBM J. Res. Develop. **32**, 306 (1988).
8 F. Guinea and J.A. Vergés, Phys. Rev. B **35**, 979 (1987).
9 D.J. Thouless and S. Kirkpatrick, J. Phys. C **14**, 235 (1981).
10 D.S. Fisher and P.A. Lee, Phys. Rev. B **23**, 6851 (1981).
11 A.D. Stone and A. Szafer, IBM J. Res. Develop. **32**, 384 (1988).
12 F. Sols, M. Macucci, U. Ravaioli, and K. Hess, to be published.

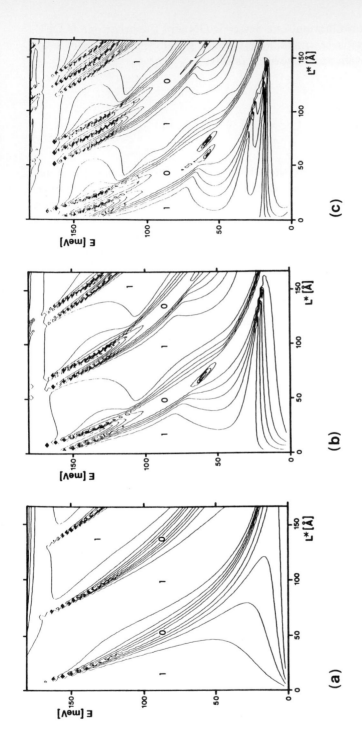

Fig. 2. Contour plots of the transmission probability $T(E,L^*)$ for several stub structures. The contour lines between probability 0 and 1 correspond to the values $T = 0.1;\ 0.3;\ 0.5;\ 0.7;\ 0.9$. In all cases, the width of the main wire is $W = 100$ Å, and the stub width is $L_z = 100$ Å. The structures considered are: (a) single stub; (b) two stubs with separation $S = 95$ Å and length difference $\Delta = 10$Å; (c) three stubs with separation $S = 95$ Å and length difference $\Delta = 10$Å between consecutive ones. L^* refers to the length of the shortest stub.

MODULATION OF THE CONDUCTANCE OF T-SHAPED ELECTRON WAVEGUIDE STRUCTURES WITH A REMOTE GATE [1]

D. C. Miller, R. K. Lake, S. Datta,
M. S. Lundstrom, M. R. Melloch, and R. Reifenberger[*]

School of Electrical Engineering
[*]Department of Physics
Purdue University
West Lafayette, Indiana

I. INTRODUCTION

Phenomena illustrating non-local effects in the "universal conductance fluctuations" of mesoscopic semiconductor structures have been observed in small Hall-bar samples with voltage probes separated by less than l_ϕ[1]. Further evidence for "waveguide-like" behavior has also been observed[2]. Non-local phenomena due to Aharonov-Bohm type effects from a ring at the end of such a structure have also been observed when a magnetic field was applied. [3]

In this paper, we report on our experiments on T-shaped 'electron waveguide' structures, where we have observed conductance oscillations when the electrostatic potential of a remote gate (above a region where no classical current flows) is changed. These are the first reported experiments on such devices, which were proposed earlier[4].

Our device is analogous to the microwave device shown in figure 1. As the distance L is changed by physically moving the plunger in the stub, the net transmission through the structure varies because of interference between the two paths shown. We would expect to see periodic oscillations in the power transmitted, with a period corresponding to moving the plunger by one-half of the wavelength of the guided wave (changing the "extra" path length by one wavelength). If the waveguide was such that only one mode was above cutoff, we could expect the transmission to go to

[1]Supported by the Office of Naval Research (Contract No. N00014-87-K-0693) and the National Science Foundation (Grant No. ECS-83-51-036).

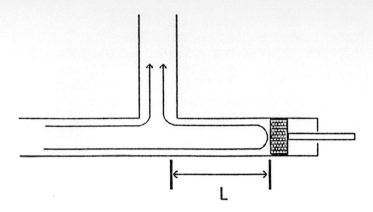

Figure 1:   A microwave waveguide analog of the device investigated. As the plunger is moved, changing the length of the shorted stub, interference between the two indicated paths modulates the amount of power transmitted periodically, with one period corresponding to a change in stub length of one-half wavelength.

zero when the extra path length was an odd number of half-wavelengths, giving 100% modulation. However, if more than one propagating mode was allowed, different modes with different wavelengths would produce a more complex transmission pattern, and we would not expect the transmission to go to zero.

We chose to use the configuration shown in figure 2(a) for our initial experiments rather than the alternative device which has the locations of the stub and drain interchanged, because we felt intuitively that in this device the two paths would be more likely to have equal amplitudes since both must turn a corner, and thus the interference effects might be easier to observe. Experiments on the alternative structure are currently in progress. We note at this point that, similar to the microwave case, large oscillations in the transmission of such a device require the number of propagating modes to be small. Also, interference effects would be reduced by inelastic scattering in the stub which would randomize the phase of the electrons.

The proposed device was simulated, with the assumptions of no inelastic scattering and hard-wall potentials, using the boundary element method to calculate the scattering matrix of a 4-way splitter with 1 port blocked and one connected to a variable length stub[5]. The result of the simulation with a single transverse subband occupied is shown in figure 2(b), and figure 2(c) shows the result with eight occupied transverse subbands. We note that, for a single mode, large (100%) oscillations are expected, and that

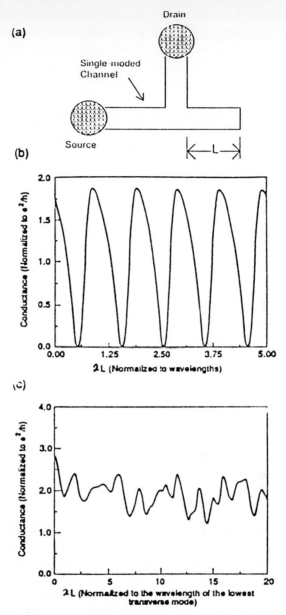

Figure 2: Numerical simulation of the device. (a.) shows the configuration simulated. (b) shows the result with a single occupied transverse electron subband. (c) shows the result with eight subbands.

with many modes, we expect to see smaller amplitude oscillations with multiple frequencies which appear similar to conductance fluctuations.

II. FABRICATION

Our devices were fabricated on a modulation doped $Al_{0.3}Ga_{0.7}As$ /GaAs heterostructure grown by molecular beam epitaxy. The structure consisted of a $1\mu m$ undoped GaAs buffer, a 200Å undoped $Al_{0.3}Ga_{0.7}As$ spacer, a 400Å doped $Al_{0.3}Ga_{0.7}As$ layer (Si doped $1\times10^{18}/cm^3$), and a 60Å thick doped GaAs cap (Si doped $1\times10^{18}/cm^3$). Measurements on standard $100\mu m$ wide Hall bars gave a carrier density of $3\times10^{11}/cm^2$ (confirmed by Shubnikov-deHaas measurements) and a mobility of $650,000cm^2/v-s$ at T=1.3K, corresponding to (two-dimensional) $L_e = 6\mu m$.

Ohmic contacts were made to the two-dimensional electron gas by evaporating Ni/Au/Ge and alloying for 5 minutes at 450°C. Fine mesas were defined using electron beam direct writing and SAL601/ER7 electron resist. The patterned width of the mesas was 0.3 to $0.4\mu m$. The resist was used as an etch mask for a shallow wet chemical etch (1:3:1000 $H_2O_2:NH_4OH:H_2O$- -etch rate $\sim$ 300 Å/minute). The etch was carried out in small steps until an unmasked region of the film stopped conducting. The resulting depth is somewhere in the doped $Al_{0.3}Ga_{0.7}As$ layer; depletion from the surface confines the underlying two-dimensional electron gas to a narrow channel.[6]

Schottky gates were then defined using PMMA and electron beam direct writing. The resulting gates are 200ÅTi/800ÅAu with a minimum linewidth of $0.2\mu m$. A scanning electron microscope image of a finished device (L=$0.6\mu m$) is shown in figure 3. A number of devices with lengths from $0.3\mu m$ to $10\mu m$ were fabricated and tested (about ten devices total).

III. RESULTS

A.C. current (f=17 Hz.) and lock-in detection were used to measure the resistance of the devices, with a typical current of 10 nA r.m.s. Devices were tested while immersed in liquid Helium either at atmospheric pressure (T=4.2K) or at reduced pressure (T $\sim$ 1.3K). Devices were illuminated briefly at T=4.2K to generate carriers in the channel, which then remained conducting in the dark. Measurements made on ring structures fabricated in the same way on a piece of the same substrate analyzed as in [7] gave estimates of the electrical width and L_ϕ of w$\sim$ $0.1\mu m$ and $L_\phi \sim$ $1\mu m$.[8] The presence of a narrow channel was confirmed by magnetoresistance measurements, which showed a characteristic large peak at B=0 and Shubnikov-deHaas oscillations at higher fields. Two ohmic contacts were

168

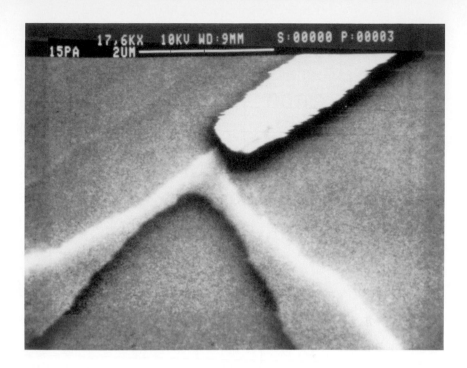

Figure 3: Scanning electron microscope image of one of the devices (L=0.6μm).

made to each end of the devices so contact resistance effects could be eliminated. Negative voltage was applied between the gate and one of the ohmic contacts used for current. Figure 4(a) shows a scan of resistance versus gate voltage for a device with L=0.3μm at T=4.2K. Figure 4(b) shows the oscillatory component after converting to conductance and subtracting a least-square fit polynomial. The r.m.s. magnitude of the oscillations is 0.13e^2/h.

To convince ourselves that these oscillations are due to quantum interference, we note that their amplitude goes down when the temperature is increased or when the measurement current is increased (causing electron heating). Figure 5 shows the effect of decreasing the temperature of various length devices on the oscillation amplitude. We see that decreasing the temperature from 4.2K to 1.3K increases the size of the oscillations, which we would not expect for a classical effect. It is clear from figure 5 that oscillations are present even in the device with a 10μm long stub. This causes us to consider alternative explanations for the origin of the oscillations.

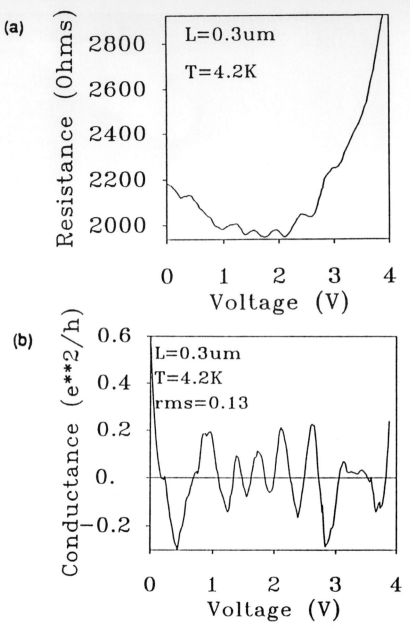

Figure 4: (a) Resistance versus gate voltage for the shortest device (L=0.3μm) at T=4.2K. (b) Oscillatory component obtained by inverting the data of (a) and subtracting a least-square fit polynomial.

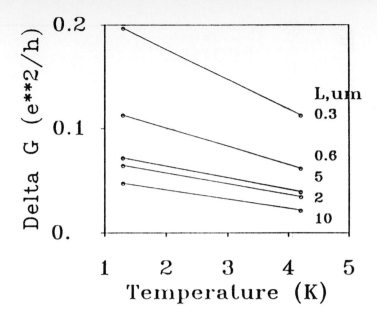

Figure 5: R.M.S. oscillation amplitudes versus temperature for devices with various lengths.

We present two possible explanations for the observed oscillations. First, it is possible that the oscillations are in fact due to interference between the two suggested paths which changes as the gate voltage makes the length of the stub shorter. An alternate explanation is that fringing fields from the gate are changing the potential in the region where classical current is flowing, and we are simply observing universal conductance fluctuations, as in a device with the gate above the conducting channel[9]. Since there is a positive background resistance (the resistance of the device increases with applied voltage) we conclude that the potential in the narrow channel is being changed, and thus we should expect to observe universal conductance fluctuations. We can separate the two effects as follows: If we measure devices with increasing L, we expect that for $2L > L_\phi \sim 1\mu$m, the oscillations due to interference from the changing stub length will disappear, since there is no longer coherence in the stub. Universal conductance fluctuations, on the other hand, will still be present (if fringing fields still reach the classical channel, evidenced by the background increase in resistance.) We would expect that more voltage would be required for a given change in potential in longer devices, since moving the gate further away should decrease the fraction of electric field lines which terminate on the

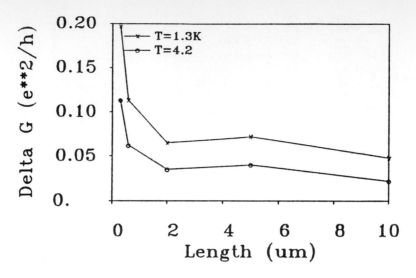

Figure 6: R.M.S oscillation amplitudes versus length between gate and bend at two different temperatures.

classically conducting region of the device, but we would expect the amplitude of oscillations produced by this mechanism to remain approximately constant, since it depends on L_ϕ and the device configuration, neither of which has changed. Figure 6 shows how the rms amplitude of the conductance oscillations varies as the stub length is increased. From this figure, we conclude that universal conductance fluctuations due to changing the potential of the entire device through fringing fields are present in devices of all lengths, but in addition, the shortest devices also show larger oscillations which may be due to interference from the changing stub length.

In order to interpret the period of the oscillations in the short device, we need to find a relationship between the applied gate voltage and the extent of the depletion region. We note that at a certain gate voltage (V_t) the resistance of the device suddenly increases sharply to above the limit of our resistance bridge (200KΩ). We attribute this behavior to the depletion layer extending all the way to the side probe and entirely eliminating the conducting path. We then assume a linear relation between voltage and depletion width, which is consistant with numerical solution of the Poisson equation for a narrow n-GaAs wire[10] and with quantized conductance steps observed in a split-gate structure[11].

Using this relation, we take the Fourier power spectrum of the oscillation data, and convert the x-axis from V^{-1} to L^{-1}. This plot is shown in figure 7 for the oscillation data in figure 4(b). The circles show values

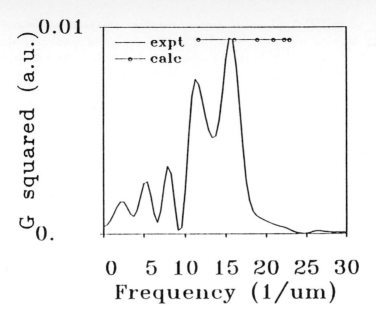

Figure 7: Fourier power spectrum of the data from figure 4(b). Also shown are wavelengths calculated for a hard-wall potential well of width 150 nm with a carrier density of $3\times10^{11}/cm^{2}$.

corresponding to the Fermi wavelengths calculated assuming a hard wall potential well with a carrier density of $3\times10^{11}/cm^{2}$ and a width of 150 nm. In view of the uncertainty in the exact value of V_t to use, and also in L (since the side probe has finite width), the agreement is reasonable.

IV. SUMMARY AND CONCLUSIONS

We have fabricated narrow T-shaped electron waveguide structures on a modulation-doped $Al_{0.3}Ga_{0.7}As/GaAs$ heterostructure, with a Schottky gate over the (classically) unimportant stub. Devices of various lengths between the end of the stub and the classical conduction path from $0.3\mu m$ to $10\mu m$ have been tested. Magnetoresistance measurements and Aharonov-Bohm oscillations observed in rings fabricated using the same substrate and process lead us to believe that the electron channel is in fact very narrow. The resistances of the various devices have been measured as the gate voltage is varied at low temperatures. In addition to universal conductance fluctuations presumably produced by fringing fields from the gate which

change the potential of the entire device, we have observed larger oscillations in the shortest devices (L~ 0.3μm). The period of these oscillations
agrees qualitatively with the expected period for oscillations due to quantum
interference from changing the length of the resonant stub by depleting it
with the gate. Additional work is needed to confirm the origin of the oscillations further. Innovative device configurations may enable us to shield the
device from fringing fields and help separate the two mechanisms.

ACKNOWLEDGEMENTS

Thanks are due to H.R. Frohne and M.J. McLennan for providing
figure 2.

REFERENCES

1. G. Timp, A.M. Chang, P. Mankiewich, R. Behringer, J. E. Cunningham, T. Y. Chang, and R.E. Howard, Phys. Rev. Lett. 59,732 (1987)

2. G. Timp, H.U. Baranger, P. deVegvar, J.E. Cunningham, R.E.
Howard, R. Behringer, and P.M. Mankiewich, Phys. Rev. Lett.
60,2081 (1988)

3. C.P. Umbach, P. Santhanam, C. van Haesendonck, and R.A. Webb,
Appl. Phys. Lett. 50,1289 (1987)

4. F. Sols, M. Macucci, U. Ravaioli, and Karl Hess, Appl. Phys. Lett.
54,350 (1989); S. Datta, SSDM (Tokyo,1988); S. Datta, Int. Conf. on
Microstructures, Microdevices, and Superlattices (Trieste, 1988)

5. H.R. Frohne, PhD Thesis, Purdue University (1988)

6. K. Owusu-Sekyere, A.M. Chang, and T.Y. Chang, Appl. Phys. Lett.
52,1246 (1988)

7. G. Timp, A.M. Chang, P. DeVegvar, R. E. Howard, R Behringer, J.E.
Cunningham, and P. Mankeiwich, Surface Science 196,68 (1988)

8. R.K. Lake, MS Thesis, Purdue University (1988)

9. W.J. Skocpol, P.M. Mankiewich, R.E. Howard, L.D. Jackel, D.M.
Tennant, and A.D. Stone, Phys. Rev. Lett. 56,2865 (1986); S.B.
Kaplan, Surface Science 196, 93 (1988)

10. P.M. Rodhe, A. Rouhani-Kalleh, and T.G. Andersson, Semicon. Sci.
and Tech. 3,823 (1988)

11. B.J. van Wees, H. van Houten, C.W.J. Beenakker, J.G. Williamson,
L.P. Kouwenhoven, D. van der Marel, and C.T. Foxon, Phys. Rev.
Lett. 60, 848 (1988)

ANOMALOUS DRAIN CONDUCTANCE IN QUASI-ONE-DIMENSIONAL AlGaAs/GaAs QUANTUM WIRE TRANSISTORS FABRICATED BY FOCUSED ION BEAM IMPLANTATION[1]

Toshiro Hiramoto[2]
Takahide Odagiri
Kazuhiko Hirakawa

Institute of Industrial Science,
University of Tokyo
Tokyo 106, Japan

Yasuhiro Iye

Institute for Solid State Physics
University of Tokyo
Tokyo 106, Japan

Toshiaki Ikoma

Institute of Industrial Science
University of Tokyo
Tokyo 106, Japan

---

[1]Supported in part by the Grant-in-Aid for Special Project Research "Alloy Semiconductor Physics and Electronics" from the Ministry of Education, Science, and Culture, Japan, by the Joint Research Project of the Institute for Solid State Physics, University of Tokyo, by Casio Science Promotion Foundation, by the Mazda Foundation, and also by the Industry-University Joint Research Program "Mesoscopic Electronics"

[2]Supported in part by the Japan Society for the Promotion of Science.

175

1.   INTRODUCTION

Recently, the quantum interference effect of electron waves in semiconductor quantum wires has attracted much attention from the technological point of view. Several quantum interference devices have been proposed (1-3). Most of these devices make use of the interference of electron waves in AlGaAs/GaAs quantum wires or rings which are controlled by the gate electric field. However, the gate-voltage dependence (i. e., the Fermi-energy dependence) of the electron transport in AlGaAs/GaAs quantum wires has been still unclear.

In this paper, we study the electron transport properties in AlGaAs/GaAs quantum wires fabricated by focused-ion-beam implantation. Drain conductance of the transistors is precisely measured by the ac and dc measurements as functions of magnetic field, gate-voltage, drain-voltage, and temperature. The origins of anomalous behaviors in the drain conductance are discussed.

2.   FABRICATION

Figure 1 shows a schematic cross-sectional view of an AlGaAs/GaAs quantum wire transistor. The starting material was a selectively doped AlGaAs/GaAs heterostructure grown by molecular beam epitaxy (MBE). Al was evaporated onto the channel to form gate electrodes. Then, a focused Si-ion-beam with diameter of less than 0.1 μm was implanted at 200 keV through the Al film into two regions. The ions penetrate the Al film and electrons in the implanted GaAs quantum well were thus depleted by damage. A very narrow wire structure was formed between the two implanted regions (4). The electron density in the wire can be precisely controlled by the gate voltage. The length of the channel is 10 μm.

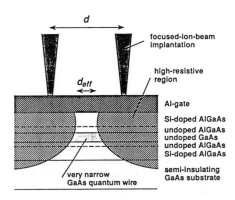

Fig. 1. A schematic cross-sectional view of an AlGaAs/GaAs quantum wire transistor.

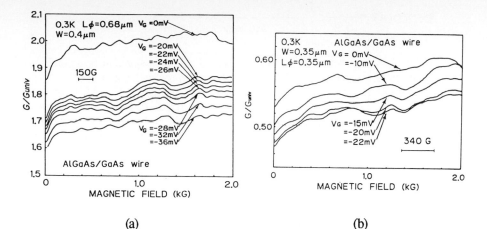

(a)　　　　　　　　　　　　　　　　(b)

Fig. 2. The ac drain conductances of two AlGaAs/GaAs quantum wire transistors at 0.3 K as a function of magnetic field. (a) Transistor #12 with W = 400 nm. (b) Transistor #22 with W = 350 nm. Conductances are normalized to $G_{univ} = e^2/h$, where e is the electron charge and h the Planck constant.

Three quantum wire transistors (#12, #22, and #36) were examined in this study. These transistors exhibited positive magnetoconductance due to the weak localization (5) at low magnetic field at 4.2 K. The phase coherence length L_ϕ and the channel width W at zero-bias ($V_g = 0$ V) were estimated by fitting the modified one-dimensional localization theory (6) to the magnetoconductance data (4). W and L_ϕ at 4.2 K were: $W = 400$ nm, $L_\phi = 680$ nm for the transistor #12; $W = 350$ nm, $L_\phi = 350$ nm for #22; and $W = 150$ nm, $L_\phi = 150$ nm for #36, respectively.

3.　　CONDUCTANCE FLUCTUATIONS

Figures 2 (a) and 2 (b) show magnetic-field dependence of the ac drain conductance of the two AlGaAs/GaAs quantum wire transistors (#12 and #22) with different channel widths measured at 0.3 K. Random oscillation patterns are observed in all the magnetoconductance spectra. This is the well known "universal conductance fluctuations (UCF)" due to the interference of electron waves (7,8). The average period of the random oscillations is apparently shorter in the transistor #12 ($W = 400$ nm) than in #22 ($W = 350$ nm). This is caused by the difference in the correlation field B_c (8) in the two wires.

The conductance fluctuation patterns of the transistors at several gate voltages V_g (i. e., at several Fermi energies E_F) are also shown in Fig. 2. It is very interesting to see how the fluctuation patterns change with V_g. In #12, the

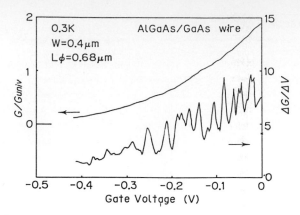

Fig. 3. Drain conductance of the gated AlGaAs/GaAs wire transistor #12 (W = 350 nm) at 0.3 K as a function of gate voltage. The derivative of the conductance is also shown.

pattern completely changes when V_g is changed from 0 to -20 mV. However, the pattern hardly changes even when V_g changes from -20 to -22 mV. When V_g is changed to -24,-26, -28 meV, and so on, the pattern changes gradually, and finally at V_g = -36 mV the pattern is quite different from that at V_g= -20 mV.

This behavior is clear in Fig. 3, which shows the ac drain conductance of the wide transistor #12 as a function of V_g at 0.3 K. The derivative of the conductance is also shown. The conductance also fluctuates randomly with V_g, that is, with the Fermi energy E_F. Since the fluctuation amplitude is almost the same as that with magnetic field, these fluctuations are due to UCF with E_F. The average period of the oscillations is about 20 mV, which corresponds to the change in the Fermi energy ΔE_F of about 1 meV.

In the theory of UCF (8), the correlation energy E_c is given by,

$$E_c = \frac{\pi^2 \hbar D}{(\min (L, L_\phi))^2} .$$ (1)

Within a bandwidth E_c, the electron states are spatially correlated and all the electrons contribute to the same fluctuation pattern (9). However, if ΔE_F exceeds E_c, the electrons are no longer correlated and the pattern changes. E_c corresponds to the typical scale of the spacing between the fluctuations in E_F (8), and we can obtain E_c from the period in Fig. 3; E_c ~ 1 meV. This is the first experimental determination of E_c in GaAs quantum wires. This value is equivalent with a temperature of ~10 K, and is much larger than in metals. This suggests that the reduction of the quantum interference by the energy averaging in semiconductors is much less than in metals, and that the quantum interference can be observed at up to 10 K. Estimated E_c from Eq. (1) is 0.6 meV, in good agreement with the experiment.

Figures 4 (a) and (b) show dc drain conductance G_d and transconductance g_m of the transistor #22 as a function of V_g at 4.2 K at several drain voltages V_{ds}. G_d decreases with decreasing V_g, but the slope of the G_d-V_g curve suddenly changes at V_g just above the threshold voltage V_{th} (i. e., G_d suddenly drops to zero) for only small V_{ds} (< 7 mV). This causes anomalous increase in g_m just above V_{th} for small V_{ds}, as shown in Fig. 4 (b). Moreover, V_{th} shifts from -0.3 to -0.24 V as V_{ds} decreases. This anomalous behavior is not observed in the transistor #12 ($W =$ 400 nm), but only in narrower transistors, #22 and #36 ($W =$ 350 and 150 nm).

This behavior is well explained by the strong localization effect (10). When V_g is sufficiently larger than V_{th}, the electron states are extended across the whole sample. However, when V_g is decreased close to V_{th} and E_F becomes comparable to the fluctuating potential, the electron states are localized and the mobility μ becomes zero, resulting in the conductivity to vanish.

For $V_{ds} =$ 240 mV, when the electrons are heated by the high electric field, G_d increases gradually when V_g exceeds V_{th} (-0.3 V). When V_{ds} is 7.0 or 0.6 mV, G_d does not appear until $V_g = $ -0.24 V. In the range of V_g from -0.3 to -0.24 V, electrons are induced at low V_{ds} but are localized, resulting in zero conductance. When V_g exceeds -0.24 V, G_d increases suddenly because of the delocalization.

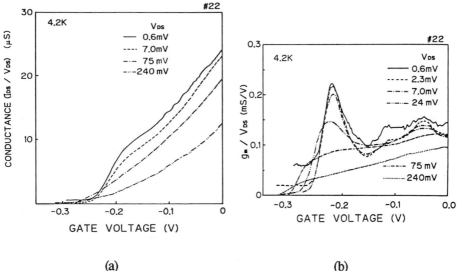

(a)   (b)

Fig. 4.   (a) dc drain conductance and (b) transconductance of the AlGaAs/ GaAs wire transistor #22 (W = 350 nm) at 4.2 K as a function of gate-voltage.

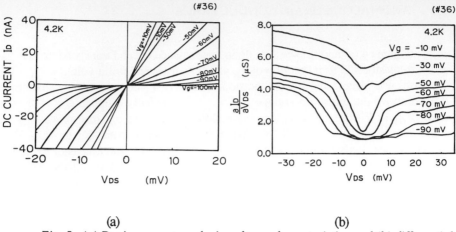

(a)   (b)

Fig. 5. (a) Drain current vs. drain voltage characteristics and (b) differential drain conductance ($\partial I_d / \partial V_{ds}$) of the AlGaAs/GaAs quantum wire transistor #36 as a function of V_{ds} at 4.2 K at several gate voltages.

In order to confirm the effect of the strong localization, dc drain current I_d is measured as a function of V_{ds}. Figure 5 (a) shows I_d-V_{ds} characteristics of the transistor #36 at 4.2 K at several gate-voltages. At V_g much higher than V_{th}, the I_d-V_{ds} curves are linear as is expected in the metallic regime. At V_{ds} close to V_{th}, however, the I_d-V_{ds} curves deviate from the ohmic behavior and a reduction of I_d is observed for small V_{ds}. This phenomenon is clearer in Fig. 5 (b), which shows the differential drain conductance ($\partial I_d / \partial V_{ds}$) of the transistor #36 as a function of V_{ds}. A significant reduction of the differential conductance is observed when both V_g and V_{ds} are small. The range of V_{ds} where the differential conductance is reduced is larger for smaller V_g. In these ranges of V_{ds}, the electron states are considered to be localized. In the strongly localized regime, the electron transport is governed by the variable range hopping (11) and resonant tunneling (12). The anomalous oscillations in the differential conductance in Fig. 5 (b) are considered to result from these kinds of electron transport. For larger V_{ds}, the electrons obtain high energy and become hot, resulting in the delocalization and the normal I_d-V_{ds} characteristics.

Now we focus on the anomalous increase in g_m in Fig. 4 (b). g_m in the linear region is given by

$$g_m / V_{ds} \quad = \frac{eW}{L} \left(\mu \frac{\partial n_s}{\partial V_g} + n_s \frac{\partial \mu}{\partial V_g} \right), \tag{2}$$

180

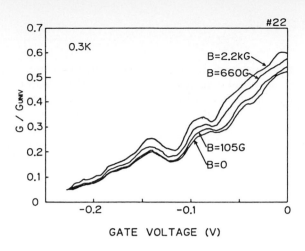

Fig. 6. The ac drain conductance of the AlGaAs/GaAs quantum wire transistor #22 at 0.3 K as a function of the gate-voltage.

where n_s is the sheet carrier concentration. Usually, g_m is simply given by $g_m = (W/L)\mu CV_{ds}$, ignoring the change in μ with V_g (the second term in bracket in Eq. (2)). This tempts us to think that the increase in g_m is due to the increased mobility (13). However, Eq. (2) is the complete form of g_m. When the electron states are strongly localized, n_s is finite but $\mu = 0$, resulting in $g_m = 0$. When V_g exceeds V_{th}, n_s does not change very much but μ drastically increases because of the delocalization. Thus, $(\partial\mu/\partial V_g)$ becomes very large temporarily, leading to a high peak in g_m. After complete delocalization, μ weakly depends on V_g and g_m shows an almost constant value corresponding to $(W/L)\mu CV_{ds}$. Therefore, the anomalous increase in g_m in Fig. 4 (b) is not due to the mobility enhancement in a one-dimensional quantum wire predicted by Sakaki (14), but due to the strong localization.

Figure 6 shows the gate-voltage dependence of the ac drain conductance in the narrow transistor #22 ($W = 350$ nm) at 0.3 K. I_d is set to 5 nA. The conductance anomalously oscillates. g_m becomes negative by this oscillation. This oscillation is not UCF because the oscillation is hardly changed even when magnetic field larger than B_c is applied (B_c is 340 G in #22), as shown in Fig. 6. This kind of oscillation can be observed only at low temperatures (≤ 1.3 K), only for small I_d (≤ 5 nA), and only in narrow transistors (#22 and #36). One possible explanation for the oscillation is the observation of the one-dimensional quantization of electrons. Although the channel width is rather wide ($W = 350$ nm) at $V_g = 0$ V, the width can be locally very narrow for smaller V_g. However, the origin of the oscillation is not clear.

5. CONCLUSIONS

The electron transport in AlGaAs/GaAs quantum wire transistors was investigated. The universal conductance fluctuation effect with the Fermi energy was observed in wide quantum wire transistors. The correlation energy was ~1 meV, which confirms the quantum interference at 10 K. The strong localization effect was found to affect the device characteristics of very narrow quantum wire transistors. It causes the anomalous increase in transconductance and the nonlinearity in current-voltage characteristics. Anomalous oscillations in drain conductance were also observed in very narrow transistors at very low temperatures.

ACKNOWLEDGEMENTS

The authors would like to thank Prof. H. Sakaki for supplying the hetero-structures. They also thank Y. Suzuki and A. Momose for technical assistance.

REFERENCES

1. S. Datta, M. R. Melloch, S. Bandyopadhyay, and M. S. Lundstrom, Appl. Phys. Lett. **48**, 487 (1986).
2. S. Datta, Extended Abstracts of 20th Int. Conf. on Solid State Devices and Materials, Tokyo, Japan, 1988, pp. 491-494.
3. M. Yamamoto and K. Hohkawa, Extended Abstracts of 20th Int. Conf. on Solid State Devices and Materials, Tokyo, Japan, 1988, pp. 495-498.
4. T. Hiramoto, K. Hirakawa, Y. Iye, and T. Ikoma, to be published in Appl. Phys. Lett.
5. G. Bergmann, Phys. Rep. **107**, 1 (1984).
6. C. W. Beenakker and H. van Houten, Phys. Rev. B**38**, 3232 (1988).
7. C. P. Umbach, S. Washburn, R. B. Laibowitz, and R. A. Webb, Phys. Rev. B**30**, 4048 (1984).
8. P. A. Lee and A. D. Stone, Phys. Rev. Lett. **55**, 1622 (1985).
9. S. Washburn and R. A. Webb, Adv. Phys. **35**, 375 (1986).
10. N. F. Mott, Rev. Mod. Phys, **50**, 203 (1978).
11. P. A. Lee, Phys. Rev. Lett. **53**, 2042 (1984).
12. M. Ya. Azbel and P. Soven, Phys. Rev. B**27**, 831 (1983).
13. M. Okada, T. Ohshima, M. Matsuda, N. Yokoyama, and A. Shibatomi, Extended Abstracts of 20th Int. Conf. on Solid State Devices and Materials, Tokyo, Japan, 1988, pp. 503-506.
14. H. Sakaki, Jpn. J. Appl. Phys. **19**, L735 (1980).

QUANTUM DEVICES BASED ON PHASE COHERENT LATERAL QUANTUM TRANSPORT[1]

S. Bandyopadhyay, G. H. Bernstein and W. Porod

Department of Electrical and Computer Engineering

University of Notre Dame

Notre Dame, Indiana 46556

Phase-coherent lateral transport phenomena hold the promise for many new types of "quantum devices" with vastly improved performance over conventional devices. In this paper, we address the performance of *lateral quantum devices* in which current flows parallel to the interfaces of a heterostructure. In particular, we focus on two specific devices, namely the Aharonov-Bohm interferometer and the recently proposed Quantum Diffraction Transistor, which have ultrafast *extrinsic* switching speed and a tremendous potential for *multi-functionality*.

I. INTRODUCTION

Over the past few years, numerous novel electronic devices have been proposed or demonstrated whose operations rely entirely on quantum transport phenomena. The most widely studied member of this class is the resonant tunneling diode in which electron transport occurs perpendicular to the interfaces of a heterostructure and the current depends on the interference of waves multiply reflected by heterobarriers. More recently, a different genre of devices has emerged (which we refer to as *lateral quantum devices*) in which current flows parallel to heterointerfaces. The inherent advantage of these devices is that the current levels in them can be much higher, which translates into a significant advantage in switching speed when the device is used in an integrated circuit. In "vertical quantum devices", such as the resonant tunneling diode, the current levels are typically low since the current is predominantly due to tunneling through large potential barriers caused by band-edge discontinuities. In contrast, there are no large potential

---

[1]Supported by the Air Force Office of Scientific Research under grant no. AFOSR-88-0096 and by IBM Faculty Development Awards.

183

barriers in lateral quantum devices so that the current levels are generally much higher.

In an integrated circuit chip, the switching speed of a device is determined not so much by the intrinsic speed of the device, but rather by the time it takes to charge and discharge the interconnect and device capacitances. This time depends on the current that can be supplied to the capacitances and the (threshold) voltage levels to which the capacitances must be charged. Roughly speaking, the extrinsic switching time τ_s is given by

$$\tau_s = \frac{C_t V_t}{I} , \tag{1}$$

where C_t is the total circuit capacitance, V_t is the threshold voltage and I is the current[2].

It is advantageous to employ such quantum devices in integrated circuit chips that have very low threshold voltages and can carry relatively large currents so that the extrinsic switching speed is high. Lateral quantum devices, whose operations depend on phase-coherent *lateral* transport, are superior in this respect. They not only exhibit larger current carrying capability, but can also have very small threshold voltages[3]. Consequently, the extrinsic switching speed of lateral quantum devices is usually much higher than that of vertical quantum devices.

In the following Sections, we discuss the performance of two different classes of lateral quantum devices. They are the Aharonov-Bohm interferometer and the recently proposed Quantum Diffraction Field Effect Transistor [1].

A. The Aharonov-Bohm interferometer

In the Aharonov-Bohm interferometer, electrons in two contiguous paths are made to interfere by an external electrostatic potential which modulates the current. If the interferometer is *two-dimensional*, which means that each path is a two-dimensional structure (viz. a quantum well), then the current can be made arbitrarily large by increasing the transverse width of the structure. This may result in certain advantages, but not necessarily in the extrinsic switching speed, since increasing the width of the structure to increase current also increases the circuit capacitance. Besides, a more crucial drawback of two-dimensional interferometers is that they do not perform sufficiently well for device applications unless transport is truly ballistic [2]. Ballistic transport is not easy to achieve in devices with

---

[2] This limitation on the switching speed does not arise if an electronic device can be switched optically. An intriguing scheme for switching an Aharonov-Bohm interferometer optically (based on virtual charge polarization), instead of electronically, has been proposed by M. Yamanishi. (Proc. of the 4th. Intl. Conf. on Superlattices, Microstructures and Microdevices).

[3] An example is the Aharonov-Bohm interferometer (see Ref. 3).

present-day capability. It therefore behooves us to consider realistic disordered structures and examine device performance in the *diffusive* regime.

In diffusive transport, two-dimensional interferometers do not work well but one-dimensional interferometers (in which the interfering paths are quantum wires) work sufficiently well [2]. The primary reason for this is that in 2-d interferometers, there is a two-fold ensemble averaging - over the electron's energy E and the transverse wavevector k_t - whereas in 1-d interferometers, the averaging over k_t is absent. The latter averaging has disastrous results when elastic scattering is operative. Therefore, for real device applications, 1-d interferometers appear to be the inevitable choice, at least for the present.

In Fig. 1, we show the current modulation (due to the electrostatic Aharonov-Bohm effect) in a disordered 1-d GaAs interferometer in the weak localization regime at 77 K. The length of the structure is 1000 Å , the carrier concentration is 1.55 x 10^6 / cm and the impurity concentration is 10^5 / cm. The model for this calculation is the same as that employed in Ref. 4. The voltage over the structure is 36 meV which is the threshold for polar optical phonon emission (onset of strong inelastic scattering) in GaAs. Again, a $\sim$ 70 % modulation of the conductance is found at LN$_2$ temperature, which may be good enough for device applications.

From Fig. 1, we find that the maximum value of the current is 2.1 μA. This is very large for single-moded quantum devices whose typical cross-sectional area is 100 Å$\times$ 100 Å. The current level of 2.1 μA translates into an effective current density of more than 10^6 A/cm^2 which is about an order of magnitude higher than what can be achieved in resonant tunneling diodes. In addition, *the threshold voltage for switching* of such devices is also very low. The threshold voltage is $\sim$ 7 mV. Therefore, using Equation (1) and assuming that the total circuit capacitance is about 1 fF[4], we find that the extrinsic switching time τ_s is $\sim$ 3 ps. (The intrinsic switching speed of the device is the transit time of electrons which in this case is $\sim$ 230 fs). The 3 ps switching speed is comparable to that of the fastest GaAs and Silicon devices or even Josephson junctions.

Apart from a fast switching speed, Aharonov-Bohm quantum devices have the additional advantage of having high transconductance (even for nanometer feature sizes) which is advantageous for analog applications. For our prototypical structure, the maximum absolute transconductance was 0.45 mS. This is comparable to the highest transconductance that one could obtain with a 100 nm wide GaAs MODFET whose transconductance would rarely exceed 1 S/mm. For such a MODFET with a feature size of 1000 Å, the absolute transconductance will be 0.1 mS which is slightly lower than the transconductance obtainable with Aharonov-Bohm devices.

The most striking feature of the Aharonov-Bohm device however is its ex-

---

[4]The estimate of 1 fF is optimistic, but certainly realizable. Capacitances of 10^{-17} F have been obtained in "Coulomb Blockade" experiments.

tremely low power-delay product. This quantity is approximately given by

$$PDP \approx C_t V_t^2 \qquad (2)$$

which gives a value of 5 x 10^{-20} Joules for a device with C_t = 1fF and V_t = 7 meV. This power delay product is a few orders of magnitude lower than what could be obtained with even Josephson junctions.

Needless to say, the above performance figures that we have calculated are theoretical projections and one must wait for the realization of actual prototype devices to see if the predicted performance is approached. But more importantly, these devices have another intriguing characteristic; they exhibit multi-functionality. For instance, they can be used to realize unipolar complementary operations, single transistor static latches and single-stage differential amplifiers [3]. It is this multi-functionality that is the most attractive feature of quantum devices, especially in view of its impact on alternate architectures such as neural networks or cellular automata.

B. The Quantum Diffraction Field Effect Transistor

Another lateral quantum device that promises extraordinary multi-functionality s the recently proposed Quantum Diffraction Field Effect Transistor (QUADFET[1]. ts schematic is depicted in Fig. 2.

The QUADFET is basically a MODFET where the gate has a narrow slit efined by electron-beam exposure. Electrons incident from the source can diffract hrough the slit, and the diffraction pattern is viewed as the currents collected at arious fingers in a "drain" consisting of closely spaced fingers. Just like in an rdinary diffraction experiment, the diffraction pattern (and hence the currents ollected at the fingers) can be changed by modulating the width of the slit. An

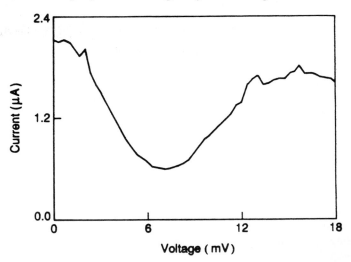

Fig. 1. The transfer characteristic of a disordered *1-d A-B interferometer*

analog voltage applied between the gate pads constricts the slit by extending the depletion layer surrounding the slit. This changes the slit width and alters the diffraction pattern so that the current levels in the fingers are changed. The current level in each finger can be made to represent a "bit" so that one can effectively transduce an analog signal between the gate pads into a bit pattern or a digital signal at the drain fingers. This suggests the operation of an A/D converter. The potential multi-functionality of this device exists in the replacement of an entire A/D circuit by a single transistor.

The problem with the QUADFET however is that it is inherently two-dimensional and therefore requires strictly ballistic transport[5]. In that sense, this device is somewhat futuristic, but assuming that future technology will routinely yield ballistic structures, we can examine the performance of this device, particularly to see if it can be operated at sufficiently high temperatures.

The condition for obtaining a minimum in the diffraction pattern, at any finger in the drain of the QUADFET, is expressed as

$$\frac{ak_F}{2\pi} sin\theta = n + \frac{1}{2} \tag{3}$$

where a is the slit-width.

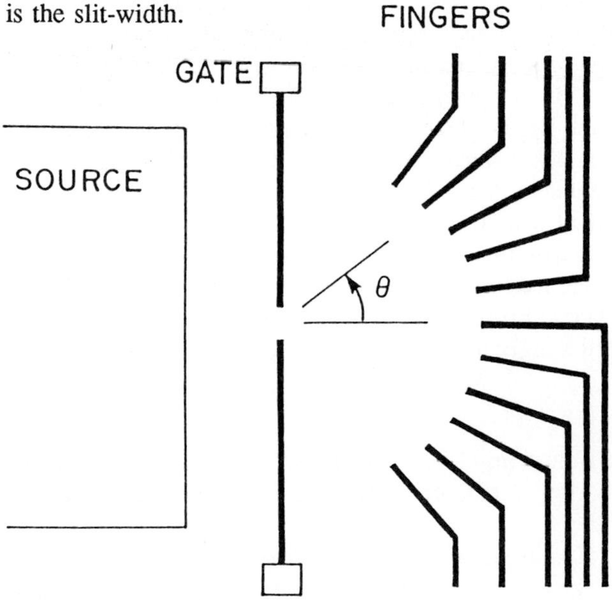

Fig. 2. Schematic of a Quantum Diffraction Field Effect Transistor

[5]The QUADFET requires ballistic transport for the same reason that a 2-d Aharonov-Bohm interferometer requires ballistic transport, namely the ensemble averaging over the electron's transverse wavevector.

In a 2-d structure, the spread in the transverse wavevector introduces a spread in θ, and the thermal spread in energy introduces a spread in the Fermi wavevector k_F, which together tend to make a minimum shallow. For a minimum to be clearly discernible, we require

$$\Delta\left(\frac{ak_F}{2\pi}sin\theta\right) = \frac{ak_F}{2\pi}\int_0^1 d(sin\theta) + \frac{a}{2\pi}sin\theta\int dk_F < \frac{1}{2} \qquad (4)$$

or

$$\frac{ak_F}{2\pi} + \frac{a}{2\pi}\frac{m^*kT}{\hbar^2 k_F} < \frac{1}{2}, \qquad (5)$$

where kT is the thermal spread in the Fermi energy and m^* is the effective mass of electrons. We have replaced $sin\theta$ by its maximum value of unity.

The minimum value of the left-hand-side is obtained when

$$E_F = \frac{kT}{2}; \quad or \quad n_s = \frac{2\pi kT m^*}{h^2}, \qquad (6)$$

where n_s is the carrier concentration. The surprising fact is that for a given carrier concentration, there is an optimum operating temperature.

It turns out coincidentally that in GaAs, for a typical carrier concentration of $\sim 10^{11}$ cm^{-2}, the optimum temperature is ~ 77 K! Of course, for 77 K operation, it is necessary that the device dimensions be smaller than the mean-free-path at LN$_2$ temperature, but this is already not too far outside the capability of x-ray or electron-beam lithography. Also, for this temperature and carrier concentration, Equation (4) is satisfied with a slit-width $a < 200$ Å and this is achievable with present lithographic capabilities. It therefore appears that the QUADFET will become a viable device for electronic applications in the future.

In conclusion, lateral quantum devices are still in their infancy. But they are likely to play an increasingly important role in electronic circuits of the future.

REFERENCES

1. Kriman, A. M., Bernstein, G. H., Haukness, B. S., and Ferry, D. K. (1988). Presented at the 4th. Intl. Conf. on Superlattices, Microstructures and Microdevices.
2. Bandyopadhyay, S., and Porod, W. (1988). Appl. Phys. Lett. 53, 2323.
3. Bandyopadhyay, S., Datta, S., and Melloch, M. R. (1986). Superlattices and Microstructures. 2, 539.
4. Cahay, M., Bandyopadhyay, S., and Grubin, H. L. (1989), Proc. Intl. Symp. on Nanostructure Physics and Fabrication (this issue).

FABRICATION AND ROOM TEMPERATURE OPERATION OF A RESONANT TUNNELING TRANSISTOR WITH A PSEUDOMORPHIC InGaAs BASE*

U. K. Reddy, G. I. Haddad, I. Mehdi and R. K. Mains

Center for High-Frequency Microelectronics
Department of Electrical Engineering
and Computer Science
The University of Michigan
Ann Arbor, Michigan, USA

I. INTRODUCTION

The use of double-barrier quantum well (DBQW) tunneling structures for obtaining negative resistance in the current-voltage characteristics of two terminal devices is now well established [1]. It is tempting to consider the three terminal operation of these structures by making contact to the quantum well layer and thus utilizing the small time scales involved in the tunneling process for high-speed applications. Although, several three-terminal structures utilizing resonant-tunneling in DBQWs have been fabricated, most employ the DBQW structure in one terminal of a more conventional transistor [2–4]. A truly resonant-tunneling transistor (RTT) consists of a traditional DBQW structure with an additional low-resistance contact made to the quantum well region. Such a three-terminal device was proposed independently by Schulman and Waldner [5] and Haddad et al. [6]. Both proposed a modified double-barrier structure in which the first level is confined below the conduction band edge of the emitter and collector regions. This is achieved by using a narrow bandgap material in the quantum well region. The existence of the confined level in the quantum well (base) region permits storing of charge and thus allowing a low-resistance contact to be made to the thin base region. Haddad et al. [6] made a further improvement to the structure by proposing a stepped barrier on the collector side to reduce the device capacitance and leakage current into the collector by better

*This work is supported by the Army Research Office under contract DAAL03-87-K-0007

confining the charge stored in the first confined state. The base contact is then used to modify the electrostatic potential in the base region permitting the independent tuning of the device between on and off resonance. The charge transport in such a device occurs by resonant tunneling into the next higher level in the quantum well. Since this level is also the first excited state in the well the advantages are increased current densities and reduced electron scattering in the base region.

II. EXPERIMENTAL

The structures grown by MBE (shown in Fig. 1(a) along with the band diagram in Fig. 1(b)) were processed into three terminal devices using conventional photolithography. The base layer was reached by electrically monitoring the etching process. Ohmic contacts were made by evaporating Ni,Ge,Au Ti and Au layers in an e-beam evaporator at a pressure of $\sim 2 \times 10^{-7}$ Torr and were then annealed in a furnace at 450° for 60 seconds in flowing nitrogen.

III. RESULTS AND DISCUSSION

The two terminal current-voltage characteristic of the RTT when operated as a conventional resonant tunneling diode is shown in fig. 2(a). The negative differential resistance (NDR) is clearly seen with a peak current density of 1.8×10^3 A/cm^2. Although the peak to valley ratio is small (~ 1.1) it should be remembered that the measurements were done at room temperature and the devices are not symmetric double barrier structures. The asymetry also explains the absence of NDR in the negative voltage direction. In order to determine the operational gain of the device the common emitter characteristics of the RTT were measured and are shown in fig. 2(b). It is seen that control is achieved through the base contact but with no gain. Also, large base voltages are needed to modulate the collector current. A calculation of the unilateral gain from the measurement of y- parameters also indicated this poor gain.

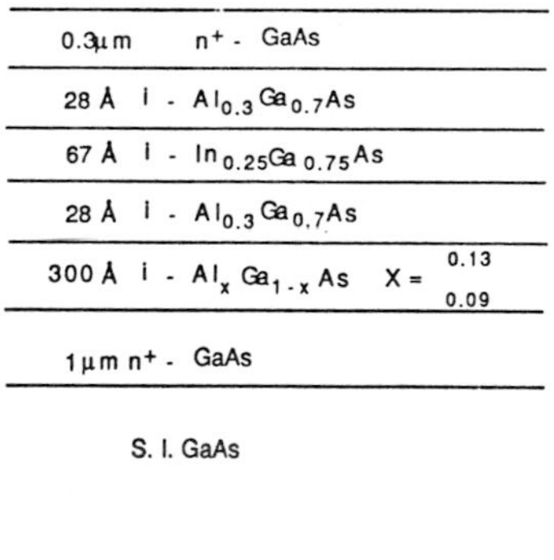

0.3μm n+ - GaAs	
28 Å i - Al$_{0.3}$Ga$_{0.7}$As	
67 Å i - In$_{0.25}$Ga$_{0.75}$As	
28 Å i - Al$_{0.3}$Ga$_{0.7}$As	
300 Å i - Al$_x$Ga$_{1-x}$As X =	0.13 0.09
1μm n+ - GaAs	

S. I. GaAs

Figure 1(a) Cross sectional view of the resonant tunneling structure

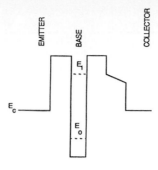

Figure 1(b) Energy band diagram of the resonant tunneling structure

In order to investigate the inability of our RTT devices to show any significant gain more measurements were made to understand the operation of the device. The measurement of emitter voltages for different emitter currents showed that for a constant emitter current the emitter voltage is not constant but varies with the base-collector voltage. This clearly indicates that the base region is not effective in shielding the emitter from the collector potential. This is further substantiated from the common emitter curves of fig. 2(b) where the collector currents show no saturation.

The lack of collector current saturation in fig. 4 and the poor gain in our devices can be understood as follows. The external base contact in our RTT is situated at a finite distance ($\sim$ 5 μm in our devices) from the intrinsic base region. Thus, the base layer which was initially conducting, is now completely depeleted due to surface Fermi level pinning. This results in a large voltage drop across the external base contact and the intrinsic base. This large voltage drop across the extrinsic base region not only kills the gain of the RTT, the surface depletion renders the base ineffective in shielding the emitter region from the electrostatic potential in the collector region.

A major design issue of the RTT, we believe, is surface depletion of the extrinsic base region. This problem, although commonly present in all vertical structures, is particularly serious in RTTs because of the extremely small dimensions involved. A satisfactory solution to this problem is very important in order to make these devices viable. We propose the introduction of a doped layer between the base contact and the quantum well (base) layer with a proper doping level so that all the depeltion occurs in this layer leaving the electrons in the base unaffected. Secondly, the quantum well can be doped to retain some extra charge. Finally, by making use of self-aligned technology the base contact can be placed close to the intrinsic base (within a few Debye lengths) largely overcoming the surface depletion. Further

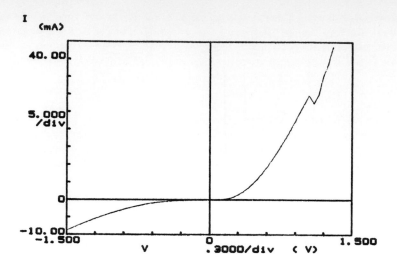

Figure 2(a) Emitter-collector current-voltage characteristic of RTT with the base open

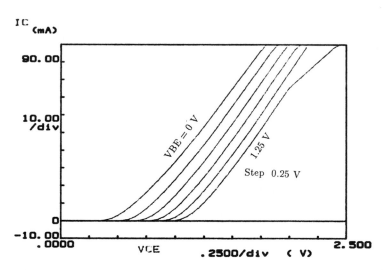

Figure 2(b) Common emitter cuttent-voltage characteristics of the RTT for different base voltages

192

experiments incorporating these design changes in our RTTs are underway in our laboratory. We also believe that operation at liquid nitorgen temperatures will improve the performance particularly in the negative transconductance region and where this device is expected to operate in the particular mode of interest here.

IV. CONCLUSIONS

We report the fabrication of a resonant tunneling transistor by actually making contact to the thin quantum well region. The base is made conducting by using narrow band gap $In_{0.25}Ga_{0.75}As$ in the quantum well where the first level is confined below the conduction band edge of the heavily doped regions and electrons tunnel resonantly through the second level. The initial results indicate that surface depletion in the extrinsic base layer can seriously affect the device performance. Several improvements to improve the device performance have been proposed.

ACKNOWLEDGMENTS

The authors would like to thank J. Oh, J. Pamulapati and Professor P. Bhattacharya for providing the materials and P. McCleer, J. East and S. Tiwari for useful comments.

REFERENCES

1. Capasso, F., and Mohammed K., and Cho, A. Y. (1986). *IEEE J. Quantum Electronics*, QE-22, 1853.

2. Bonnefoi, A. R., and McGill T. C., and Burnham, R. D. (1985). *IEEE Electron Device Letters*, 6, 636.

3. Woodward, T. K., and McGill, T. C., and Burhnham R. D. (1987). *Appl. Phys. Lett.*, 50, 451.

4. Capasso, F. and Sen, S. and Cho A. Y. (1987). *Appl. Phys. Lett.*, 51, 526.

5. Schulman, J. N. and Waldner, M. (1988). *J. Appl. Phys.*, 63. 2859.

6. Haddad, G. I., and Mains, R. K. and Reddy, U. K., and East, J. R. (1988). 4th Int. Conf. on Superlattices and Microstructures, Trieste, Italy.

OVERSHOOT SATURATION IN ULTRA-SHORT CHANNEL FETS DUE TO MINIMUM ACCELERATION LENGTHS[1]

J. M. Ryan, J. Han, A. M. Kriman, and D. K. Ferry
Center for Solid State Electronics Research, Arizona State University, Tempe, AZ 85287-6206

P. Newman
Electronics Technology and Devices Laboratory, Ft. Monmouth, NJ 07703

I. INTRODUCTION

Most simple models of semiconductor devices are built upon the gradual channel approximation, in which a one-dimensional problem is solved for the potential along the channel length. In the simplest of these models, the mobility is assumed to be constant, and the velocity rises along the channel as the electric field rises (the density drops in going from the source to the drain, which leads to the rise in field). This leads to the general behavior that the drain current is proportional to $(V_{GS} - V_{DS})^2$. The important first modification to this behavior, and one seen in most submicron devices, is that the velocity cannot continue to increase, but is limited to the saturation value, typically 10^7 cm/sec. The current density through the device is then limited to a value given by the product of the saturation velocity and the carrier density at the position in the device where velocity saturation sets in. Beyond this point, both the velocity and total current through the device remains constant, and this point acts as the injection point for the saturated region. Thus, typically, we find that

$$J = nev = C_o v_{sat}(V_{GS} - V_{sat}) \; , \tag{1}$$

where v_{sat} is the saturation velocity and V_{sat} is the voltage in the channel at the saturation point -- the position at which the velocity rises to v_{sat}. Equation (1) is written for a MOSFET, but the equivalent behavior is found in MESFETs. We also note from (1) that the transconductance no longer increases for further decreases in the gate length, since the gate length dependence has disappeared from the current equation. If we are to obtain any further increase in performance by down-scaling the device size, we must go sufficiently far to see increases in the saturation velocity due to *non-stationary transport*. In this latter case, the gate length is sufficiently short that electrons transit the entire gate region without ever stabilizing to the saturation velocity. Due to *velocity overshoot*, the effective velocity can be greater than the saturation value (1).

The existence of velocity overshoot in FETs can be used to enhance their characteristics, if the devices can be fabricated small enough to utilize this effect. The velocity overshoot can also be used to increase the frequency response of the MESFETs if the drain can collect the electrons within a few inelestic mean free paths of the gate. For this case, the electrons will spend a larger portion of their transit time in the overshoot region, and the transit time will decrease, thus increasing the transconductance. From the wide range of reported transconductances in the literature for small-gate MESFETs, a comparison is difficult. Although variations in doping and active layer thicknesses can be determined, differences due to fabrication techniques are difficult to accurately estimate. By fabricating MESFETs with varying gate lengths and doping, this uncertainty can be eliminated.

II. FABRICATION

All exposures were accomplished using electron-beam lithography in a system described earlier (2). The heavily doped wafer had an active layer thickness of 50nm, and the lightly doped wafer had a depth of 200nm. The fabrication process required four mask levels; mesa isolation, ohmics contacts, contact pads and gate metalization.

---

[1] Work supported by the Army Research Office

The exposures were done with an acceleration voltage of 40kV, and except for the mesa mask, the resist used was 950,000 PMMA positive resist. The thickness of the resist varied from 120nm to 600nm, and was developed in a 3:7 solution of cellosolve:methanol. For the mesa mask, Shipley SAL 601-ER7 negative resist was used. It was spun to obtain a thickness of 500nm and then baked for 30 minutes at 90 C. Exposure was done at a dose of 8.5 x 10^{-6} Coul/cm^2 and followed by a 15 minute post-bake at 90 C. The development was done in Shipley SAL MF-622 for 2.5 minutes. The ohmic contacts were AuGe:Ni:Au, and the contact resistance was determined by a transmission line test pattern. A typical finished ultra-submicron device is shown in Fig. 1.

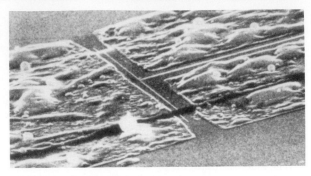

Fig. 1. The ultra-short gate length MESFETs are generally of the structure shown here. The gate widths are 20 µm. The normal Au-Ge-Ni contacts appear in their usual muddled form.

III. dc MEASUREMENTS

The devices were tested for their dc characteristics using an HP 4145B semiconductor parameter analyzer. Figure 2 displays the I_d versus V_{ds} curve for a L_g=30nm, heavily-doped device. Several of the devices exhibit bulk negative differential conductivity, known as the Gunn effect. This occurs at an electric field of 6.6 kV/cm, which is reasonable if the voltage drop of the contacts is taken into account.

The extrinsic transconductance of each device was also directly measured. The intrinsic transconductance was calculated using

$$g_m = g_m'/(1-g_m'R_s) \qquad (2)$$

where g_m' is the measured transconductance, g_m is the intrinsic transconductance and R_s is the contact resistance of the source obtained from the transmission line pattern. The transconductance versus gate length has been plotted in Fig. 3 for both the heavily- and lightly-doped devices. The transconductance decreases as the gate length decreases for gates larger than 55nm. This is due to the decreasing L_g/a ratio causing the depletion region under the gate to be less like a parallel plate capacitor and to be more influenced by the circular fringing regions. As the gate lengths decrease below 55nm, the transconductance increases. This is caused by velocity overshoot. The electrons in the channel are spending an increasing portion of their transit time with a higher velocity. The devices show a limit to the velocity overshoot effect at 37 nm. For devices with gate lengths less than this, the carriers see an acceleration length too short to experience full velocity overshoot.

The current in the channel can be written as

$$I = qN_dv(x)hZ \quad , \qquad (3)$$

where q is the electric charge, N_d is the doping, $v(x)$ is the carrier velocity, h is the channel opening and Z is the channel width. Hauser [3] has expressed the channel opening as

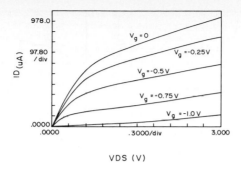

VDS (V)

Fig. 2 The characteristic curves of drain current as a function of source-drain potential for a typical ultra-short gate MESFET.

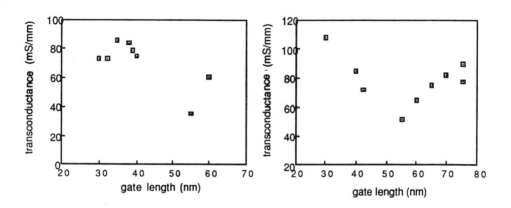

Fig. 3. The transconductance observed in the ultra-short gate MESFETs fabricated via the program described in the text. The heavily doped material (1.5×10^{18} cm^{-3}) is shown on the right, while the lightly-doped material (2×10^{17} cm^{-3}) is on the left.

$$h = a[1-((V_d-V_g)/V_p))^{1/2}] \quad , \tag{4}$$

where a is the epi layer thickness and V_p the pinchoff voltage. Thus by plotting I_d versus $((V_d-V_g)/V_p)^{1/2}$, the average transit velocity can be determined. Figure 4 shows a plot of v_s versus gate length for a fixed V_{ds}. It shows the average transit velocity for gate lengths larger then 55nm (those which did not show overshoot in the transconductance), an average transit velocity approximately equal to the high field velocity. As the gate lengths decrease below 55nm, the transit velocity increases until it saturates at a gate length of nearly 40nm. This supports the previous transconductance data. The lower high field velocity for the heavily doped devices is only 8.0×10^6 cm/s and rises only 15%, while the lightly doped devices show a high field velocity of 1×10^7 cm/s and rise nearly 40%. This can be attributed to the increased scattering due to the increased doping.

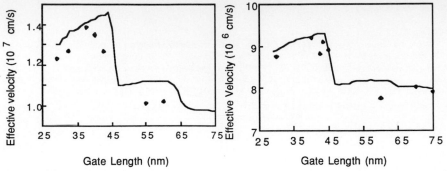

Fig. 4. The inferred values for the effective saturation velocity in (a-left) lightly-doped, and (b-right) heavily doped material. The simple computer simulations are also shown for comparison.

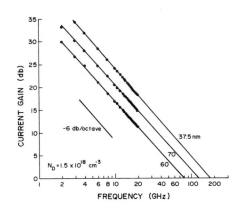

Fig 5. The current extrapolates to a value that can be used to infer the effective saturated velocity.

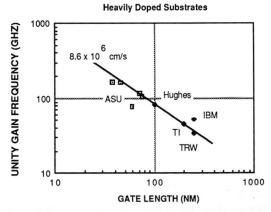

Fig. 6. The cutoff frequency is shown as a function of the gate length of the MESFET.

IV. ac MEASUREMENTS

The ac measurements were done on a Cascade Microtech Model 42 microwave probe station coupled to an HP 8510 network analyzer and an 8350B sweep oscillator, with an HP 8518A S-parameter test set. The devices were tested at frequencies ranging from 2 to 20 GHz. Figure 5 shows the current gain for several devices as a function of frequency. Extrapolating the gain at a rate of -6 dB/octave gives the minimum estimated value of the gain-bandwidth frequency. Figure 6 plots f_t as a function of gate length and compares them with other epitaxial GaAs MESFETs reported. The gain-bandwidth frequency can be approximated by the expression

$$f_t = v_s/2\pi L_g \quad .$$ (5)

The slope of the plot can be used to determine the transit velocity. Drawing the line through our best results gives a value of 8.6 x 10^6 cm/s. This compares with the value previously obtained for the heavily doped average transit velocity. These results show the simple expression for f_t still holds for gate lengths down to 40nm. No rise was observed in f_t below 55nm, corresponding to the onset of velocity overshoot, but the smallest devices did fall below the line suggesting the limited acceleration length might be coming to effect.

V. SIMULATION

The saturation of the overshoot velocity was investigated utilizing a tranport model based upon a parameterized velocity autocorrelation function (3). This took into account both the energy and momentum relaxation rates, since the motion of the carriers in an applied electric field involves the acceleration due to the field and the relaxation due to collisions. The relaxation process is retarded due to its time variation, and appears as the presence of a convolution integral of these two time varying quantities.

The model has been modified by the introduction of an initial velocity that represents the heating of the carriers due to the barriers introduced by the ohmic contact. The resulting average transit velocity has also been plotted in Fig. 4 as a function of gate length. It shows that this simple approach yields results that match the data reasonably well, and confirms that the overshoot is limited by having a minimum acceleration length.

VI. SUMMARY

We have presented data on ultra-short gate-length MESFETs which indicates that velocity overshoot is occuring in these devices. This behavior continues until the gate length is less than approximately 37 nm. Below this length, the carriers are heated in the parasitic source resistance to a sufficient degree that they no longer have sufficient gate length to accelerate to the full overshoot velocity.

REFERENCES

1. Ferry, D. K. (1982). In *Advances in Electronics and Electron Physics*," (C. Marton, Ed.), vol. 58, 311. Academic Press, New York.
2. Bernstein, G., and Ferry, D. K. (1988). IEEE Trans. Electron Dev. **35**, 887.
3. Hauser, J. R. (1967). Solid-State Electron. 10, 577.

TAILORING TRANSPORT PROPERTIES BY WAVEFUNCTION ENGINEERING IN QUANTUM WELLS AND ITS DEVICE APPLICATIONS[1]

Suyog Bhobe, Wolfgang Porod, and Supriyo Bandyopadhyay
Department of Electrical and Computer Engineering
University of Notre Dame

David J. Kirkner
Department of Civil Engineering
University of Notre Dame

We investigate a semi-classical mesoscopic phenomenon in which the dependence of a system's macroscopic transport properties on the microscopic details of the electronic wavefunction is exploited to realize an ultrafast switching transistor. The conductance of a quantum well with a selectively-doped region depends on the precise nature of the wavefunction in the well which can be altered by an external field that "pushes" the wavefunction in and out of the doped region. This modulates the conductance of the well (by few orders of magnitude at liquid helium temperature) on timescales of the order of 100 femtoseconds. We have investigated this phenomenon using a combination of self-consistent Schrödinger Equation solution and ensemble Monte Carlo simulation to model transient electronic transport in the well.

I. INTRODUCTION

It is generally believed that mesoscopic "quantum devices", whose operations rely on quantum mechanical phenomena, will be much faster than classical devices such as an ordinary field-effect-transistor. The reason for this is that classical devices are switched by moving carriers in and out of the device so that the switching time is limited by the transit time of carriers. Quantum devices, on the other hand, do not usually require infusion and extraction of carriers. Typically, they are switched by inducing constructive or destructive interference of electrons

---

[1]This work was supported by the Air Force Office of Scientific Research.

201

which does not require physical movement of charges. Therefore, the switching speed of quantum devices is not likely to be limited by the transit time of carriers[2].

Although quantum devices have this inherent advantage, they also have certain disadvantages. Devices that are based on phase-coherent phenomena (such as interference) must have dimensions smaller than the phase-breaking length of electrons which is given by

$$L_\phi = \sqrt{D\tau_{in}} \, , \qquad (1)$$

where τ_{in} is the inelastic scattering time and D is the diffusion coefficient of electrons which depends on the elastic scattering time or "mobility".

It is evident that quantum devices will have the following drawbacks. Firstly, they must be operated well below room temperature so that τ_{in} is sufficiently large. Room temperature operation would require such small feature sizes that the lithographic capabilities necessary for delineating them are presently unavailable. Secondly, the material for the devices must be sufficiently "clean" since D depends on the elastic scattering time and hence the mobility μ. This is quite critical in two- or three-dimensional structures in which elastic scattering is far more frequent than inelastic scattering at cryogenic temperatures. Only in one-dimensional structures, the cleanness of the material is not that critical because of the drastic suppression of elastic scattering by one-dimensional confinement[3]. But one-dimensional structures (quantum wires) are not easy to fabricate and their current carrying capability is inherently low which makes them inappropriate for many applications.

There is however at least one semi-classical device that combines the advantages of both quantum devices (fast switching speed not limited by the transit time) and classical devices (no requirement of phase coherence and associated complications). The principle behind the operation of this device is very simple. The conductance of a two-dimensional structure such as a quantum well is given by

$$G\left(in \ (\Omega/\square)^{-1}\right) = qn_s\mu \qquad (2)$$

Instead of modulating G by modulating n_s (as is done traditionally), one can modulate it by changing μ. The advantage is that μ can be changed on timescales of the order of the momentum relaxation time so that the switching speed of such a device is not limited by the transit time.

---

[2]There are exceptions however. An example is the electrostatic Aharonov-Bohm interferometer in which the switching speed is in fact limited by the transit time of carriers.

[3]This does not mean that one-dimensional structures can be arbitrarily "dirty" since many quantum interference effects may not survive in the strong localization regime.

The important question now is how to modulate μ. The mobility depends on the scattering rates of electrons which (even in the semiclassical formalism of Fermi's Golden Rule) depends explicitly on the electronic states (wavefunctions) in the system. By applying an external field, the wavefunction in a quantum well can be altered - it can be skewed and pushed in and out of a selectively doped region within the well - which modulates the scattering rates and the mobility. Such wavefunction engineering is essentially similar to the quantum-confined Franz-Keldysh effect. When the wavefunction resides mostly the doped region, the mobility is low, otherwise it is high. If we neglect the time required for skewing the wavefunction (which is very small) then the switching time of such a device is essentially the momentum relaxation time which can be less than a picosecond.

Such a device, termed a velocity modulation transistor (VMT), has been proposed by Hamaguchi and his co-workers [1]. In this paper, we analyze this device and evaluate the magnitude of the conductance modulation as well as the switching time.

II. THEORY AND COMPUTATION

The electronic states of a quasi two-dimensional system, such as a quantum well, have been studied widely in the literature [2]. We have calculated the wavefunctions inside a 500 Å GaAs-AlGaAs quantum well, whose right-half [250 - 500 Å] is doped with impurities. The wavefunctions are obtained by solving self-consistently the Schrödinger and Poisson equations using the Finite Element Method. The details of the calculation are presented in Ref. 3. We assume a carrier concentration of 5 x 10^{11} cm^{-2} and an impurity concentration of 2 x 10^{17} cm^{-3} in the selectively doped region. The wavefunctions and the energy levels are shown in Fig. 1.

The scattering rate of two-dimensionally confined electrons in the well is obtained from Fermi's Golden Rule and is given by

$$\frac{1}{\tau(k)} = \frac{N_{eff} m^* e^4}{4\hbar^3 \epsilon^2 \lambda \sqrt{\lambda^2 + 4k^2}} , \tag{3}$$

where k is the Fermi wavevector $\left(k = \sqrt{2\pi n_s}\right)$, n_s is the two-dimensional electron concentration, m^* is the effective mass of electrons, ϵ is the dielectric constant and λ is the screening constant which was taken as 300 Å for GaAs.

The term in the above equation that depends explicitly on the precise details of the wavefunction is the so-called effective impurity density N_{eff} which is related to the "effective" number of impurities that interact *in situ* with the electrons. This quantity is obtained as

$$N_{eff} = \sum_m \int_{doped\ region} N_D(x) |\psi_m(x)|^2 dx , \tag{4}$$

where $\psi_m(x)$ is the wavefunction of the mth. subband in the well. The integration is performed over the selectively doped regions of the quantum well. This "effective" density was calculated for various electric fields applied transverse to the well interfaces that skew the wavefunctions away from the doped region by different amounts. It is easy to see now that the scattering rate can be modified by altering $\psi_m(x)$ by a transverse field (or equivalently a "gate voltage") which alters N_{eff}.

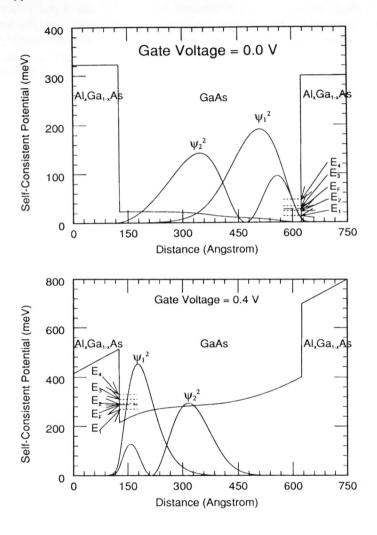

Fig. 1. The self-consistent potential and the wavefunctions in the two lowest occupied subbands in the selectively-doped GaAs-AlGaAs well in the absence (top figure) and presence (bottom figure) of a gate voltage.

The scattering rates obtained from Equations (3) and (4) for various gate-voltage-dependent $\psi_m(x)$ were used in a two-dimensional Monte Carlo simulation to model transient electronic transport at 4.2 K. From the simulation results, we extracted the momentum decay characteristics of an ensemble of electrons injected parallel to the interfaces of the well with the Fermi velocity. We included only the impurity scattering mechanism in the simulation and neglected all other kinds of scattering. The impurity scattering rate was found to be sufficiently high to be dominant at 4.2 K. Fig. 2. shows the decay of the initial momentum with time at various gate voltages. From the decay characteristics, we evaluated the momentum relaxation time τ_m by defining τ_m to be the time that elapses before the ensemble average momentum decays to $\frac{1}{e}$ times its initial value. From the momentum relaxation time, we calculate the effective "mobility" in the quantum well using $\mu = e\tau_m/m^*$, and from this we obtain the conductance G using Equation (2).

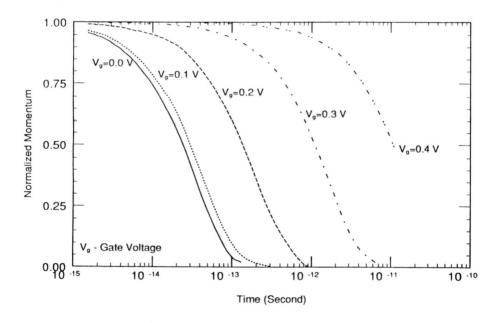

Fig. 2. The decay of the initial momentum of an electron injected parallel to the well interfaces at various gate voltages. The results are valid for a temperature of 4.2 K. These curves were obtained from ensemble Monte Carlo simulation.

Gate voltage (V)	Scattering rate $(\sec^{-1})$	Conductance $((\Omega/\square)^{-1})$
0.0	3.8×10^{13}	5.5×10^{-5}
0.1	3.0×10^{13}	6.9×10^{-5}
0.2	6.5×10^{12}	3.2×10^{-4}
0.3	8.0×10^{11}	2.6×10^{-3}
0.4	8.3×10^{10}	2.5×10^{-2}

Table 1: Scattering rates and conductance vs. gate voltage

Table I above lists the conductances for various gate voltages. Note that the conductance can be modulated over more than three orders of magnitude by varying the gate bias over a range of only 400 millivolts at 4.2 K. The modulation will certainly decrease significantly at higher temperatures due to phonon scattering, but the rather small threshold voltage of 0.4 V still indicates a large transconductance and also a very low power-delay product. In addition, the switching speed, being of the order of the momentum relaxation time, is about 100 fs which is comparable to the switching speeds of quantum devices presently extant. In these respects, the performance of this device is quite comparable to the performance of ultra-high performance quantum devices.

III. CONCLUSION

In this paper, we have explored a semi-classical device whose performance is comparable to those of mesoscopic quantum devices but whose fabrication is much easier. It is an interesting example of a mesoscopic phenomenon where a macroscopic property, namely the conductance, depends on the microscopic details of the electronic wavefunction. The advantage here is that there is no requirement of phase coherence so that the device could operate at elevated temperatures and the demand on the material quality is not stringent.

V. REFERENCES

[1] C. Hamaguchi, K. Miyatsuji, and H. Hihara, *Jpn. J. Appl. Phys.* **23**, L132 (1984). A similar concept has been proposed by H. Sakaki, *Jpn. J. Appl. Phys.* **21**, 381 (1982).
[2] See, for example, F. Stern, *Physical Review* **B 5**, 4891 (1972).
[3] S. Bhobe, W. Porod and S. Bandyopadhyay, *Surface and Interface Analysis*, (1989) (in press).

InP BASED RESONANT TUNNELING DIODES FOR MILLIMETER-WAVE POWER GENERATION [1]

I. Mehdi
G. I. Haddad

Center for High-Frequency Microelectronics
Department of Electrical Engineering and Computer Science
The University of Michigan
Ann Arbor, Michigan 48109

I   Introduction

Ever since the first demonstrated application of resonant tunneling diodes(RTDs) by Sollner et al.[1] there has been a growing interest[2-11] in improving the negative differential resistance (NDR) characteristics of the double-barrier resonant tunneling structures (RTS). The two important performance criteria for device applications used at present are the current peak-to-valley ratio (PVR) and the peak current density. However, how these criteria of performance relate to actual device operation has not been explained previously. We have recently derived and presented[12] a figure-of-merit expression for these devices being used as high frequency sources. Using this figure-of-merit as the performance criterion we report experimental results for lattice matched and pseudomorphic InP based double barrier structures with high figure-of-merit numbers. Various ways of improving the figure-of-merit number are incorporated in a pseudomorphic InP based InGaAs/InAlAs deep quantum well structure with a long drift layer that has a higher figure-of-merit than conventionally designed resonant tunneling structures.

II   Figure-of-merit expression for RTDs

To calculate the power output from resonant tunneling devices, first the device area must be selected. To find the area we assume that the device is matched to a minimum achieveable circuit resistance of 1-Ω and the series resistance of the device is negligible. This is an approximation and may be difficult to realize at very high frequencies.

[1]This work was supported by the U.S. Army Research Office under the URI program contract no. DAAL03-87-K-0007.

However, if a larger load or series resistance is present, then the output power will scale inversely proportional to the load resistance. It is further assumed that the device is biased in the negative resistance region of the I(V) curve and the negative resistance is linear. In the high frequency limit, as shown in ref. 12, one may write the rf power as:

$$P_{rf} \approx \frac{J_p{}^2 \left(\frac{J_v}{J_p} - 1\right)^2}{8\omega^2 C^2} \tag{1}$$

where J_p is the peak current density, J_v is the valley current density and C is the device capacitance per unit area. The figure-of-merit(FM) is then defined as

$$FM = \frac{J_p \left(1 - \frac{J_v}{J_p}\right)}{C} \tag{2}$$

and it has the units of Volts/sec. The peak current can be measured experimentally at DC. The measurement of the capacitance, however, for a given structure remains nontrivial. A reasonable conservative guess for the device capacitance can be given by assuming a parallel plate capacitor model, $C = \epsilon/d$, where d is the depletion region width and can be approximated to be the width of the double barrier structure and any intrinsic spacer or drift layers. A word of caution must be added here. This method of calculating the capacitance is an approximation at best since it does not include the effects of charge depletion and doping concentration.

III   Results and Discussion

The choice of the InP based heterostructure material system is based on the advantages it presents for the high frequency operation of the resonant tunneling devices. Inata et al.[8] have shown that a lighter effective electron mass in the barrier improves the NDR of resonant tunneling devices. $In_{0.52}Al_{0.48}As$ has an effective mass of $0.075m_o$ compared to $0.092m_o$ for $Al_{.30}Ga_{.70}As$ and this reduction in the effective mass improves the peak current density.

Another device parameter shown to have an effect on the current PVR is the barrier potential height[12]. For the lattice matched GaAs/AlGaAs system the barrier height is approximately 0.23 eV. An increase in the Al composition of the barrier above a certain fraction and a certain barrier thickness, results in the barrier material becoming an indirect band gap semiconductor which places some restrictions on the design of $GaAs/Al_xGa_{1-x}As$ RTSs. On the other hand, in the $In_{0.53}Ga_{0.47}As/In_{0.52}Al_{0.48}As$ RTSs a relatively large potential barrier($\approx 0.53eV$) can be obtained with no such restrictions on the barrier or well widths because both materials are direct band-gap semiconductors. Thus performance for such a material system can be optimized.

$In_{0.53}Ga_{0.47}As$ has a higher mobility than GaAs which could become an important factor for high frequency RTSs using a transit-time drift region[13]. A practical advantage of the InGaAs material system is that it is relatively easier to make ohmic contacts to InGaAs than GaAs. In fact, by growing a heavily doped InGaAs layer at the point of ohmic contacts, low contact resistance non-alloyed ohmic contacts can be obtained. Non-alloyed contacts are advantageous for resonant tunneling structures because they do not promote diffusion of impurities into the well region where they can enhance impurity scattering and thus reduce the current PVR. The low contact resistance is important since one of the device parameters effecting the maximum oscillation frequency of practical oscillators is the minimum achievable contact resistance[13].

Using the FM as the performance criterion some of the device structures reported in the literature can be analyzed. This is done in Table 1 (300 K) and Table 2 (77 K).

No.	Material System	Capacitance 10^{-7} F/cm^2	J_p kA/cm^2	P/V ratio	Figure-of-merit 10^{10} V/s	Ref
1	GaAs/AlGaAs	5.31	7.77	2.2	0.80	3
2	GaAs/AlGaAs	5.31	6.1	3.9	0.85	3
3	GaAs/AlAs	5.31	1.3	3.6	0.18	3
4	GaAs/AlAs	12.3	40	3.5	2.32	4
5	GaAs/AlAs	8.43	15	3	1.8	5
6	GaAs/AlAs	8.27	10	1.3	0.28	5
7	InGaAs/InAlAs	6.80	63	4.1	7.0	6
8	InGaAs/InAlAs	6.12	22	2.3	2.0	8
9	InGaAs/InAlAs	1.08	0.32	7.1	0.25	9
10	InGaAs/InAlAs	6.14	11	7	1.54	this work
11	InGaAs/InAlAs ($In_{0.70}Ga_{0.30}As$ well)	0.90	19.8	2	11.0	this work
12	InGaAs/AlAs	8.74	23	14	2.44	10
13	InGaAs/AlAs (InAs well)	6.32	6	30	0.92	11
14	InGaAs/AlAs	7.80	32	23	3.92	11
15	InGaAs/AlAs	5.56	15	24	2.59	this work

Table 1: Summary of the figure-of-merit calculation for some of the published device structures and the results presented here for operation at 300 K.

Also included are results that have been recently obtained by us using lattice matched and pseudomorphic InGaAs/InAlAs resonant tunneling structures.

The lattice matched and pseudomorphic heterostructures were grown on n$^+$ InP substrates. The lattice matched $In_{0.53}Ga_{0.47}As/In_{0.52}Al_{0.48}As$ device structure consists of 41.5 Å barriers with a 61.5 Å well. The barriers are enclosed by 15 Å spacer layers and the doping outside the spacer layers in 2×10^{18}. Mesa diodes were fabricated using conventional lithography techniques. Non alloyed ohmic contacts of Ni/Ge/Au/Ti/Au were used. The device structure for the pseudomorphic InGaAs/AlAs RTS consists of 23.2 Å barriers surroynding a 41.5 Å well. The barriers are enclosed by 50 Å spacer layers and the doping outside of the spacer layers is 2×10^{18}. Mesa diodes were fabricated using the procedure described above.

As can be seen from the results in Table 1 and Table 2, a large current PVR does not guarantee a large figure-of-merit number. Although we have reported results with some of the highest current PVRs, the corresponding figure-of-merit is not noteworthy. It is important to realize that in order to increase the FM the peak current density of the structures must be increased and the device capacitance per unit-area must be reduced.

One way to increase the current density is by reducing the barrier width. However, if the barrier widths are reduced past a certain thickness then the barriers do not provide sufficient confinement of the quasi-bound state thus substantially reducing the PVR. To avoid this problem we have fabricated a RTS where the well material is $In_{0.70}Ga_{0.30}As$, the barriers are $In_{0.52}Al_{0.48}As$ and the material outside the barriers is $In_{0.53}Ga_{0.47}As$. To increase the current density the barriers are only 30 Å wide. The well is 70 Å wide. The pseudomorphic well width is limited by critical length considerations. By using a deep quantum well the conduction is through the second quasi-bound state and thus a higher

No.	Material System	Capacitance 10^{-7} F/cm^2	J_p kA/cm^2	P/V ratio	Figure-of-merit 10^{10} V/s	Ref
1	GaAs/AlGaAs	2.98	4	6.5	1.13	2
2	GaAs/AlGaAs	5.16	20	3.5	2.77	2
3	GaAs/AlGaAs	5.31	7.77	7.0	1.25	3
4	GaAs/AlGaAs	5.31	8.6	14.3	1.51	3
5	GaAs/AlAs	5.31	6.4	21.7	1.15	3
6	GaAs/AlAs	8.27	10	2.7	0.76	5
7	InGaAs/InAlAs	6.80	55	11.4	7.4	7
8	InGaAs/InAlAs	6.12	22	11.7	3.29	8
9	InGaAs/InAlAs	6.74	100	6.3	12.5	8
10	InGaAs/InAlAs	1.08	0.41	39	0.37	9
11	InGaAs/InAlAs	6.14	10	21	1.55	this work
12	InGaAs/InAlAs ($In_{0.70}Ga_{0.30}As$ well)	0.90	18.5	6.3	17.3	this work
13	InGaAs/AlAs	8.74	23	35	2.55	10
14	InGaAs/AlAs (InAs well)	6.32	6	63	1.93	11
15	InGaAs/AlAs	5.56	15	51	2.64	this work

Table 2: Summary of the figure-of-merit calculation for some of the published device structures and the results presented here for operation at 77 K.

current density can be expected compared to conducting through the ground state in a conventionally designed structure. To decrease the per unit-area capacitance of the device a relatively long drift layer (1000 Å) after the second barrier has been introduced. The resonant tunneling diodes were fabricated as described earlier. The contacts however were alloyed at 450 degrees for 30 seconds. It is important to realize that with such a long drift layer transit-time effects can become dominant at very high frequencies [13]. Assuming carrier velocity of 10^7, the transit angle for a 1000 Å drift layer at 100 GHz would be 0.2π. This can be neglected at 100 GHz but will become significant at higher frequencies. The figure-of-merit, which is higher for this structure than conventionally designed structures, is presented in Table 1 and Table 2. Note that this structure has not been optimized yet, but rather it establishes the trend.

IV Conclusion

Using the figure-of-merit criterion for resonant tunneling diodes we have reported experimental results for lattice matched and pseudomorphic InGaAs/InAlAs resonant tunneling diodes with high current PVR and high figure-of-merit numbers. Finally, we have reported results for a pseudomorphic InGaAs/InAlAs deep qauntum well structure that was designed to increase the FM number and it has a higher figure-of-merit than conventionally designed resonant tunneling structures.

Acknowledgment: We thank Drs J. Oh and Bhattacharya for material growth.

References

1. T. C. L. G. Sollner, W. D. Goodhue, P. E. Tannenwald, C. D. Parker, and D. D. Peck, Appl. Phys. Lett., Vol. 43, September 1983, pp.588-590.

2. S. Muto, T. Inata, H. Ohnishi, N. Yokiyama,and S. Hiyamizu, Jpn. J. Appl. Phys., Vol. 25,July 1986, pp.L577-L579.

3. C. I. Huang, M. J. Paulus, C. A. Bozada, S. C. Dudley, K. R. Evans, C. E. Stutz, R. L. Hones, M. E. Cheney, Appl. Phys. Lett., Vol. 51,July 1987, pp. 121-123.

4. W. D. Goodhue, T. C. L. G. Sollner, H. Q. Lee, E. R. Brown, and B. A. Vojak, Appl. Phys. Lett., Vol. 49, October 1986, pp.1086-1088.

5. M. Tsuchiya and H. Sakaki, Jpn. J. Appl. Phys., Vol. 25, March 1986, pp. L185-L187.

6. Y. Sugiyama, T. Inata, S. Muto, Y. Nakata, and S. Hiyamizu, Appl. Phys. Lett. , Vol. 52, January 1988, pp. 314-316.

7. S. Muto, T. Inata, Y. Sugiyama, Y. Nakata, T. Fujii, and S. Hiyamizu, Jpn. J. Appl. Phys., Vol. 26, March 1987, pp. L220-L222.

8. T. Inata, S. Muto, Y. Nakata, T. Fujii, H. Ohniski, and S. Hiyamizu, Jpn. J. Appl. Phys., Vol. 25, December 1986,pp. L983-L985.

9. T. Inata, S. Muto, Y. Nakata, S. Sasa, T. Fujii, and S. Hiyamizu, Jpn. J. Appl. Phys., Vol. 26, August 1987, pp. L1332-L1334.

10. A. A. Lakhani, R. C. Potter, D. Beyea, H. H. Hier, E. Hempfling, L. Aina, J. M. O'Connor, Electronics Letters, Vol. 24, 4th February 1988, pp.153-154.

11. T. P. E. Broekaert, W. Lee, and C. G. Fonstad, Appl. Phys. Lett., Vol. 53, October 1988, pp. 1545-1547.

12. R. K. Mains, I. Mehdi, and G. I. Haddad, presented at the 13th International Infrared and Millimeter Waves Conference, December 1988, Hawaii. Also to be published in the International Journal of Infrared and Millimeter Waves, April 1989.

13. V. P. Kesan, D. P. Neikirk, P. A. Blakey, B. G. Streetman, and T. D. Linton, Jr., IEEE Trans. Elec. Dev., Vol. 35, April 1988, pp. 405-413.

FIELD/SCATTERING INTERACTION QUANTIZATION
IN HIGH-FIELD QUANTUM TRANSPORT[1]

R.Bertoncini
A.M.Kriman
D.K.Ferry

Center for Solid State Electronics Research
Arizona State University
Tempe, Arizona 85287-6206

I. INTRODUCTION

Rapid development in the area of submicron devices has provided a renewed interest in a theory of electronic transport that goes beyond the semiclassical Boltzmann equation (1). This theory is based on the validity of the adiabatic approximation and on perturbation theory. The electrons are represented by nearly stationary, free-particle-like states of momentum **k**, and scattering is treated as weak and infrequent and the external fields are considered as weak and slowly varying in both space and time scales.

In ultrasmall semiconductor components, however, both electric fields and scattering rates can reach very high values and the spatio-temporal variations of the phenomena involved begin to be comparable to those of the microscopic interactions. In these regimes, the limits of applicability of the Boltzmann theory must be surpassed and transport must be considered from a fully quantum mechanical viewpoint since neglecting the uncertainty relations can now lead to erroneous results (2).

Two important quantum effects that appear as consequences of the position-momentum uncertainty relation and the interference of various perturbations are collisional broadening (CB) and the intra-collisional field effect (ICFE) (2) and they should be taken into account in a proper treatment of quantum transport.

The formulation of a scheme capable of describing non-linear transport phenomena has been a long-standing and debated problem theoretically, and most formalisms have been restricted to gradient expansions in the external fields (3,4). Some models to include collisional broadening within a Monte Carlo approach have been proposed (5,6), but they are far from satisfactory since they do not have any solid first-principles justification. On the other hand, only a formalism has been proposed for the ICFE (7).

Several approaches are based upon the Kadanoff-Baym-Keldysh (KBK) non-equilibrium Green's function technique (8). This technique allows, at least in principle, a non-perturbative calculation by including the field in the "unperturbed" (absence of scattering) Hamiltonian. However, current applications of the KBK method have been limited to weak fields and/or slowly varying systems (9) and have not been applied to problems beyond quadratic response.

In this work, we calculate the spectral density function $A(\mathbf{k},\omega)$, within the KBK formalism. Through $A(\mathbf{k},\omega)$, which represents the probability distribution for finding an electron with energy $\hbar\omega$ in a state of momentum $\hbar\mathbf{k}$ we are able to account for both CB and ICFE in a relatively simple and rigorous way. We demonstrate the technique by considering a problem in which non-polar optical phonon scattering and a high electric field are simultaneously applied to the semiconductor. We show that the field, acting in conjunction with the phonons, leads to an effective quantization of the energy (motion) in the direction of the electric field.

---

[1]Work supported by the Office of Naval Research.

II. THE AIRY TRANSFORM REPRESENTATION

The spectral function can be defined (10) as the anti-commutator

$$A(r,t;r',t') = \left\langle \left\{ \Psi(r,t), \Psi^+(r',t') \right\} \right\rangle \quad , \tag{1}$$

where Ψ^+ and Ψ are the creation and annihilation field operators for fermions and the bracket <...> indicates the non-equilibrium expectation value. The equivalent definition (11)

$$A(\mathbf{k},\omega) = -2 \operatorname{Im} [G^r(\mathbf{k},\omega)] \tag{2}$$

can be used to calculate the spectral density in reciprocal space $(\mathbf{k},\omega)$ through the knowledge of the non-equilibrium retarded Green's function (3)

$$G^r(r,t;r',t') = -\frac{i}{\hbar}\theta(t-t')\left\langle \left\{ \Psi(r,t),\Psi^+(r',t') \right\} \right\rangle \quad . \tag{3}$$

G^r is determined by solving the appropriate Dyson's equation, which can be formally written as

$$G^r = G_0^r + G_0^r \Sigma^r G^r \quad , \tag{4}$$

where G_0^r is the retarded free-particle propagator and the self-energy Σ^r describes the interaction mechanisms as well as the external fields.

In order to describe high-field effects we must be able to handle the difficulties associated with an electron system that is no longer translationally invariant in the direction of the applied electric field (1). This implies, as a consequence, that momentum along the field is no longer a good quantum number and that Dyson's equation cannot be diagonalized in the basis of the unperturbed, free-particle Hamiltonian. Low-field formulations can ignore this, since the field is treated as a perturbation of a homogeneous system.

In the attempt to overcome these difficulties, many authors (12,13,14), have represented the electric field E by a vector potential $A(t) = -\int E(t')dt'$. This breaks time-translational invariance, which is equally important in the proof of conservation laws. A kind of translation symmetry *does* persist in the presence of a constant field: a real-space translation along the field, combined with a shift of the *energy*, preserves the form of the problem (13). However, describing the system in terms of shifted coordinates, like $(z-\hbar\omega/eE)$, does not de-convolve Dyson's equation. The introduction of the Wigner coordinates, used in many earlier attempts to go beyond small perturbative fields, does not make the problem easier, but actually confuses it by artificially introducing the extra center-of-mass time T in a scalar potential gauge and the extra center-of-mass coordinate $\mathbf{R}$ if a vector potential is used.

Therefore, we have used a different approach. In order to treat the electric field (represented by a scalar potential $\phi = eEz$) exactly, we work in the Hilbert space defined by the normalized eigenfunctions of an electron in a uniform electric field, i.e. plane waves on the plane normal to the field and Airy functions of the first kind along the direction of the field. This allows us to define a coordinate system $(\mathbf{k}_\perp, s)$ (where $\mathbf{k}_\perp = k_x, k_y$ and s is the argument of the Airy function) for Fourier-transforming to momentum in the transverse directions and to Airy coordinates s along the field. The transformation that connects the two coordinate systems (x,y,z) and $(\mathbf{k}_\perp, s)$ is defined by the integral operation

$$g(\mathbf{k}_\perp, s) = \int \frac{d\rho}{2\pi} \int \frac{dz}{L} e^{i\mathbf{k}_\perp \cdot \rho} Ai\left(\frac{z-s}{L} \right) f(\rho z) \tag{5}$$

where $f(\rho,z)$ is an arbitrary function, ρ is the transverse position vector and $L=(\hbar^2/2meE)^{1/3}$. The Airy transform variable s has a physical interpretation as the turning point in z of a non-coupled electron with energy $\varepsilon(\mathbf{k}_\perp,s)=\hbar^2 k_\perp^2/2m+eEs$. In this space, a function diagonal in both $\mathbf{k}_\perp$ and s variables is translationally invariant in the transverse direction, but not along the z-direction. This is a very appealing property since it implies the possibility of dealing with diagonal functions without requiring an assumption of translational invariance along the direction of the applied field.

We take G_o^r in Eqn. (4) to be the Green's function for an electron in the presence of the electric field, but without scattering and denote it by G_E^r. This approach has been used previously (15), but in our formalism, the unperturbed, field-dependent Green's function can be written down quickly as

$$G_E^r(k_\perp,s,t\text{-}t') = -\,i\,\theta(t\text{-}t')\exp[\,-\frac{i}{\hbar}\epsilon_{k_\perp,s}(t\text{-}t')\,]\ .\tag{6}$$

Dyson's equation in $(k_\perp,s)$-space reads :

$$G^r(k_\perp,s,s';\omega) = G_E^r(k_\perp,s;\omega)\,\delta(s\text{ - }s')$$

$$+\,G_E^r(k_\perp,s,\omega)\int ds_2\,\Sigma^r(k_\perp,s,s_2;\omega)G^r(k_\perp,s_2,s';\omega)\ .\tag{7}$$

For the self-energy Σ^r we use the approximation (11,16)

$$\Sigma^r(r,t;r't') = iD_o^>(r,t;r't')G^r(r,t;r',t')\ ,\tag{8}$$

which is valid for non-degenerate systems, where $D_o^>$ is the equilibrium phonon correlation function

$$D_o^>(r,t;r',t') = \left\langle \phi(r,t)\phi^+(r',t')\right\rangle\ ,\tag{9}$$

with ϕ and ϕ^+ the phonon-field operators.

We consider non-polar optical processes and assume only one-phonon scattering with the approximation (17) $G^r = G_E^r$ in Eqn. (8). Then $\Sigma^r(k_\perp,s,s_2;\omega)$ is highly peaked about $s=s_2$. This makes a diagonal approximation for Σ^r appropriate, and we obtain

$$\Sigma^r(s,\omega) = \frac{2\pi}{\hbar}|V|^2\sum_{\eta=\pm 1}(N_o + \frac{\eta+1}{2})F(s,\omega)$$

$$\text{Re}[F(s,\omega)] = \frac{2\pi}{\sqrt{2}}\,\frac{m^{3/2}}{\hbar^2}\,\Theta^{1/2}[Ai'(\zeta)Bi'(\zeta) - \zeta Ai(\zeta)Bi(\zeta) + \frac{\sqrt{\zeta}}{\pi}\theta(\zeta)]\tag{10}$$

$$\text{Im}[F(s,\omega)] = \frac{2\pi}{\sqrt{2}}\,\frac{m^{3/2}}{\hbar^2}\,\Theta^{1/2}\Big[Ai'^2(\zeta) - \zeta Ai^2(\zeta)\Big]\ ,$$

where $\Theta = [3(\hbar eE)^2/2m]^{1/3}$, $\zeta = [eEs\text{-}\hbar(\omega-\eta\omega_o)]/\Theta$.

III. RESULTS

Fig.1 shows the real part of the self-energy as a function of the argument ζ. The oscillatory nature of the self-energy has not been seen previously in other treatment of the high-field behavior and is a consequence of the non-perturbative inclusion of the electric field in the problem. These oscillation imply the presence of regions in which the electron energy is lowered and regions where the energy is raised. These regions alternate with one another and the zero-crossings in the figure represent the quantized energies

215

which the quasi-particle energy concentrates. The zero-crossings occur asymptotically where $\zeta = [3\pi(2n+1)/8]^{2/3}$. Because of the irrational factor in $\Theta = 3^{1/3}eEL$, the oscillations are incommensurate with those occurring in the phonon-decoupled problem.

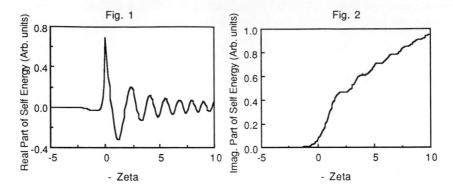

Fig. 1. The real part of the self-energy in the presence of both optical phonon scattering and high accelerating electric fields, in the Airy coordinates, defined in the text.

Fig. 2. The imaginary part of the self-energy in the Airy coordinate, which is defined in the text.

In Fig.2 we plot the results for the imaginary part of the self-energy, which is related to the scattering rate Γ as (5)

$$\Gamma(k_\perp,s,\omega) = \frac{2}{\hbar} \, \text{Im}[\Sigma^r(k_\perp,s,\omega)] \qquad . \qquad (11)$$

The presence of the ICFE is found to generate a tail in the scattering rate for $\zeta < 0$ and a series of damped oscillations at $\zeta > 0$ associated with the oscillations in the real part of the self-energy and reflecting the influence of the quantized levels through the Airy functions. The existence of such a tail smooths out the sharp threshold in energy of the scattering rate, making possible transitions which cannot occur in the absence of the field. The appearance of step-like oscillations, on the other hand, signals the onset of additional densities of final states corresponding to the subbands generated by each quantized level.

With this model for the self-energy, Dyson's equation (7) is a multiplicative equation and can be immediately solved for the retarded Green's function :

$$\text{Re}[G^r(k_\perp,s,\omega)] = \frac{\hbar\omega - \epsilon_{k_\perp,s} - \text{Re}\Sigma^r(s,\omega)}{[\hbar\omega - \epsilon_{k_\perp,s} - \text{Re}\Sigma^r(s,\omega)]^2 + [\text{Im}\Sigma^r(s,\omega)]^2} \qquad (12)$$

$$\text{Im}[G^r(k_\perp,s,\omega)] = \frac{\text{Im}\Sigma^r(s,\omega)}{[\hbar\omega - \epsilon_{k_\perp,s} - \text{Re}\Sigma^r(s,\omega)]^2 + [\text{Im}\Sigma^r(s,\omega)]^2} \qquad . \qquad (13)$$

The spectral density found in (14), using the definition (2), is plotted in Fig. 3. It is positive definite and satisfies the normal sum rules (18). In Airy coordinates, it can exhibit an unusual double peak near the zero point (where the limiting δ-function occurs, which is the semiclassical turning point). This double-peak structure suggests that there is a length scale associated with the motion along the electric field direction. We conjecture that this motion might therefore be more appropriately treated in terms of "hopping " transport in the z-direction between states described by discrete values of the Airy coordinate.

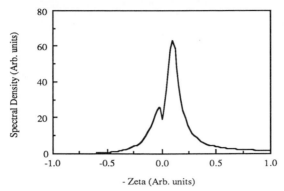

Fig. 3. The spectral density that arises from the Green's function that incorporates both the accelerating field and the self-energy shown in Figs. 1,2. The double peak structure is thought to incorporate the preferred local shift in z-z' that accompanies the interaction of the field and the scattering.

IV. SUMMARY

We have introduced a treatment in which the momentum coordinate representation along the field direction is replaced by a representation in terms of Airy coordinates. This yields the mathematical advantage of simplified Dyson's equations involving fewer coordinates. We have applied this to the case of non-polar optical phonon scattering in semiconductors in high electric fields. The results provide new insight into the coupled electron-phonon system in these high fields.

REFERENCES

1. For a recent review on the subject see Reggiani, L. (1985). Physica . 134, 123.
2. Barker, J.R., and Ferry, D.K. (1980). Sol. State Electron. 23, 531.
3. Kadanoff, L.P., and Baym, G. (1962). "Quantum Statistical Mechanics." W.J.Benjamin, New York.
4. Wilkins, J.W. (1972). Phys. Rev. B6, 3189.
5. Fischetti, M.V., and DiMaria, D.J. (1985). Phys. Rev. Lett. 42, 2475.
6. Porod, W., and Ferry, D.K. (1979). Physica 134B, 137.
7. Barker, J.R., and Ferry, D.K. (1979). Phys. Rev. Letters 42, 1779.
8. A review of these techniques is Rammer, J., and Smith, H. (1986). Rev. Mod. Phys. 58, 323.
9. Mahan, G.D. (1984). Phys. Rep. 110, 321. and references contained therein.
10. Davies, J.H., and Wilkins, J.W. (1988). Phys.Rev. 38, 1667.
11. Langreth, D.C. (1976). In "Linear and Non-linear Electron Transport in Solids" (J.T.Devreese and E. vanBoren, ed.) Plenum, New York.
12. Jauho, A.P., and Wilkins, J.W. (1984). Phys. Rev. B29, 1919.
13. Krieger, J.B., and Iafrate G.J. (1986). Phys. Rev. B33, 5494. and (1987). Phys.Rev. B35, 9644.
14. Reggiani, L., Lugli, P., and Jauho, A.P. (1987). Phys. Rev. B36, 6602.
15. Khan, F.S., Davies, J.H., and Wilkins, J.W. (1987). Phys. Rev. B36 , 2578.
16. Langreth, D.C., and Wilkins, J.W. (1972). Phys. Rev. B6, 3189.
17. Fetter, A.L., and Walecka, J.D. (1971). "Quantum Theory of Many-Particle Systems." McGrew-Hill, New York.
18. Mahan, G.D. (1981). "Many-Particle Physics." Plenum, New York.

Chapter 5

Equilibrium and Nonequilibrium Response in Nanoelectronic Structures

NONEQUILIBRIUM ELECTRON DYNAMICS IN SMALL SEMICONDUCTOR STRUCTURES

A. F. J. Levi
S. Schmitt-Rink

AT&T Bell Laboratories
Murray Hill, New Jersey 07974

I. INTRODUCTION

Electrons in thermal equilibrium with a semiconductor lattice of temperature T have average excess kinetic energy $E \sim k_B T$, where k_B is Boltzmann's constant. Electron motion is controlled by diffusive processes and typical diffusion velocities are around $v_{diff} \simeq 5 \times 10^6$ cms^{-1}. A non-equilibrium electron can have excess kinetic energy significantly higher than the ambient thermal energy (e.g. $E \gtrsim 10 \, k_B T$) and electron motion is no longer diffusive. In a typical direct band gap III-V semiconductor this can lead to conduction band electron velocities approaching 10^8 cms^{-1}. Thus, extreme nonequilibrium electron transport allows for more than an order of magnitude increase in velocity compared to the diffusive regime. Clearly, understanding nonequilibrium transport in semiconductors has technological significance by, for example, assisting in the design of very high speed transistors. There are also fundamental issues, such as the interplay between quantum mechanical reflection (elastic scattering) and dissipative processes (inelastic scattering), which are important to understand. In this note we explore qualitative changes in the physics of device operation when dimensions are comparable to an electron mean free path.

Figure 1 shows a schematic band diagram of a N-p-n heterojunction bipolar transistor (HBT) of base width Z_b and collector depletion region width Z_c. A bias voltage V_{be} applied between emitter and base causes electrons to flow across the abrupt emitter/base junction. An extreme nonequilibrium distribution of electrons is injected into the base with excess kinetic energy E and a large forward component of momentum. The mean free path of these carriers in the base region is determined by inelastic scattering with the coupled excitations of the p-type majority carriers/optical phonons and elastic scattering from ionized impurities.

A typical HBT operates by transforming a low impedance current source (the emitter) into a high impedance current source (the collector). Electrons injected from the forward biased emitter/base diode traverse the base and are accelerated in the electric field of the reverse biased base/collector diode. Irreversible scattering processes, such as optical phonon emission, take place in the collector depletion region, the electron loses energy, remaining close in energy to the

NANOSTRUCTURE PHYSICS
AND FABRICATION

221

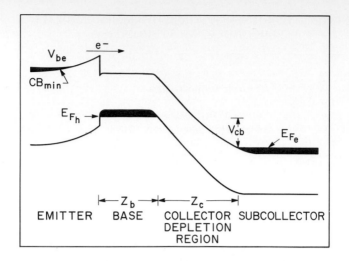

Fig. 1   Schematic band diagram of a N-p-n HBT under bias.
Base width Z_b, collector transit region width Z_c,
conduction band minimum CB_{min} and electron (hole)
quasi-Fermi level E_{F_e} (E_{F_h}) are indicated.

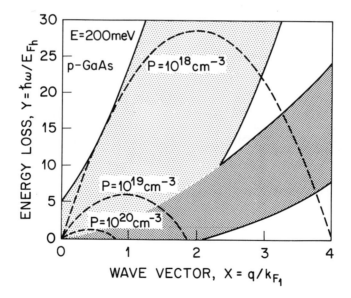

Fig. 2   Dispersion relation for single particle excitations in p-
type GaAs. Broken lines are parabolas of integration for
three different p-type carrier concentrations and for
conduction band electron injection energy E = 200 meV.

conduction band minimum, and reverse current flow against the electric field in the depletion region is small. In this way the transformation to high impedance occurs over the length of the collector depletion region. The situation is qualitatively different for a transistor in which at least a portion of current flow through the base and collector is via coherent "ballistic" transport. In such a transistor the impedance transformation for coherent electrons takes place in the degenerately doped n^+ subcollector contact where the electron suffers scattering events. The probability amplitude of the coherent electron wave function, $|\psi|^2$, is damped at the injector and collector contact where electrons are scattered into and out of this state. Under these operating conditions it is clear that scattering at the n^+ subcollector contact could influence the emitter current resulting in a fundamental change in the impedance transforming properties of the transistor.

II. EFFECT OF REDUCING Z_b ON BASE/TRANSPORT

Extreme nonequilibrium electron transport in the base and collector depletion region has been observed experimentally in AlGaAs/GaAs HBTs in which Z_b and Z_c are comparable to the electron mean free path (1). However, in a useful device a reduction in Z_b requires an increase in p, the majority charge carrier density in the base, to keep the base sheet resistance $R_{b\square}$ acceptably small. For example, GaAs with a p-type carrier concentration of $p = 2 \times 10^{20}$ cm^{-3} has a mobility $\mu \simeq 50$ cm^{2}V^{-1}s^{-1} (2), so that a sheet resistance of $R_{b\square} \lesssim 200$ $\Omega_\square$ requires $Z_b \gtrsim 300\text{Å}$. Now consider the effect these parameters have on nonequilibrium electron transport in the base. A conduction band electron of energy E above the conduction band minimum CB$_{min}$ is injected into the p-type base. This electron may scatter inelastically, losing energy $Y = \hbar\omega/E_{F_h}$ and changing momentum by $X = q/k_{F_1}$, where E_{F_h} is the Fermi energy of the majority p-type carriers in the base and k_{F_1} is the Fermi wave vector of the heavy-hole band. To calculate the total inelastic scattering rate $1/\tau_{in}$ for a fixed injection energy E, we need to integrate over the spectral weight of the coupled majority carrier/optical phonon excitations in p-type GaAs (3). Within effective mass theory, energy and momentum conservation results in a parabola of integration which, for the indicated p-type carrier concentration, leads to the broken lines plotted in Fig. 2. Calculation of $1/\tau_{in}$ requires integration of the spectral weight within the parabola. It is clear from the figure that small values of p involve an integration over a large portion of the heavy-hole intraband single particle excitations (dark shaded region in Fig. 2), which carry most of the inelastic scattering strength. With increasing p a reduced portion of phase space is integrated. Although, for large values of p, the maximum scattering strength increases, over the small region of phase space in which the integration takes place the scattering strength can decrease. A consequence of this is that, with increasing carrier concentration, $1/\tau_{in}$ increases, reaches a maximum and then decreases. This fact is illustrated in Fig. 3(a), in which results of calculating $1/\tau_{in}$ for GaAs as a function of p are given for three different values of E. For low values of $p \lesssim 10^{17}$ cm^{-3} optical-phonon scattering dominates. For $p \sim 2 \times 10^{19}$ cm^{-3} the scattering rate reaches a maximum and at very high values of $p \gtrsim 5 \times 10^{20}$ cm^{-3}, $1/\tau_{in}$ decreases, becoming less than the bare optical-phonon scattering rate because of screening effects. Thus, because with decreasing Z_b we must increase p, a fundamental limit to device performance, $1/\tau_{in}$, can decrease. In addition, since the physics

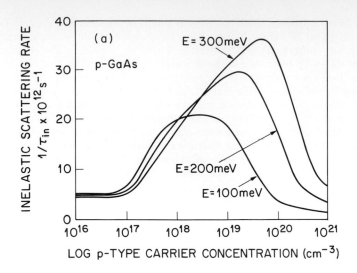

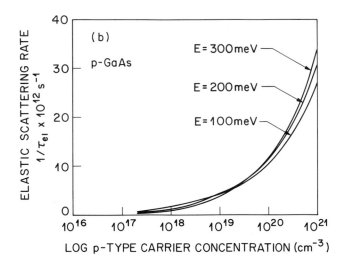

Fig. 3   (a)   Total inelastic scattering rate $1/\tau_{in}$ as function of p-type carrier concentration in GaAs for three indicated values of electron energy E.

(b)   Total elastic electron scattering rate $1/\tau_{el}$ for p-GaAs. Parameters used in calculations were effective heavy-hole mass $= 0.5m_0$, light-hole mass $= 0.082m_0$, conduction band effective electron mass $= 0.07m_0$, high-frequency dielectric constant $\epsilon_\infty = 10.91$, longitudinal optical-phonon energy $\hbar\omega_{LO} = 36.5$ meV, transverse optical-phonon energy $\hbar\omega_{TO} = 33.8$ meV, and T = 0.

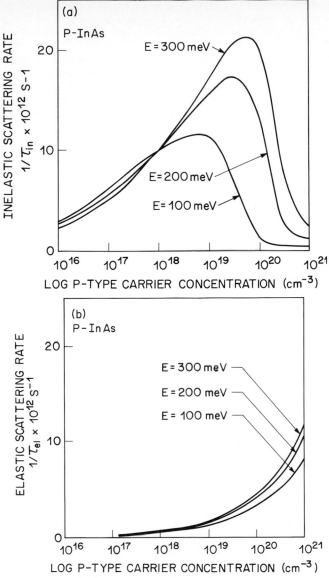

Fig. 4   (a) Total inelastic scattering rate $1/\tau_{in}$ as function of
p-type carrier concentration in InAs for three indicated
values of electron energy E.

(b) Total elastic electron scattering rate $1/\tau_{el}$ for p-
InAs. Parameters used in calculations were effective
heavy-hole mass = $0.41m_0$, light-hole mass = $0.025m_0$,
conduction band effective electron mass = $0.021m_0$,
high-frequency dielectric constant ϵ_∞ = 11.8,
longitudinal optical-phonon energy $\hbar\omega_{LO}$ = 30.2 meV,
transverse optical-phonon energy $\hbar\omega_{TO}$ = 27.1 meV, and
T = 0.

225

underlying the decrease in $1/\tau_{in}$ at high p is based on a phase space argument, the results are general and apply to other material systems such as $Ga_{0.47}In_{0.53}As$ and InAs. In Fig. 4(a) we show results of calculating $1/\tau_{in}$ as a function of p for InAs.

If the high carrier concentration in the base is created using randomly positioned impurities, then it is necessary to consider the elastic scattering rate due to those ionized impurities, $1/\tau_{el}$. The results of calculating $1/\tau_{el}$ as a function of p for three values of E are shown for GaAs in Fig. 3(b) and InAs in Fig. 4(b). It is clear from the figures that $1/\tau_{el}$ is the dominant scattering process for $p \gtrsim 5 \times 10^{20}$ cm^{-3}. Rather than attempt to optimize $1/\tau_{el}$ and $1/\tau_{in}$, it is better to eliminate the contribution from $1/\tau_{el}$ by placing impurities in a periodic superlattice (4). One possible approach towards achieving this might utilize atomic layer epitaxy techniques such as delta doping.

III. FEEDBACK BETWEEN ELASTIC AND INELASTIC SCATTERING

As Z_b and Z_c are reduced in length coherent electron transport between emitter and n^+ subcollector contact becomes possible. In this situation, local equilibrium concepts such as diffusive transport become invalid and we must consider the combined effects of dissipation and elastic scattering from the conduction band profile. To correctly describe this process requires care. For example, in a quantum mechanical problem involving both elastic and inelastic scattering channels one has to ensure correct normalization of the particle wave function. If the probability of finding the particle in the initial state is unity then, after interaction, the sum over all possible final state probabilities must also be unity. This unitary condition leads directly to a feedback mechanism by which inelastic scattering processes influence the probability of elastic scattering (5). Clearly this feedback mechanism, which is beyond the scope of simple perturbation theory, could be of importance in the design of very small device structures.

Rather than solve the problem for the complicated structure illustrated in Fig. 1 we have so far considered the simpler problem of a single conduction band electron with initial kinetic energy E impinging on a potential barrier of energy V_0 and width δ, see Fig. 5. Confined within the barrier are Einstein phonons of frequency ω to which the electron couples. The electron may be inelastically scattered inside the barrier, and is either transmitted or reflected emerging with energy $E' = E \pm n\hbar\omega$, where n is an integer. We have solved this problem for a tight-binding Hamiltonian by employing a continued fraction expansion (5).

Electron-phonon coupling is usually small so that changes in transmission due to phonons can only be seen on a difference scale. In Fig. 6 we show results of calculating the change in transmission coefficient as a function of electron energy, E, for the case T=0, $\hbar\omega = 0.2t$, and $V_0 = 1.6t$ where 4t is the conduction band width. For low electron energy, $E < -2t + \hbar\omega$, no real phonons can be emitted but fluctuations can lower the potential barrier increasing the total elastic current. Inelastic transmission sets in at the threshold for 1-phonon emission $E = -2t + \hbar\omega$. The elastic (0-phonon) transmission decreases at that point because electron probability is transferred into the inelastic (1-phonon) channel. Notice that the transmission coefficient exhibits singularities (cusps and infinite slopes) at the threshold energies for phonon emission, $E = -2t + n\hbar\omega$, $n \geq 1$.

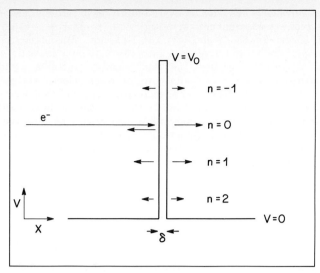

Fig. 5   Schematic diagram of an electron of initial energy E
approaching a potential barrier of average energy V_0 and
width δ. Inelastically scattered electrons can be
transmitted or reflected at the barrier and emerge with
energy $E' = (E \pm n\hbar\omega)$, where n is an integer and ω is the
phonon frequency.

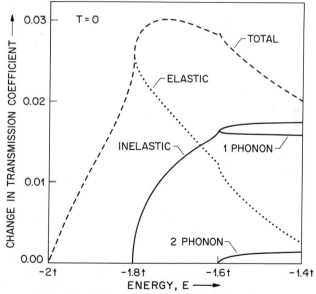

Fig. 6   Change in total (broken curve), elastic (dotted curve),
and inelastic (solid curve) transmission coefficient
(relative to static-barrier case) for $V_0 = 1.6t$, $V_1 = 0.4t$,
$\hbar\omega = 0.2t$, $T = 0$. The conduction bandwidth is 4t and
the phonon amplitude is V_1.

This interesting threshold behavior is well known from atomic and nuclear reactions and is a direct consequence of unitarity (6,7). At $E = -2t + 2\hbar\omega$, 2-phonon emission becomes possible and now the 1-phonon transmission decreases because electron probability is transferred into the 2-phonon inelastic channel. Elastic current decreases because inelastic current increases and total transmission decreases because of the admixture of lower velocity states. In our example for $T = 0$ (Fig. 6) we find a change in total transmission with infinite slope at the 1-phonon threshold and a small cusp at the 2-phonon threshold. This translates respectively into a peak and a dispersive signal in the second derivative of a typical tunnel junction's current-voltage characteristic. Experimentally, 1-phonon peaks are observed and the measured 2-phonon structure appears to have just such a dispersive signature (8).

IV. CONCLUSIONS

Reducing vertical dimensions in electronic devices such as heterojunction bipolar transistors results in coherent electron transport through the active region of the device. This fact raises new questions concerning the physics of device operation, outside the range of conventional theories. Clearly, if we are to make useful high speed devices with small active regions we will require a better understanding of nonequilibrium electron dynamics in the relevant structures, in particular the interplay between irreversible processes and scattering off the various band discontinuities.

REFERENCES

1. Berthold, K., Levi, A. F. J., Walker, J., and Malik, R. J. (1988). Appl. Phys. Lett. *52*, 2247 and Berthold, K., Levi, A. F. J., Walker, J., and Malik, R. J. (1989). Appl. Phys. Lett. *54*, 813.
2. Madelung, O., Schultz, H., and Weiss, H. (Eds): "Landolt-Bornstein tables" (Springer, Berlin, 1982), Vol. III/17a and references therein.
3. Levi, A. F. J. (1988). Electronics Lett. *24*, 1273.
4. Levi, A. F. J., McCall, S. L., and Platzman, P. M. (1989). Appl. Phys. Lett. *54*, 940.
5. Gelfand, B. Y., Schmitt-Rink, S., and Levi, A. F. J. (1989). Phys. Rev. Lett. *62*, 1683.
6. Landau, L. D., and Lifshitz, E. M.: "Quantum Mechanics" (Pergamon, Oxford, 1981).
7. Breit, G. (1952). Phys. Rev. *107*, 1612.
8. Bowser, W. M., and Weinberg, W. H. (1977). Surf. Science *64*, 377.

BALLISTIC TRANSPORT IN THE VERTICAL AND HORIZONTAL DOMAINS

M. Heiblum

IBM Research Division, T. J. Watson Research Center,
Yorktown Heights, NY, 10598

The use of hot electrons for ballistic transport studies has proven to be extremely useful since it provides a simple way to separate the ballistic ensemble (which is hot and directional) from the non-ballistic one (which is cooler and less directional). This idea was implemented to verify unambiguously that ballistic transport is indeed possible in semiconductors. The realization of ballistic transport led also to many other spin-offs enhancing the understanding of the transport. In my talk I describe recent results utilizing Tunnelling Hot Electron Transfer Amplifier (THETA) devices as amplifier devices and as energy spectrometers; they have served as an almost ideal mini-laboratory for studying hot electron ballistic effects. Since by the time these proceedings will be published all the material covered in my talk will be published, I'll use this space to only summarize my talk briefly, and supply the relevant references.

In principle, our device is composed of a double barrier structure and three conductive layers in between. Contacts are formed to the conductive layers and biasing is applied across each barrier. The first barrier is used as a tunnel injector of a quasi monoenergetic hot electron beam moving forward, and the second (a thicker one) to do energy spectroscopy of the arriving electrons (1). Spectroscopy measurements were used to prove, for the first time, unambiguously, the existence of ballistic electron transport in a semiconductor (2). This was done in a vertical type device where conductive and barrier layers layers were deposited sequentially by MBE (doped GaAs and undoped AlGaAs, respectively). With further improvements ballistic fractions greater than 90% have been observed in vertical GaAs devices (3). Similar experiments were subsequently carried out on hot hole devices. Ballistic transport of holes was also observed after separating the light holes from the heavy hole population and injecting the light ones into the structure. In the p type devices light hole ballistic fractions smaller than 10% have been measured to traverse heavily doped 30 nm wide GaAs layers (4).

At low temperatures, as the energy of the ballistic electrons increases, a clear threshold can be observed when their energy exceeds 36 meV, the longitudinal optical phonon energy in GaAs. (5) Electrons having this energy can emit a phonon, be removed from the ballistic distribution, and never surmount the collector barrier. As the energy of the ballistic electrons increases, the velocity of the electrons increases and overall scattering cross - sections decrease, resulting in a longer mean free path. However, when the injection energy increases above the L - Γ valleys separation, electrons can scatter into the L valleys, which have a high density of states to scatter into, and the mean free path abruptly decreases (6). In order to increase the mean free path further, pseudomorphic layers of InGaAs had replaced GaAs in the active layer of the device. In these layers the L - Γ separation can be larger by about 100 meV than in GaAs, enabling thus to inject electrons at higher energies. Indeed current gain in these modified THETA devices is the highest ever reported in hot electron devices (about 40 at 4.2 K). (7)

229

The carriers that did not scatter, the ballistic ones, maintain their phase and strongly interfere among themselves after reflecting from potential band discontinuities as they cross interfaces. Electrons and holes injected from a bulk emitter, through a tunnel barrier, into a quasi two dimensional layer (a potential well) exhibit strong resonances as the injection energy crosses the confined levels in the well. Virtual levels are also observed. These are resonances due to reflections of electrons, with energies greater than the potential heights confining the wells, from the abrupt band discontinuities forming the wells. (8) These resonances were analyzed to determine the mass of the carriers and the non - parabolicity of the energy band.

More favorable conditions for ballistic transport occur in the plane of high mobility 2DEG where no heterointerfaces and dopant atoms are in the current path. We have recently fabricated lateral hot electron structures where the ballistic hot electrons are injected and propagate in 2DEG. The two potential barriers were induced with negatively biased thin metal gates (some 50 nm long) deposited above the surface of a high mobility 2DEG. After establishing lateral tunnelling and hot electron injection through these induced potential barriers in the plane, we have observed lateral ballistic transport of hot electron beams and found ballistic mean free paths longer than 0.5 micron. (9) With some modifications, lateral THETA devices with current gain as high as 100 was measured at 4.2 K. Note that this gain was measured in a device with an active layer wider than 200 nm in the dimension of transport, compared to the vertical devices where the active layers were only 20 nm wide (10).

I'd like to acknowledge my collaborators who have participated in this work and coauthored our joint papers. The work was partly supported by DARPA and administered by ONR, contract #N00014-87-C-0709.

REFERENCES

1. M. Heiblum, Solid - St. Electron. **24**, 343 (1981).
2. M. Heiblum, M. I. Nathan, D. C. Thomas, and C. M. Knoedler, Phys. Rev. Lett. **55**, 2200 (1985).
3. M. Heiblum, I. M. Anderson, and C. M. Knoedler, Appl. Phys. Lett. **49**, 207 (1986).
4. M. Heiblum, K. Seo, H. P. Meier, and T. W. Hickmott, Phys. Rev. Lett. **60**, 828 (1988).
5. M. Heiblum, D. Galbi, and M. Weckwerth, Phys. Rev. Lett. **62**, 1057 (1989).
6. M. Heiblum, E. Calleja, I. M. Anderson, W. P. Dumke, C. M. Knoedler, and L. Osterling, Phys. Rev. Lett. **56**, 2854 (1986); I. Hase, H. Kawai, S. Imanaga, K. Kaneko, and W. Watanabe, International Workshop on Future Electron Devices: Superlattice Devices, (Japan, 1987), Conference Procedings, p. 63.
7. K. Seo, M. Heiblum, C. M. Knoedler, J. Oh, J. Pamulapati, and P. Bhattacharya, Electron Dev. Lett. **10**, 73 (1989).
8. M. Heiblum, M. V. Fischetti, W. P. Dumke, D. J. Frank, I. M. Anderson, C. M. Knoedler, and L. Osterling, Phys. Rev. Lett. **58**, 816 (1987).
9. A. Palevski, M. Heiblum, C. P. Umbach, C. M. Knoedler, A. N. Broers, and R. H. Koch, Phys. Rev. Lett. **62**, (1989).
10. C. P. Umbach, A. Palevski, and M. Heiblum, Submitted to Appl. Phys. Lett.

QUANTUM KINETIC THEORY OF NANOELECTRONIC DEVICES[1]

William R. Frensley

Texas Instruments Incorporated
Dallas, Texas 75265

1. INTRODUCTION

One of the principal features of nanometer-scale structures is the appearance of quantum-mechanical effects in such phenomena as electron transport. Many of these effects, including the quantum Hall effect [1], the Aharonov-Bohm effect [2], and universal conductance fluctuations [3], are observed in systems which are kept very near to thermal equilibrium (that is, in the linear response regime). Such effects are observed only at cryogenic temperatures and only with sophisticated signal-detection schemes (such as a lock-in amplifier). In marked contrast to such phenomena are the quantum-interference effects seen in some semiconductor nanostructures, such as the quantum-well resonant-tunneling diode (RTD) [4,5]. These effects are observed when the system is driven far from equilibrium by an external bias voltage and they persist to well above room temperature. They could, if one so desired, be observed using only a low-precision voltmeter and a milliammeter. There is thus a persuasive reason for studying the physics of nanostructures driven far from equilibrium. We shall see that under such circumstances pure-state quantum mechanics is not an adequate description and must be replaced with a quantum kinetic theory.

The RTD is made of very thin heteroepitaxial layers designed such that the conduction-band profile includes a double-barrier structure. Multiple reflections of the electron wavefunction from these barriers gives rise to resonances in the transmission probability in the same way that such resonances occur in an optical Fabry-Perot interferometer. The structure and behavior of the RTD is summarized in Fig. 1. The $I(V)$ curve of a good experimental

---

[1]Supported in part by the Office of Naval Research and the Defense Advanced Research Projects Agency.

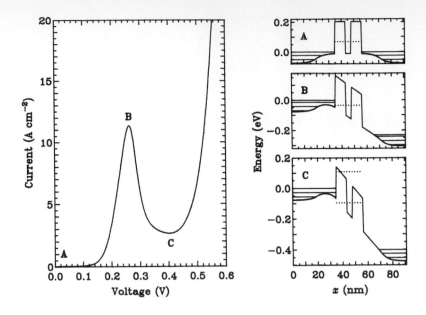

Fig. 1. Structure and behavior of the resonant-tunneling diode.

device measured at 77K is shown to the left and the conduction-band profiles at voltages corresponding to the labeled points are shown to the right. (Occupied electron states are indicated by horizontal hatching and the resonant levels by dotted lines.) In equilibrium, **A**, the chemical potentials are equal and no current flows. At such a voltage that the resonant level lines up in energy with the occupied states in the cathode, **B**, the resonant-tunneling current reaches a peak. At higher voltages the resonant level is pulled below the lowest occupied state in the cathode, **C**, the resonant-tunneling current ceases, leading to a negative differential resistance between **B** and **C**.

Why are quantum interference effects so much more prominent in the resonant-tunneling diode than those that occur in metallic nanostructures? The different topologies of the electron trajectories is a part of the reason. Aharonov-Bohm and similar devices are analogous to the Michelson interferometer, which is considerably less robust than the Fabry-Perot configuration. However, a much more fundamental reason is that the quantum effects in the RTD are observed far from thermodynamic equilibrium (*i.e.*, with a large chemical potential difference across the diode), where a much wider variety of phenomena are possible than in the linear response regime. The present work will be concerned with semiconductor nanostructures driven far from equilibrium. We shall assume that the electrons in such a structure interact only via their mean field (the Hartree potential), and that the Hartree potential and thus the spatial electron density is a physical observable. One might

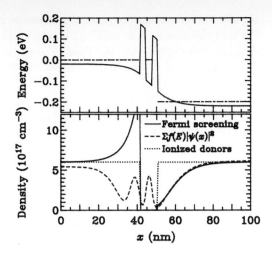

Fig. 2. Fermi-screened potential and resulting charge densities.

expect that under these circumstances elementary quantum mechanics would provide an adequate theory. In fact it is not adequate and we will see how the inadequacies appear in an attempt to evaluate the self-consistent potential. The problem can be traced to the *a priori* assumption that the single-particle density matrix is diagonal in the eigenbasis of the Hamiltonian. Removing this assumption leads one to a quantum kinetic theory.

2.   SCHROEDINGER EQUATION AND SELF-CONSISTENCY

In the absence of any contrary results we would expect that the self-consistent potential for a resonant-tunneling diode under bias would resemble that obtained from the Thomas-Fermi screening theory, such as that shown in Fig. 2. Generalizing the Thomas-Fermi theory to the finite temperature case, we obtain the Fermi-screened potential $V(x)$ by solving

$$- \epsilon \nabla^2 V = q \left\{ 2(m^*/2\pi\hbar^2\beta)^{3/2} \mathcal{F}_{1/2}[\beta(\mu - qV)] - N_d \right\}, \qquad (1)$$

where $\mathcal{F}_{1/2}$ is the Fermi-Dirac integral, μ is the chemical potential, and N_d is the density of ionized donors. The potential of Fig. 2 was obtained by solving (1) with the appropriate chemical potential in each of the thick contact layers and neglecting any charge density in the barriers or well. As one would expect the potential drop is accomodated by an accumulation layer on the left-hand side of the structure and a depletion layer on the right-hand side. The electron density implied by the Fermi-screening approximation is shown by the solid line. In this and all subsequent computations the temperature was taken to be 300 K.

Now, let us take the Fermi-screened potential and insert it into Schroedinger's equation. We find the eigenstates $\psi_l(E, x)$ incident from the left with unit amplitude and $\psi_r(E, x)$ incident from the right. Using these solutions we find the total electron density, which we will express as the diagonal elements of the density matrix:

$$
\begin{aligned}
\rho(x, x') &= \int_{V_l}^{\infty} \frac{dE}{2\pi\hbar v_l(E)} f(E - \mu_l)\psi_l(E, x)\psi_l^*(E, x') \\
&+ \int_{V_r}^{\infty} \frac{dE}{2\pi\hbar v_r(E)} f(E - \mu_r)\psi_r(E, x)\psi_r^*(E, x'),
\end{aligned}
\tag{2}
$$

where $V_{l,r}$ are the asymptotic potentials to the left and right, $v_{l,r}(E)$ is the velocity of an electron of energy E at the respective boundary. Here f is the Fermi-Dirac distribution function integrated over the transverse momenta: $f(E) = (m^*/\pi\hbar^2\beta)\ln(1 + e^{-\beta E})$. The density resulting from this calculation is shown by the dashed line in Fig. 2, and it displays several interesting features. One is the small peak centered in the quantum well. This is the charge build-up due to resonant tunneling. Another interesting feature is the asymptotic values of the electron density, lower than the donor density on the left and slightly higher than the donor density on the right. This is due to the resonant transmission of a part of the electron distribution incident from the left. For most energies the electrons are reflected and thus for each incident k there will be an equal density of electrons in the left-hand contact with $-k$. However these electrons will be absent for states in the transmission resonance. We can estimate the deficit as $j/qv_l(E_0)$ where j is the tunneling current density and E_0 is the resonant energy. Numerical calculations verify that this value is equal to the electron deficit in the left-hand electrode. Similarly, the resonant-tunneling electrons contribute an excess density to the right-hand electrode which is well approximated by $j/qv_r(E_0)$. These effects are really artifacts of the assumption of purely ballistic transport inherent in Schroedinger's equation and will not persist if inelastic scattering processes are present.

The most important feature of the electron density inferred from Schroedinger's equation is the reduction in the electron density in the region where screening theory predicts an accumulation layer ($20\,\text{nm} \leq x \leq 40\,\text{nm}$). This decreased density is readily explained: The current density $\langle\psi|j|\psi\rangle$ from each scattering eigenstate is independent of x. As such a state propagates into a region of decreasing potential, its speed of propagation must increase and thus, to maintain the current constant, its amplitude must decrease.

Where, then, does the increased electron density in the screening theory come from? It comes from electron occupation of a greater number of states, rather than from increased amplitude in any given state. The potential "notch" permits states with nonnegligible amplitude on the left-hand side of the barrier and with energies below V_l. These states are connected via

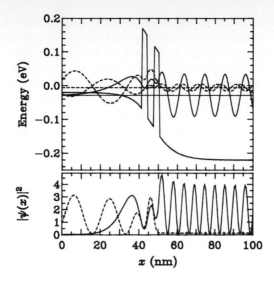

Fig. 3. Fermi-screened potential and resulting wavefunctions.

tunneling to the propagating states incident from the right. The discrepancy between screening and scattering theory is thus explained: Screening theory assumes that the notch states are in thermal equilibrium with the left-hand side of the structure while scattering theory assumes that they are in equilibrium with the right-hand side. Actually, the situation will be governed by the rates of two competing processes: (i) inelastic scattering of electrons from propagating states into the notch states and (ii) tunneling out of the notch states into the right-hand electrode [6]. Again, inelastic processes are the key to obtaining a physically reasonable distribution of electrons.

To evaluate the effects of inelastic processes one naturally turns first to the Fermi golden rule. Actually, the more complete expression underlying this picture is the Pauli master equation, which assumes that the density matrix is diagonal in the eigenbasis of the Hamiltonian [7]. In the RTD this means that the form (2) is assumed, with the distribution functions f replaced by probabilities P_i to be determined. The master equation is then

$$dP_i/dt = \sum_j [W_{ij} P_j(t) - W_{ji} P_i(t)], \qquad (3)$$

where the W_{ij} are the golden-rule transition rates. When applied to the RTD problem, this picture leads to a violation of the continuity equation. To observe this, let us look at the eigenstates derived from the Fermi-screened potential. Fig. 3 shows two typical states, one with $E_1 > V_l$ (dashed line, both real and imaginary parts shown) and the other in the "notch" with

$E_2 < V_l$ (solid line). The spatial distributions of these states are shown in the lower part of the figure. Now, the Pauli master equation (3) implies that a transition process like phonon scattering will transfer electron density at some rate from the state at E_1 to the state at E_2. Because of the very different spatial distributions of the these states, this must involve a current flow with nonzero divergence. However each of the states has a uniform current density and thus the diagonal density matrix describes a state with no current divergence. Using such a model, which violates continuity, in a self-consistent calculation is a very dangerous procedure. To satisfy the continuity equation, we must allow off-diagonal elements of the density matrix. This leads us to a quantum kinetic theory.

3. QUANTUM KINETIC THEORY

A quantum kinetic theory is expressed in terms of the single-particle density matrix $\rho(x, x')$, or a mathematically equivalent object such as the Wigner distribution function. The time evolution of ρ is described by a kinetic equation, which should include the quantum-interference effects described by Schroedinger's equation, and may include the effects of inelastic processes such as those described by the Pauli master equation. The density matrix is directly obtained by solving this kinetic equation; thus the natural basis states (the eigenvectors of ρ) are obtained from the kinetic theory as a result, rather than inserted as an *a priori* assumption.

A quantum kinetic theory appropriate for nanostructure devices such as the RTD has been developed over the past few years [8,9]. Due to the openness of any device with respect to electron flow, the theory is most naturally expressed in terms of the Wigner distribution function $f(x, k)$. The kinetic equation has the form

$$\partial f/\partial t = \mathcal{L}f/i\hbar + \mathcal{C}f, \tag{4}$$

where $\mathcal{L}$ is the Liouville superoperator describing the ballistic motion and $\mathcal{C}$ is a collision superoperator including dissipative interactions. The effects of the electrical connections to the device are described by open-system boundary conditions which are time-irreversible and modify $\mathcal{L}$ so as to render it non-hermitian [8,10]. The kinetic equation without the collision term has been successfully used to calculate the dc, large-signal transient, and small-signal ac response of the RTD [8,11].

The form of the kinetic equation (4) is Markovian, which means that the time evolution depends only upon the state of the system at present, and not upon its prior history. Non-Markovian behavior can result when information about the state of the system is stored in degrees of freedom which are not explicitly included in the kinetic model. These could include the occupation numbers and phases of phonon modes, electron correlations in the contacts, or two-or-more-electron correlation effects. Thus the Markovian kinetic equation

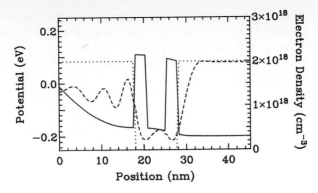

Fig. 4. Potential and density from self-consistent Wigner function with no scattering.

is only a first approximation, but we will see that at least it does not display the unphysical features that we observed in the scattering theory.

The form of the collision superoperator $\mathcal{C}$ is constrained by the requirement of continuity. An instantaneous scattering process cannot move an electron from one position to another, so $\mathcal{C}$ must contain a factor of $\delta(x - x')$. That is, $\mathcal{C}$ must be local in its effect upon the electron distribution, but it can be nonlocal in the sense that $\mathcal{C}(x, k; x, k')$ can depend upon the impurity distribution or phonon modes at other values of x. We would expect on physical grounds that $\mathcal{C}$ should be of the form of a master operator, such as appears in the Pauli equation (3). Thus $\mathcal{C}$ should have the form:

$$\mathcal{C}(x, k; x', k') = \delta(x - x') \left[W_{kk'} - \delta(k - k') \int dk'' \, W_{k''k} \right], \qquad (5)$$

where $W_{kk'}$ is the transition rate from k' to k. Published analyses imply that, at least at some level of approximation, the $W_{kk'}$ are equal to the golden rule transition rates between plane wave states [12,13]. Note that the kinetic approach avoids making assumptions about which states are participating in a transition by considering all possible transitions between complete sets of states.

The influence of inelastic processes on the self-consistent potential and electron density of the RTD using the kinetic theory are illustrated in Figs. 4–6. The kinetic equation (4) for steady state was solved self-consistently with Poisson's equation for the Hartree potential, using a multidimensional Newton technique. A device with a somewhat larger doping level in the contacts than that of Figs. 2–3 was assumed, to reduce the screening length and thus the size of this rather time-consuming computation.

If no collision term is included, the self-consistent solution is that shown

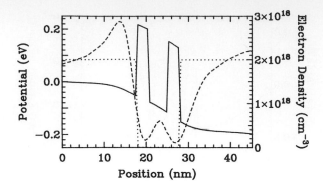

Fig. 5. Potential and density from self-consistent Wigner function with simple relaxation, $\tau = 100\,\mathrm{fs}$.

in Fig. 4. The solid line represents the potential, the dashed line is the resulting electron density, and the dotted line shows the ionized donor distribution. The problem we first observed in the Schroedinger model is also apparent here: No accumulation layer has formed on the left side of the barriers and the electron density is in fact depleted. This leads to an upward curving of the potential on the left side, which is certainly unphysical. Because the boundary in this case is artificial, occuring in contacting layers which are much thicker in the experimental devices, its position should have a negligible effect on the form of the solution. This can only happen if the electric field approaches a small value at the boundary.

The electric field is much better behaved if inelastic processes are included in the calculation. Fig. 5 shows the self-consistent solution with a simple relaxation-time collision operator ($W_{kk'} = f_0(k)/\tau$, where $f_0(k)$ is a normalized Maxwellian distribution) with $\tau = 100\,\mathrm{fs}$. In this case, the shape of the potential and charge distribution is much closer to that obtained from screening theory (Fig. 2). However, this relaxation rate is really too large for GaAs devices.

A more realistic model of the scattering processes is used in the calculation of Fig. 6. In this case $\mathcal{C}$ was constructed using golden-rule transition rates for longitudinal optical (LO) and acoustic phonons in GaAs. It is apparent that the phonon scattering is trying to form an accumulation layer on the left-hand side, but the scattering rate is not large enough to enforce the screening-theory picture. Such a situation may well occur in structures with thin barriers and thus whose tunneling rate exceeds the inelastic scattering rate. However, the present calculation is not yet entirely satisfactory. If the electric field extended deep into the contact layers, it would certainly affect the shape of the distribution function which those layers supply to the device

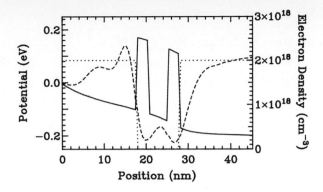

Fig. 6. Potential and density from self-consistent Wigner function with LO and acoustic phonon scattering.

(by displacing the peak of a Maxwellian distribution, for example). Such effects would tend to increase the density on the left and decrease it on the right, which would result in a potential closer to the screening theory result. Simulations which include a simple model of this effect do indeed show a marked reduction of the electric field at the boundaries [14].

4. SUMMARY

The quantum effects observed in nanostructures that are close to thermal equilibrium are quantitatively weak effects, but they can be adequately modeled by elementary quantum theory because the form of the density matrix in equilibrium, $\rho \propto e^{-\beta H}$, assures us that the electrons actually occupy the eigenstates of the Hamiltonian. The quantum effects can be much larger in nanostructures that are driven far from equilibrium, but then we have no assurance that ρ is diagonal in the eigenbasis. The inadequacies of pure-state quantum mechanics become particularly apparent when one tries to include self-consistency in an analysis of the resonant-tunneling diode. Such an analysis requires a satisfactory treatment of inelastic processes, which cannot be obtained in terms of pure quantum states.

A satisfactory treatment of inelastic processes can be formulated at a kinetic level. When both self-consistency and inelastic scattering are included in such a model, we find that the shape of the self-consistent potential is qualitatively changed by modifications to the rate of inelastic scattering.

REFERENCES

[1] D. R. Yennie, Rev. Mod. Phys. **59**, 781 (1987).

[2] A. G. Aronov and Yu. V. Sharvin, Rev. Mod. Phys. **59**, 755 (1987).

[3] P. A. Lee and T. V. Ramarkishnan, Rev. Mod. Phys. **57**, 287 (1985).

[4] L. L. Chang, L. Esaki, and R. Tsu, Appl. Phys. Lett. **24**, 593 (1974).

[5] T. C. L. G. Sollner, W. D. Goodhue, P. E. Tannenwald, C. D. Parker, and D. D. Peck, Appl. Phys. Lett. **43**, 588 (1983).

[6] N. S. Wingreen and J. W. Wilkins, Bull. Am. Phys. Soc. Ser. II **32**, 833 (1987).

[7] H. J. Kreuzer, *Nonequilibrium Thermodynamics and its Statistical Foundations*, (Oxford Univ. Press, New York, 1981), ch. 10.

[8] W. R. Frensley, Phys. Rev. B **36**, 1570 (1987).

[9] U. Ravaioli, M. A. Osman, W. Pötz, N. Kluksdahl, and D. K. Ferry, Physica **134B**, 36 (1985).

[10] W. R. Frensley, "Boundary Conditions for Open Quantum Systems Driven Far from Equilibrium," (unpublished).

[11] W. R. Frensley, Superlattices and Microstructures **4**, 497 (1988).

[12] I. B. Levinson, Zh. Eksp. Teor. Fiz. **57**, 660 (1969) [Sov. Phys —JETP **30**, 362 (1970)].

[13] J. Lin and L. C. Chu, J. Appl. Phys. **57**, 1373 (1985).

[14] R. K. Mains and G. I. Haddad, "Numerical Considerations in the Wigner Function Modeling of Resonant-Tunneling Diodes," (unpublished).

QUANTUM TRANSPORT WITH DISSIPATION: LINEAR AND NON-LINEAR RESPONSE [1]

Supriyo Datta
Michael J. McLennan

School of Electrical Engineering
Purdue University
West Lafayette, Indiana

I. INTRODUCTION

Much of our understanding of electron transport in solids is based on the Boltzmann transport equation (BTE). Despite its impressive successes, it cannot describe transport phenomena in which the wave nature of electrons plays a crucial role. An important topic of current research is to develop a quantum transport formalism that can be used to describe current flow in mesoscopic structures where wave interference effects are significant. In quantum transport theory the semiclassical distribution function $f(\mathbf{r};k;t)$ is replaced by an appropriate distribution function derived from the field correlation function (1).

$$G^<(\mathbf{r}_1,\mathbf{r}_2;t_1,t_2) = i<\psi^\dagger(\mathbf{r}_2;t_2)\psi(\mathbf{r}_1;t_1)> \tag{1.1}$$

For example, the Wigner distribution function is obtained by transforming to center-of-mass $\mathbf{r} = (\mathbf{r}_1 + \mathbf{r}_2)/2$, $t = (t_1 + t_2)/2$ and relative coordinates, and then Fourier transforming with respect to the relative coordinate $(\mathbf{r}_1 - \mathbf{r}_2 \rightarrow k, t_1 - t_2 \rightarrow E)$.

We will present a steady state quantum kinetic equation in terms of the electron density per unit energy $n(\mathbf{r};E)$. We emphasize that $n(\mathbf{r};E)$ is not a semiclassical concept but a well-defined quantum mechanical quantity proportional to the Wigner function averaged over k and t.

$$n(\mathbf{r};E) = -i\int dt\int dk \ G^<(\mathbf{r};k;E,t) \tag{1.2}$$

_____

[1] Supported by the Semiconductor Research Corporation (Contract No. 88-SJ-089) and the National Science Foundation (Grant No. ECS-83-51-036).

The averaging over t is simply due to our restriction to steady state. On the other hand, the averaging over k is made possible by assuming a special form for the inelastic scattering. Averaging over k is equivalent to setting $r_1 - r_2 = 0$ in the correlation function $G^<(r_1,r_2;t_1,t_2)$ so that

$$n(r;E) \sim -i \int dt_1 \int dt_2 e^{-iE(t_1-t_2)/\hbar} G^<(r,r;t_1,t_2) \qquad (1.3)$$

In general, it is not possible to write a quantum transport equation involving only the electron density $n(r;E)$; spatial correlations of the wavefunction represented by the off-diagonal terms $G(r_1,r_2;t_1,t_2)$, $r_1 \neq r_2$, must also be taken into account. The simplification presented in this paper is made possible by the assumption that inelastic scattering is caused by a distribution of independent oscillators, each of which interacts with the electrons through a *delta potential*. We also assume that inelastic scattering processes are weak and infrequent, just as one does in deriving Fermi's golden rule; however, the elastic processes are treated exactly. This model closely approximates a laboratory sample with magnetic impurities, or impurities with internal degrees of freedom. For other types of inelastic scattering the model may not be realistic; however, we believe that it should still be possible to describe much of the essential physics of dissipation in quantum transport. Physically, it is easy to see why the above assumption leads to a simple transport equation that does not involve spatial correlations of the wavefunction. In the "golden rule" approximation, each scatterer acts independently. Since we have assumed a delta interaction potential, an inelastic scattering event only involves the wavefunction at a particular point and is insensitive to spatial correlations.

This simplification is important for two reasons. Firstly, the number of independent variables is reduced from $(r_1;r_2;E)$ (or equivalently, $(r;k;E)$) to $(r;E)$. Secondly, the transport equation involves only positive quantities, so that it is easy to make intuitive approximations. Monte Carlo analysis based on a probabilistic interpretation should also be possible. We believe that this is a consequence of the fact that each inelastic scattering event in our model can be viewed as a quantum measurement of the position and energy of the electron. Every time an electron is inelastically scattered it leaves one of the oscillators in an excited state, and energy is dissipated into the surroundings as the oscillator relaxes back to its state of thermodynamic equilibrium. An observer who monitors the states of the oscillators will see a series of flashes with different energies from different spatial locations and can, in principle, deduce the electron density $n(r;E)$ from the observations. Our transport equation is thus formulated in terms of a variable that is actually measured by the inelastic scattering process rather than a conceptual quantity from which observable quantities can be deduced.

II. THE MODEL

We consider any arbitrary structure in which the propagation of electrons is described by the following one-electron effective-mass Hamiltonian.

$$H_0 = \frac{[\mathbf{p} - e\mathbf{A}(\mathbf{r})]^2}{2m} + eV(\mathbf{r}) \tag{2.1}$$

The vector and scalar potentials $\mathbf{A}(\mathbf{r})$ and $V(\mathbf{r})$ include the self-consistent Hartree potential computed from the Poisson equation, as well as external fields and sources of *elastic* scattering such as impurities, defects, boundaries, etc. For the inelastic scattering we assume a reservoir of independent oscillators labeled by the index m,

$$H_p = \sum_m \hbar\,\omega_m \left(a_m^\dagger a_m + \frac{1}{2}\right) \tag{2.2}$$

where $a_m^\dagger$ and a_m are the creation and annihilation operators for oscillator m. We assume that each oscillator interacts with the electrons through a delta-potential, so that the interaction Hamiltonian H' can be written as

$$H' = \sum_m U\,\delta(\mathbf{r} - \mathbf{r}_m)(a_m^\dagger + a_m) \tag{2.3}$$

Note that we have assumed the interaction strength U to be a constant. There is no loss of generality since the strength of inelastic scattering can be adjusted through the density of scatterers per unit volume per unit energy, described by some function $J_0(\mathbf{r};\hbar\,\omega)$. The summation over m is eventually replaced by an integral.

$$\sum_m \;\Rightarrow\; \int d\mathbf{r} \int d(\hbar\,\omega) J_0(\mathbf{r};\hbar\,\omega) \tag{2.4}$$

III. INELASTIC SCATTERING TIME

It can be shown (2) that if we treat the interaction H' in the 'golden rule' approximation, then the damped quasi-particle propagator is obtained from the 'Schrödinger' equation with an additional imaginary potential.

$$\left[E - H_0 + \frac{i\hbar}{2\tau_i(\mathbf{r};E)}\right] G(\mathbf{r},\mathbf{r}';E) = \delta(\mathbf{r} - \mathbf{r}') \tag{3.1}$$

We emphasize that the imaginary potential in eq. (3.1) is not a phenomendogical quantity. It is an optical potential (3) that is derived rigorously from microscopic theory by evaluating the self-energy. In general, this potential is non-local; but it reduces to a local potential for our model with point inelastic scatterers because the imaginary part of the self-energy (the real part is neglected here) is a delta function in the 'golden rule' approximation.

$$\Rightarrow \quad \Sigma(\mathbf{r}_1,\mathbf{r}_2;E) = -\frac{i\hbar}{2\tau_i(\mathbf{r}_1;E)}\delta(\mathbf{r}_1 - \mathbf{r}_2) \tag{3.2}$$

where the inelastic scattering time $\tau_i(\mathbf{r};E)$ is given by

$$\tau_i^{-1}(\mathbf{r};E) = \int dE' \; S(\mathbf{r};E',E) \tag{3.3a}$$

$$S(\mathbf{r};E',E) = \frac{2\pi}{\hbar}N_0(\mathbf{r};E') \; F(\mathbf{r};E'-E) \; (1-f(\mathbf{r};E')) \tag{3.3b}$$

where $\quad F(\mathbf{r};E-E') = |U|^2 J_0(\mathbf{r};\hbar\,\omega) \begin{cases} N(\hbar\,\omega) & , \text{ if } E - E' = \hbar\,\omega > 0 \\ (N(\hbar\,\omega)+1) & , \text{ if } E' - E = \hbar\,\omega > 0 \end{cases}$.

$N_0(\mathbf{r};E)$ is the electronic density of states given by $\sum_M |\phi_M(\mathbf{r})|^2\delta(E - \epsilon_M)$ where $\phi_M(\mathbf{r})$ are the eigenfunctions of H_0 with eigenvalues ϵ_M; the 'distribution function' $f(\mathbf{r};E)$ is defined by $n(\mathbf{r};E) = N_0(\mathbf{r};E)f(\mathbf{r};E)$. Assuming that the oscillators are in thermal equilibrium, the average number of "phonons" $N(\hbar\,\omega)$ in each oscillator is related to its frequency ω by the Bose-Einstein factor. $S(\mathbf{r};E',E)$ tells us the rate at which electrons are scattered at a point $\mathbf{r}$ from an energy E to an energy E'. Eq. (3.3b) has the appearance of a 'golden rule'; however, unlike the usual golden rule we are using the position representation instead of an eigenstate representation.

The factor $(1 - f(\mathbf{r};E'))$ in eq. (3.3b) is inserted heuristically to account for the exclusion principle; in our one-electron model it does not appear automatically. We are currently developing a formal derivation of the transport equation in the second quantized picture using the Keldysh formulation of non-equilibrium statistical mechanics.

It is well-known that an imaginary potential of the type in eq. (3.1) causes particles to decay with a lifetime τ_i. This decay corresponds to the inelastic scattering rate per unit volume per unit (initial) energy at a point $(i_S(\mathbf{r};E)/e)$

$$i_S(\mathbf{r};E)/e = n(\mathbf{r};E)/\tau_i(\mathbf{r};E) \tag{3.4}$$

Note that, in general, with a non-local optical potential the inelastic scattering rate would depend not only on the local electron density but also on the spatial correlations of the wavefunction. The simple result in eq. (3.4) is a consequence of eq. (3.1) which follows from our assumption of independent point inelastic scatterers. This leads to a simple picture of quantum transport as a process of diffusion in $\mathbf{r}$ and E (Fig. 1). The transport equation we will derive in Section IV is based on this picture. We emphasize that the use of **r** and E simultaneously does not violate the uncertainty principle. As shown in eq. (1.3) the energy spectrum is derived from the temporal correlations of the wavefunction at a point **r** and bears no relationship to **k** which has to do with the spatial correlations. We are not using conjugate variables like **r** and **k** or E and t simultaneously.

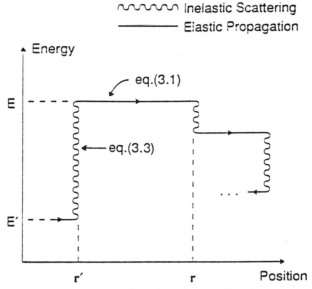

Fig. 1: Quantum transport can be viewed as a diffusion process in (**r**;E).

IV. TRANSPORT EQUATION

We will now show that the scattering current must satisfy an integral equation of the form

$$i_S(\mathbf{r};E) = \int d\mathbf{r}' \int dE' K(\mathbf{r},\mathbf{r}';E,E') i_S(\mathbf{r}';E') + I(\mathbf{r};E) \qquad (4.1)$$

where the kernel $K(\mathbf{r},\mathbf{r}';E,E')$ is given by

$$K(\mathbf{r},\mathbf{r}';E,E') = \frac{|G(\mathbf{r},\mathbf{r}';E)|^2}{\tau_i(\mathbf{r};E)} F(\mathbf{r}';E - E') \tau_i(\mathbf{r}';E')(1 - f(\mathbf{r}';E))$$

Note that the kernel $K(\mathbf{r},\mathbf{r}';E,E')$ is proportional to the square of the Green function of the 'Schrödinger' equation, and thus contains all quantum interference effects due to elastic scatterers, boundaries, etc. It is interesting to note that the BTE without the diffusion and field terms,

$$\frac{d}{dt}f_k = -\frac{f_k}{\tau_k} + \sum_{k'}S_{k,k'}(1-f_k)f_{k'} \qquad (4.2)$$

can also be written in a form similar to eq. (4.1), assuming steady state $(df_k/dt = 0)$ and defining a scattering current $(i_S)_k/e \equiv f_k/\tau_k$.

$$(i_S)_k = \sum_{k'}K_{k,k'}(i_S)_{k'} \qquad (4.3)$$

where $K_{k,k'} \equiv S_{k,k'}(1-f_k)\tau_{k'}$. Eq. (4.2) is derived assuming that the momentum eigenstates $|k>$ are energy eigenstates with definite energies E_k. For this reason, the energy variable does not appear explicitly in eq. (4.2). The fact that the position eigenstates $|r>$ are not energy eigenstates makes the derivation of eq. (4.1) more complicated than that of eq. (4.2). Indeed, were it not for our assumption of point-size inelastic scatterers, it would not be possible to write down an equation such as eq. (4.1) solely in terms of the electron density; in general, such an equation would also involve spatial correlations of the wavefunction. In general, the externally injected current $I(r;E)$ may also have spatial correlations, which we are neglecting in this treatment.

The integral equation, eq. (4.1), is illustrated in Fig. 2 with a schematic diagram showing the in-flow and out-flow at each coordinate $(r;E)$. It is

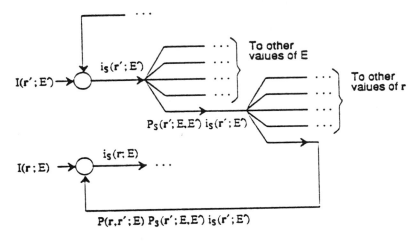

Fig. 2:   A schematic diagram illustrating the transport equation (eq. (4.1)).

apparent that the kernel $K(r,r';E,E')$ is given by the product of two factors.

$$K(\mathbf{r},\mathbf{r}';E,E') = P(\mathbf{r},\mathbf{r}';E)P_S(\mathbf{r}';E,E') \tag{4.4}$$

$P_S(\mathbf{r}';E,E')$ is the fraction of the electrons inelastically scattered at $\mathbf{r}'$ with an initial energy E', that acquire a final energy E after the scattering. $P(\mathbf{r},\mathbf{r}';E)$ is the fraction of electrons injected at $\mathbf{r}'$ with energy E (by inelastic scattering from some other energy) that suffer their *next* inelastic scattering event at $\mathbf{r}$. The kernel is the product of these two factors and can be viewed as a 'transfer function' from one inelastic scattering event at $(\mathbf{r}';E')$ to the next one at $(\mathbf{r};E)$.

The first of these factors is written down easily from eqs. (3.3a) and (3.3b).

$$P_S(\mathbf{r}';E,E') = S(\mathbf{r}';E,E')/\int dE\, S(\mathbf{r}';E,E')$$

$$= \frac{2\pi}{\hbar} N_0(\mathbf{r}';E)F(\mathbf{r}';E-E')(1-f(\mathbf{r}';E))\tau_i(\mathbf{r}';E') \tag{4.5}$$

To compute the second factor we note that inelastic scattering may be viewed as a two-step process involving a decay out of an initial energy E, followed by an injection into a different energy. In calculating $P(\mathbf{r},\mathbf{r}';E)$, the second step is irrelevant. We are simply interested in the probability that an electron injected at $\mathbf{r}'$ with energy E suffers its *very next* inelastic event at $\mathbf{r}$; the reinjection at $\mathbf{r}$ is a separate part of the problem that is already taken into account by the integral transport equation. Thus, for the purpose of calculating $P(\mathbf{r},\mathbf{r}';E)$ we can ignore the reinjection process and assume that we are dealing with decaying quasi particles. As we have seen, such particles are described in our model by the Schrödinger equation (eq. (3.1)), modified to include the optical potential $i\hbar/2\tau_i(\mathbf{r};E)$. Since we have assumed point-size inelastic scatterers, an electron is injected as a point source by the inelastic scattering process. We can thus expect $P(\mathbf{r},\mathbf{r}';E)$ to be proportional to the squared magnitude of the Green function $G(\mathbf{r},\mathbf{r}';E)$ obtained from eq. (3.1). Consider the continuity equation obeyed by the probability density $n = |G|^2$ and the probability current density $\mathbf{J} = i\hbar[(\nabla G)^*G - G^*(\nabla G)]/2m$ that we obtain from the solution to eq. (3.1). It can be shown from eq. (3.1) that

$$\frac{1}{e}\nabla \cdot \mathbf{J} + \frac{n}{\tau_i} = \frac{i}{\hbar}\delta(\mathbf{r}-\mathbf{r}')[G-G^*] \tag{4.6}$$

Integrating over all volume, using the divergence theorem and assuming that the boundaries are far away so that no current flows out of the surface, we have

$$\int d\mathbf{r} \frac{|G(\mathbf{r},\mathbf{r}';E)|^2}{\tau_i(\mathbf{r};E)} = \frac{2\pi}{\hbar} N_0(\mathbf{r}';E) \tag{4.7}$$

since $N_0(\mathbf{r};E) = -\mathrm{Im}\{G(\mathbf{r},\mathbf{r};E)\}/\pi$. The integrand on the left in eq. (4.7) is the steady-state current n/τ_i due to electrons lost from the coherent state by inelastic scattering; the term on the right is the total steady-state current injected at $\mathbf{r}'$ (Fig. 3). The ratio of these two terms is equal to the

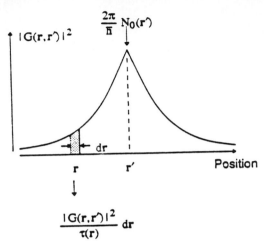

Fig. 3:   Sketch of the probability density $|G(\mathbf{r},\mathbf{r}';E)|^2$ calculated from eq. (3.1). The index E has been dropped for convenience.

probability function $P(\mathbf{r},\mathbf{r}';E)$, which is the quantity we sought.

$$P(\mathbf{r},\mathbf{r}';E) = \frac{\hbar}{2\pi} \frac{|G(\mathbf{r},\mathbf{r}';E)|^2}{N_0(\mathbf{r}';E)\tau_i(\mathbf{r};E)} \tag{4.8}$$

Substituting eqs. (4.8) and (4.5) into eq. (4.4) we obtain the expression for the kernel $K(\mathbf{r},\mathbf{r}';E,E')$ stated earlier following eq. (4.1).

Eq. (4.1) can be solved to determine the inelastic scattering current $i_S(\mathbf{r};E)$, or equivalently, the electron density $n(\mathbf{r};E)$. Given the static potentials $V(\mathbf{r})$ and $A(\mathbf{r})$ (eq. (2.1)), and a distribution of inelastic scatterers $J_0(\mathbf{r};\hbar\,\omega)$, the kernel $K(\mathbf{r},\mathbf{r}';E,E')$ can be computed. We can then proceed to solve eq. (4.1). At any node $(\mathbf{r};E)$ we have two variables: the external current $I(\mathbf{r};E)$ and the electron density $n(\mathbf{r};E)$ (or equivalently, $i_S = en/\tau_i$). At all nodes which are not connected to some external source, $I(\mathbf{r};E) = 0$, and we must solve for $n(\mathbf{r};E)$. At contact nodes, we can assume $n(\mathbf{r};E)$ to be given by a thermodynamic distribution characterized by a local chemical potential, and solve instead for $I(\mathbf{r};E)$. It should be noted that I is the current flowing into the structure and not the current density $\mathbf{J}$ within the structure; the two are related by $I = -\nabla \cdot \mathbf{J}$.

V. RELATION TO THE MULTIPROBE LANDAUER FORMULA

In this paper we have adopted a microscopic approach starting from a model Hamiltonian for the inelastic scatterers; however, our model is closely related to the Landauer picture. Since the inelastic scattering process is purely local it can be viewed as an exit into a reservoir followed by reinjection into the structure. From this point of view it would seem that distributed inelastic scattering processes can be simulated by connecting a continuous distribution of reservoirs throughout a structure. Indeed, when we simplify our transport equation to linear response assuming local thermodynamic equilibrium (2)

$$i_S(\mathbf{r};E) = \frac{N_0(\mathbf{r};E)}{\tau_i(\mathbf{r};E)} \frac{1}{e^{(E-e\mu(\mathbf{r}))/k_BT} + 1} \tag{5.1}$$

($\mu(\mathbf{r})$ is the local chemical potential) we obtain what looks like the multiprobe Landauer formula with a continuous distribution of probes (4,5).

$$I(\mathbf{r}) = \frac{e^2}{h} \int d\mathbf{r}' \left[T(\mathbf{r}',\mathbf{r})\mu(\mathbf{r}) - T(\mathbf{r},\mathbf{r}')\mu(\mathbf{r}') \right] \tag{5.2}$$

where $\quad T(\mathbf{r},\mathbf{r}') = \int dE \left(-\frac{\partial f_0}{\partial E} \right) \frac{\hbar^2 |G(\mathbf{r},\mathbf{r}';E)|^2}{\tau_i(\mathbf{r};E)\tau_i(\mathbf{r}';E)} \tag{5.3}$

$f_0(E)$ is the Fermi-Dirac function at equilibrium with a constant chemical potential μ_0. The expression for $T(\mathbf{r},\mathbf{r}')$ agrees with the result obtained from the Kubo conductivity $\sigma(\mathbf{r},\mathbf{r}')$ using the well-known relationship between σ and T (6). It can be shown that the kernel $T(\mathbf{r},\mathbf{r}')$ obeys the relations $T(\mathbf{r},\mathbf{r}')|_H = T(\mathbf{r}',\mathbf{r})|_{-H}$ and $\int d\mathbf{r}' [T(\mathbf{r}',\mathbf{r}) - T(\mathbf{r},\mathbf{r}')] = 0$ just like the corresponding coefficients in the multiprobe Landauer formula.

This paper also serves to clarify the meaning of the chemical potential $\mu(\mathbf{r})$ in quantum transport theory. The transport equation derived in this paper is formulated in terms of the electron density per unit energy $n(\mathbf{r};E)$. In order to derive the linear-response transport equation, we assume *local thermodynamic equilibrium* so that $n(\mathbf{r};E)$ can be expressed in terms of a local chemical potential $\mu(\mathbf{r})$. It is with this assumption that our transport equation simplifies to a form resembling the multiprobe Landauer formula generalized to a continuous distribution of probes. On the other hand, if the driving forces are large enough (or the inelastic scattering weak enough) then local thermodynamic equilibrium may not be maintained. It is then not appropriate to talk in terms of a local chemical potential; we should solve for the actual distribution $n(\mathbf{r};E)$ using the more general transport equation.

VI. RELATION TO THE DRIFT-DIFFUSION EQUATION

As we have pointed out earlier, the transport equation discussed in this paper (eq. (4.1)) can be viewed as describing a random diffusion process in $(\mathbf{r};E)$, where the kernel $K(\mathbf{r},\mathbf{r}';E,E')$ represents the probability of "hopping" from $(\mathbf{r}';E')$ to $(\mathbf{r};E)$. Thus, the transport process can be viewed as classical Brownian motion; the only quantum mechanical input is in computing the kernel. In specializing to linear response (eq. (5.2)), we have integrated over energy, so that we are left with a diffusion process in real space only. Eq. (6.1) can be rewritten in terms of the electron density $n(\mathbf{r})$ rather than the chemical potential $\mu(\mathbf{r})$.

$$I(\mathbf{r}) = \int d\mathbf{r}' \left[\nu(\mathbf{r}',\mathbf{r})n(\mathbf{r}) - \nu(\mathbf{r},\mathbf{r}')n(\mathbf{r}') \right] \qquad (6.1)$$

Eq. (6.1) has a simple physical interpretation. $\nu(\mathbf{r}',\mathbf{r})d\mathbf{r}'$ tells us the fraction of electrons per unit time that "hop" from $\mathbf{r}$ to $\mathbf{r}'$. The first term on the right-hand side of eq. (6.1) is the total number of electrons hopping per unit time *out* of the volume element $d\mathbf{r}$ while the second term is the number of electrons hopping per unit time *into* the volume element $d\mathbf{r}$. Quantum transport is thus much like classical Brownian motion with a distribution of hopping lengths $\nu(\mathbf{r}',\mathbf{r})$ that is determined quantum mechanically. It can be shown (2) that in a homogeneous medium with a slowly varying electron density $n(\mathbf{r})$, eq. (6.1) can be reduced to the drift-diffusion equation

$$I(\mathbf{r}) = \nabla \cdot \mathbf{J} = e(D\nabla^2 n + \mathbf{v}_d \cdot \nabla n) \qquad (6.2)$$

where $\quad \mathbf{v}_d = \int d\boldsymbol{\rho} \, \boldsymbol{\rho} \, \nu(\boldsymbol{\rho}) \quad$ (drift velocity) $\qquad (6.3a)$

$$D = \int d\boldsymbol{\rho} \, \rho^2 \, \nu(\boldsymbol{\rho}) \quad \text{(diffusivity)} \qquad (6.3b)$$

and $\boldsymbol{\rho} = \mathbf{r} - \mathbf{r}'$. We have checked that the drift velocity and diffusivity obtained from these relations agrees with well-known results in a few simple cases. For example, if we use semiclassical dynamics to compute $\nu(\boldsymbol{\rho})$ we obtain the correct semiclassical magnetoresistance. Again, when we compute numerically the ensemble-averaged diffusivity of a disordered resistor with a constant inelastic scattering time τ_i we obtain results similar to Thouless and Kirkpatrick (7).

VII. CONCLUSIONS

In this paper we have presented a simple steady state quantum transport equation that involves only the electron density per unit energy n(r;E) and not the spatial correlations of the wavefunction. Under conditions of local thermodynamic equilibrium, the electron density n(r;E) can be written in terms of a local chemical potential and the transport equation reduces to a form that resembles the multiprobe Landauer formula extended to include a continuous distribution of probes. We believe that the simplicity of this transport equation will make it feasible to obtain numerical solutions for specific microstructures, and thereby quantitatively answer some of the fundamental questions of quantum transport (8). Also, by comparing the predictions of our model with experiment, it should be possible to identify new phenomena arising from correlations between inelastic scatterers, and to shed light on the microscopic origin of irreversibility.

ACKNOWLEDGEMENTS

The authors are grateful for many helpful discussions with Paul Muzikar.

REFERENCES

1. Mahan, G. D. (1987). Phys. Rep. **145**, 251.

2. Datta, S. and McLennan, M. J. (1989). Technical Report No. TR-EE 89-12, Purdue University.

3. Koltun, D. S. and Eisenberg, J. M. (1988). **Quantum Mechanics of Many Degrees of Freedom**, Chapter 11, John Wiley & Sons, Inc.

4. Büttiker, M. (1986). Phys. Rev. Lett., 1761.

5. Imry, Y. (1986). **Directions in Condensed Matter Physics**, edited by G. Grinstein and G. Mazenko, World Scientific Press, Singapore, 1986.

6. Stone, A. D. and Szafer, A. (1988). IBM J. Res. Develop. **32**, 384; Fisher, D. S. and Lee, P. A. (1981). Phys. Rev. **B23**, 6851.

7. Thouless, D. J. and Kirkpatrick, S. (1981). J. Phys. C.: Solid State Phys. **14**, 235.

8. Landauer, R. (1988). IBM J. Res. Dev. **32**, 306.

THEORY OF QUANTUM TRANSPORT
IN LATERAL NANOSTRUCTURES

John R Barker

Nanoelectronics Research Centre
Department of Electronic and Electrical Engineering
University of Glasgow, Glasgow G12 8QQ, United Kingdom.

Abstract

A number of theoretical formalisms for modelling electron waveguide structures are reviewed where the emphasis is on reducing the effective dimensionality to take advantage of 1-D algorithms. Applications to the Aharonov-Bohm effect are described.

1. INTRODUCTION

Nanolithography is now capable of producing a wide range of low-dimensional electron guiding structures such as GaAs wires, lateral resonant tunnel devices, lateral superlattices and ring structures for observing the Aharonov-Bohm effect. Various fabrication schemes are known; and for illustration, figure 1 shows the gate patterns required to re-configure the quasi-two dimensional electron gas in a HEMT device into a number of different electron waveguide structures. Typically, such structures require a source-drain distance or effective channel length $L \ll L_e$, where L_e is the inelastic coherence length for true quantum ballistic

transport or equivalently, electron waveguide phenomena to be observable. In the present paper we discuss the theory of electron transport in the class II-IV devices shown in figure 1, for which carriers are confined to a set of laterally extended channels (for example, by squeezed gate techniques). The channels may or may not have soft walls (ie penetrable by tunnelling) and they may be subject to three-dimensional applied macro-potentials and fluctuation potentials arising, for example, from non-self-averaging configurations of remote donors (this topic is treated elsewhere: Nixon, Davies and Barker, 1989).

One-dimensional quantum ballistic transport theory, including multiply-connected geometries is quite well developed; it is ultimately based on the 1-D Schrödinger equation, which for simplicity, we shall write in the form:

$$H\Psi = (p_z^2/2m^* + V(z))\Psi = i\hbar\, \partial\Psi/\partial t \qquad (1)$$

Very powerful methods exist for handling this equation, in particular the Villars-Fesbach transformation (used for the K-G equation, Fesbach and Villars, 1958),

$$\psi^{\pm} = (1/2)\{\,\psi \pm (1/ik)\partial\psi/\partial z\} \qquad (2)$$

reduces the time-independent form of (1), where $k = (2mE)^{1/2}/\hbar$, to two coupled first-order equations, or equivalently a single equation for the spinor-like variable $\Psi \equiv (\psi^+,\psi^-)'$, which have a direct analogy with *classical transmission-line theory.* Using $\tau = -kz$ as a dimensionless space variable, the equation of motion for $\Psi(\tau)$ becomes

$$i\,\partial\Psi/\partial\tau \;=\; \begin{pmatrix} 1-u/2 & -u/2 \\ u/2 & -1+u/2 \end{pmatrix} \Psi \;\equiv\; H\,\Psi \qquad (3)$$

where $\mathbf{H} \equiv (1-u/2)\sigma_z - (u/2)\sigma_y$ is the effective "Hamiltonian" for the state Ψ,

$u = V(\tau)/E$ and σ_z and σ_y are standard Pauli spin matrices. Here, $\psi^+(z)$ (or ψ^-) may be interpreted locally (in vicinity of z_0, say) as a forward (or backward) propagating plane wave $\psi^+(z) \sim \psi^+(z_0)\exp(ik(z-z_0))$ (or $\psi^-(z_0) \exp(-ik(z-z_0))$). Evidently, $\psi = \psi^+ + \psi^-$. It is easy to demonstrate that the usual current is the conserved quantity $j = (\hbar k/m) \Psi^+ \sigma_z \Psi$. If j is normalised to $(\hbar k/m)$ we see that σ_z acts as an indefinite metric for the normalisation of Ψ. The "Hamiltonian" $\mathbf{H}$ is pseudo-Hermitian in the sense that $\mathbf{H}^+ = \sigma_z \mathbf{H} \sigma_z$. $\mathbf{H}$ generates a pseudo-unitary transformation of Ψ such that the transfer or transmission matrix $\mathbf{T}$ belongs to the $SU(1,1)$ group with the significance that $|T_{11}|^2 - |T_{12}|^2 = 1$:

$$\mathbf{T} = \begin{pmatrix} T_{11} & T_{12} \\ T_{12}{}^* & T_{11}{}^* \end{pmatrix} \tag{5}$$

$$\Psi(\tau) = \mathbf{T}(\tau)\ \Psi(0) \tag{6}$$

Numerically (or analytically), one proceeds by integrating Ψ from a point $z = z_2$ ($>z_1$) back along the negative z-axis to z_1 (ie in direction of +ve τ). The transmission matrix for a null barrier ($V = 0$) of length L is given by $T_{11} = \exp(-ikL)$, $T_{12} = 0$. If $\mathbf{T}_1, \mathbf{T}_2, \mathbf{T}_3,...$are transmission matrices for successive regions (ordered left to right) the transmission of the total range is given by the product $\mathbf{T} = \mathbf{T}_1\mathbf{T}_2\mathbf{T}_3...$In terms of the usual transmission and eflection amplitudes for a plane wave incident from the left of a potential barrier it is trivial to prove that $t(k) = 1/T_{11}$ and $r(k) = T_{12}{}^*/T_{11}$. Advanced methods for parameterising the "equations of motion" are given by Peres (1983a). The related scattering or S-matrix connects the outgoing waves to the incoming waves defined by the unitary transformation:

$$\begin{pmatrix} \psi^-(\tau) \\ \psi^+(0) \end{pmatrix} = \mathbf{S} \begin{pmatrix} \psi^-(0) \\ \psi^+(\tau) \end{pmatrix} \tag{7}$$

A group-theoretical treatment of this approach is given by Peres (1981, 1983a,1983b) and it has been applied to the determination of resistances, localisation length and chaotic band structures in 1-D disordered systems. The related invariant embedding technique (Kumar, 1985, Heinrichs,1986,1987) provides a useful technique for obtaining a Riccati equation for the reflection amplitude and hence to a closed equation for the Landauer resistance in 1-D. For general two-dimensional problems, or higher, the prospects for analytical approaches appear to be poor at first sight and numerical techniques seem to be the only way forward. But for quantum waveguide structures we can recover some of the power of the 1-D methodology at the expense of using a coupled-mode theory.

2. COUPLED-MODE FORMULATION

A range of potentially exciting devices may be constructable as *quantum electron waveguides*: metal wires, GaAs wires, squeezed Q2DEG channels (Wharam et al, 1988; Kirczenov, 1988, has described the uniform channel case to model the quantised resistance and quantum Hall effect in narrow ballistic channels; a WKB version is given by Glazman et al, 1988), ring structures (Ford et al, 1988), tapers. We may also include the problem of injecting electrons from a narrow channel into a wide contact. Until recently these have only been modelled as one-dimensional structures (we may quote the elegant work by Buttiker (1986), Gefen, Imry and Azbel(1984) as examples) with a few numerical calculations based on the 2D Schrödinger equation (section 3 of the present paper). Frohne and Datta (1988) have described an approximate numerical technique based on wavefunction matching to calculate the scattering matrix for electron transfer between 2D channel regions with different confining potentials in the transverse direction. The problem has interesting analogies with electromagnetic tapered waveguide theory (Sporleder and Unger, 1979). Indeed, the analogy is exploited by us (Barker and Laughton 1989-unpublished) in a new formalism for one-electron 2-D and 3-D quantum waveguides which generalises the one-dimensional transmission equations of section 1 to a set of one dimensional coupled-mode equations.

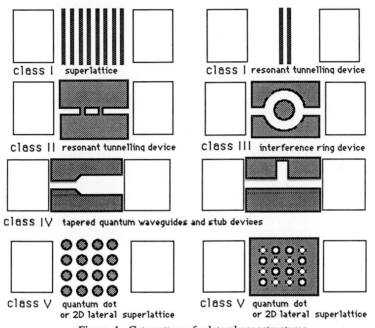

Figure 1 : Gate patterns for lateral nanostructures

Suppose that $\{\varphi_m[a_\mu(z), y]; \mu=1,2,...N\}$ are a complete orthonormal set of wavefunctions defined over y-space, where the $a_\mu(z)$ are N parameters which will generally depend on the z-spatial coordinate and which will be eventually related to the geometry of the confinement potential. As an example, let us refer to fig. 2 for a tapered 2D channel, channel radii a=a(z),b=b(z) with an internal potential V= V(z,y) and perfectly reflecting walls, we might represent the wavefunction at a location z,y as the superposition of an amplitude factor $\psi_m(z)$ multiplied by a local eigenstate $\varphi_m[a(z),b(z), y]$ of a uniform guide of width b(z)-a(z) at the same energy (evanescent modes are included):

$$\Psi = \sum_m \psi_m(z) \, \varphi_m[a(z), b(z), y] \qquad (8)$$

Substituting this expression, or its generalisation $\Psi = \Sigma \, \psi_m(z) \, \varphi_m[a_\mu(z), y]$, into the Schrödinger equation and projecting onto the z-domain by forming the scalar product (integrate over y) on the complete set $\{\varphi_m[a(z), b(z), y]\}$ we find that the amplitudes $\psi_m(z)$ obey coupled 1-D equations of the form:

$$[\partial^2/\partial z^2 + \gamma_m(z)^2]\psi_m = \Sigma \, \{A_{mn} \, \psi_n + B_{mn} \, \partial\psi_n/\partial z \, \} \tag{9a}$$

Here, $\gamma_m(z) = (2m^*[\,E - E_m]/\hbar^2)^{1/2}$ is a z-dependent propagation factor and $E_m(z)$ is the eigenvalue of state $\varphi_m[a_\mu(z), y]$. The coupling coefficients $A_{mn}(z)$, $B_{mn}(z)$ are known functions of : the local guide eigenstates φ_m; the guide parameters and their derivatives, $a_\mu(z)$, $(\partial/\partial z)a_\mu(z)$, $(\partial^2/\partial z^2)a_\mu(z)$ (in our example this dependence is via a(z) and b(z)); and the partial matrix elements $V_{mn}(z)$ of the internal potential V(z,y).

$$A_{mn} = \int dy \, \varphi_m\{(2m^*/\hbar^2)V(z,y)\varphi_n - \{\textstyle\sum_\mu \partial^2\varphi_n/\partial a_\mu^2.[\partial a_\mu/\partial z]^2 + \partial\varphi_n/\partial a_\mu.\partial^2 a_\mu/\partial z^2\}\} \tag{9b}$$

$$B_{mn} = \int dy \, \varphi_m\{-2\textstyle\sum_\mu \partial\varphi_n/\partial a_\mu.[\partial a_\mu/\partial z]\} \tag{9c}$$

$$V_{mn} = \int dy \, \varphi_m(2m^*/\hbar^2)\{V(z,y)\}\varphi_n \tag{9d}$$

$$D_{mn} = A_{mn} - V_{mn} \tag{9e}$$

Eqns. (9) are a good starting point for a Green function formulation, but in the present paper we shall only discuss the relation to the 1D theory of section 1. We note that a very similar treatment allows us to handle the time-domain or a 3- dimensional structure for which y is replaced by x,y.

Using the Villars-Fesbach transformation in the form

$$\psi_m^\pm = (1/2)\{\,\psi_m \pm (1/i\gamma_m)\partial\psi_m/\partial z\} \tag{10}$$

we find the coupled-mode equations corresponding to equation (3):

$$\partial\psi^+_m/\partial z = i\gamma_m\psi^+_m(z) + \Sigma \, \{C^{++}_{mn}(z) \, \psi^+_n(z) + C^{+-}_{mn}(z) \, \psi^-_n(z)\} \tag{11}$$

$$\partial\psi^-_m/\partial z = -i\gamma_m\psi^-_m(z) + \Sigma \, \{C^{-+}_{mn}(z) \, \psi^+_n(z) + C^{--}_{mn}(z) \, \psi^-_n(z)\} \tag{12}$$

The coefficients C are z-dependent and are related to the A, B coefficients via:

$$C^{s\,s}_{mn}(z) = s\,(1/i\gamma_m)(\,V_{mn} + D_{mn} + s\,i\,B_{mn}\,) \tag{13a}$$

$$C^{s-s}_{mn}(z) = s\,(1/i\gamma_m)(\,V_{mn} + D_{mn} - s\,i\,B_{mn}\,) \tag{13b}$$

where $s = \pm 1$. The detailed form of the C coefficients depends on the parameterisation. For example, for a hard-walled guide with $a_1 = a(z)$, $a_2 = 0$, where the guide boundary is slowly-varying in space, and $\varphi_m[a(z), y] = \sqrt{(2/a)} \sin(m\pi y/a)$, the dominant contribution from the confinement potential comes from the term proportional to $[\partial a/\partial z]$:

$$B_{mn} = (-1)^{(n\pm m)} [\partial \ln(a)/\partial z] \, 4mn/(n^2 - m^2)$$
$$= (-1)^{(n\pm m)} [\partial \ln(a)/\partial z] \, 4\sqrt{E_m}\sqrt{E_n}/(E_n - E_m) \qquad (m \neq n) \qquad (14)$$

where $E_n(z) = n^2(\hbar^2/2m^*)[\pi/a(z)]^2$ is the energy of the nth transverse mode at z. The slow decrease of $B_{mn} \sim 1/(E_n - E_m)$, as a function of the energy separation implies that a very large number of modes may be required to obtain good convergence. Interestingly, in this case the coupling coefficients only depend on the logarithmic derivative of the guide width.

We may use these results to estimate the influence of a boundary fluctuation on an otherwise uniform channel. Let $a(z) = a_0 + \delta a.\sin(k_g z)$, where $\delta a/a_0 \ll 1$ and $2\pi/k_g$ is the wavelength of the fluctuation. Choosing m,n to maximise B_{mn} we find that strong scattering occurs between modes when $|\gamma/k_g| \sim 3\delta a/a_0$, or in terms of effective wavelengths when $\lambda_g \leq (3\delta a/a_0)\lambda$. The reverse of this condition $\lambda_g \gg (3\delta a/a_0)\lambda$, where λ_g is the wavelength of the dominant Fourier components of the boundary fluctuation potential, is essentially the condition for adiabatic channel conduction in which carriers smoothly adjust to the local modes and $\psi^{\pm}_m(z) \sim \exp(\pm i \int^z dx \, \gamma_m(x))$). Note that even very weak fluctuation potentials on the boundaries, or otherwise, will induce strong scattering between degenerate states $(\gamma_m(z), E_m(z)) \rightarrow (\gamma_n(z), E_n(z))$ in favour of non-evanescent states with the lowest real propagation constant γ_n since the density of states $\rho \sim [E - E_m]^{-1/2} \sim 1/\gamma_n$.

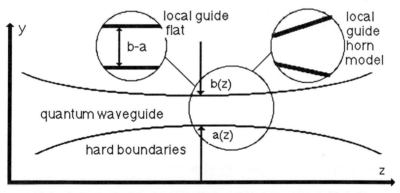

Figure 2: Example of the parameterisation of a quantum waveguide in 2D

The best choice of states $\{\varphi_m[a_\mu(z), y] \}$ depends on the problem; for example, in a slowly tapering channel we could use local states corresponding to the case of a uniform channel with the same instantaneous wall parameters. The latter may correspond to a hard wall boundary or mark the points at which the confinement potential changes abruptly or indeed to any suitable parameterisation of the confinement potential. For more severely changing confinement potentials it is better to use local states which correspond to the local horn solutions (see figure 2) which better describe the wave spreading near a local throttle region.

We have been particularly interested in a confinement potential of the form: $V(z,y) = V_i(z)$ for $y_{i-1} < y < y_i$ $(i=1,2,3)$ and V -> infinity for $y= y_0$ and $y= y_3$ (the latter two parameters may be extended to infinity). It is possible to to obtain the instantaneous transverse states exactly. The model allows the investigation of parallel channels, ring devices, tapers and soft-walled confinement potentials for which evanescent states are important. Fig. 3 shows some of the local cross sections which may be set up. Parabolic confinement potentials are also amenable.

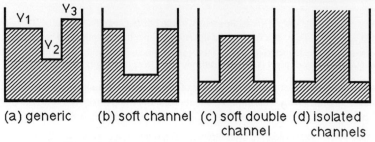

(a) generic   (b) soft channel   (c) soft double (d) isolated
                                       channel   channels

Figure 3: A parameterised model confinement potential

We are currently applying this technique to the design of quantum waveguides which can exhibit large current modulation due to interferometer action and tuning stub gating.

3. FINITE-DIFFERENCE METHOD

Recently we have devised a method for computing wavepacket propagation in two-dimensional geometries including the quantising effect of an applied magnetic field (Barker, 1989; Barker and Finch, 1989; Finch, 1989). The method uses the fact that the Hamiltonian $H= (\mathbf{p}-e\mathbf{A})^2/2m^* + e\phi$ is the generator of infinitesimal time translations, but a variant on the Cayley expansion is used. Introducing the variable $\tau = \delta t/2\hbar$ we replace the infinitesimal evolution operator $e^{i\delta tH(t)/\hbar}$ by

$$e^{-i\delta tH(t)/\hbar} \longrightarrow (1+iH_y\tau)^{-1}(1- iH_x\tau) (1+iH_x\tau)^{-1}(1-iH_y\tau) \qquad (15)$$

If we introduce the uniform time-mesh $t= s\delta t$ and the space mesh $x= n\Delta$, $y=m\Delta$ we find the discretised Schrödinger equation in the form

$$[1 + iH_x(s,nm)\tau] \, \psi(nm,s+1/2)= [1 - iH_y(s,nm)\tau] \, \psi(nm,s) \qquad (16a)$$

$$[1 + iH_y(s,nm)\tau] \, \psi(nm,s+1)= [1 - iH_x(s,nm)\tau] \, \psi(nm,s+1/2) \qquad (16b)$$

Here we write the total Hamiltonian as $H= H_x + H_y$ where H_x and H_y are the most symmetrical separation of the x and y dependent components. Two time steps are deployed. The method is analogous to the Alternating Direction Implicit method traditionally used for diffusion problems. Despite the recurrence relations the algorithm can be re-written into an easily vectorised form for computation on supercomputers.

4. AHARONOV-BOHM EFFECT

We have used the finite-difference method to compute wavefunction propagation and the magnetotransmission for a wide range of semiconductor ring structures with the aim of discovering the influence of finite channel width and fluctuation potentials on the Aharonov-Bohm effect. One dimensional theory (Gefen, Imry and Azbel, 1984) predicts perfect modulation of the transmission coefficient as a function of the enclosed magnetic flux, but experimentally the best observed modulations have been < 25%-30%. Many experimental ring structures have the geometry sketched in figure 4(a), where we notice that electrons entering from the left must make a transition from a narrow to a wide waveguide region before settling into the arms of the ring; the equivalent tapered guide is shown in fig.4(b) where it is obvious that the first and second spatial derivatives of the guide radius a(z) are large at the ring entrance and exits. The coupled mode theory of section 2 indicates that even if a wavepacket approaches from the left in a single (eg lowest) transverse mode it may be scattered into higher modes at the ring junction. This effect is confirmed numerically (fig.5a) where we see that an incident gaussian wavepacket (13.0 meV) in the ground transverse state has scattered pre-dominantly into three transverse states in the ring and maintains occupancy of those states in the exit channel. As a consequence, the Aharonov-Bohm resonance (fig.5b) is quite different from the pure 1-D result, only the central lobe of the exiting wavepacket exhibits destructive interference and the two side lobes are transmitted; the situation is closer to a two slit interference experiment where the envelope of the interference pattern remains constant in shape and is shifted slightly by the Lorentz force but the interference pattern developes a zero at the centre. At the most this can only lead to 30% current modulation. At very low incident energy ~1 meV only evanescent modes are coupled from higher states and the transmission is mono-mode. The latter situation has not yet been obtained experimentally. The computed magneto-transmission for a practical ring is shown in figure.6 and agrees qualitatively and quantitatively with published experimental data (Ford et al. 1988). The insertion of a narrow taper in the exit channel can be shown numerically to significantly improve the modulation. Couple-mode theory suggest that a design such as that in figure 4(c) will give much better transmission and current modulation because of the mode commensurability between channel and ring (fig4d).

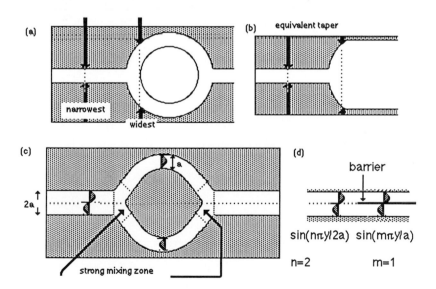

Figure 4: Aharonov-Bohm ring geometries

Fig.5a
Probability
Distribution
of incident
Gaussian packet
on a 400 nm
ring. B=0
Central exit
lobe transmits

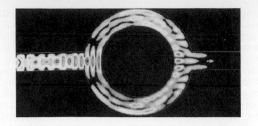

Fig.5b
Probability
Distribution
of incident
Gaussian packet
on a 400 nm
ring.
enclosed
flux = h/e
Central exit
lobe reflects.

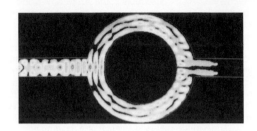

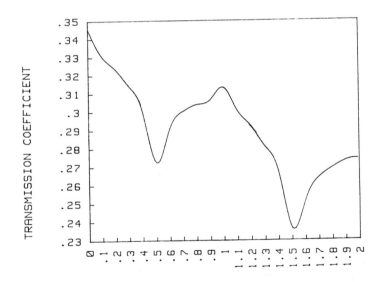

Fig. 6 Computed transmission for a 690 nm outside radius ring with 80 nm wire width

5. LATTICE-GROUP THEORETIC METHOD

Very recently a new approach has been developed by Barker and Pepin (1989, to be published) which represents the general multi-dimensional transport problem as a multiply-connected <u>one-dimensional</u> flow on a lattice . The method gives very high accuracy numerically but also allows a semi-analytical attack on the problem using group theoretic methods developed for one-dimensional potential problems. The method defines a wavefunction on a finite, continuous 1-D network spanning the n-D electron waveguide. On each branch of the network the wavefunction is one-dimensional and is propagated between network nodes by a 1-D S-matrix or T-matrix determined by 1-D algorithms. The wavefunctions along different branches are matched at the nodes by a unique S-matrix which preserves local physical continuity. In the case of two dimensions the nodal S-matrix is

$$(1/2) \quad \begin{array}{cccc} -1 & 1 & 1 & 1 \\ 1 & -1 & 1 & 1 \\ 1 & 1 & -1 & 1 \\ 1 & 1 & 1 & -1 \end{array}$$

The network equations of motion are formally solved by a S and T sub-matrix formalism which builds in the preservation of unitarity and pseudo-unitarity. The method is intrinsically more accurate than finite-difference methods with the same mesh size. The only major proviso with the method is that the wavevector k should be kept away from the band edges of

the network ie $k \ll \pi/\Delta$ where Δ is the lattice spacing (non uniform meshes may be used).A very large number of transverse modes are automatically included in the formalism and it provides an economical route to the computation-intensive 3-D models.

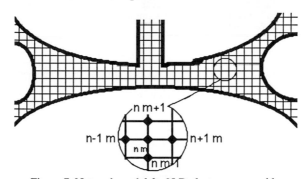

Figure 7: Network model for N-D electron waveguides.

6. FLUID MODELS

Schrödinger and Madelung developed a transformation of quantum mechanics to a fluid dynamical-like formalism which has some merit in describing the results of numerical computations for extended quantum waveguides. Let us write the wavefunction for a pure

state as $\psi = \sqrt{\rho}\, \exp(\,iS(\mathbf{r})/\hbar)$ where ρ and S are real functions of position. If we insert this expression into Schrödinger's equation using a quadratic kinetic energy for simplicity, and separate out the real and imaginary parts we get two coupled equations for the velocity field $\mathbf{v}(\mathbf{r},t) = (\nabla S)/m$ and the probability density ρ. They are just the continuity equation and an equation very similar to the Euler equation in fluid dynamics.

$$\partial\rho/\partial t + \nabla(\rho\mathbf{v}) = 0 \qquad\qquad (17)$$

$$\partial \mathbf{v}/\partial t + \mathbf{v}.\nabla \mathbf{v} + \nabla V(r)/m = (\hbar^2/2m^2) \; \nabla \{ \; \rho^{-1/2} \nabla^2 \; \rho^{1/2} \} \quad (18)$$

The RHS vanishes in the classical limit to give precisely an Euler equation - a perfectly valid construct to handle particles if we use delta functions to represent the position density and velocity field. If a vector potential is present we can use the form $\mathbf{v}(r,t) = (\nabla S - e\mathbf{A}/m)$ to pick up Lorentz force terms. An interesting point here is the significance of the classical fluid concept of circulation, defined as the loop integral of the velocity field: $\Gamma = \oint \mathbf{v}.\mathbf{dr}$. From the definition of velocity this becomes: $\Gamma = e \oint \mathbf{A}.\mathbf{dr}/m$ which is proportional to the non-integrable phase factor occurring in the Aharonov-Bohm effect.

7. CONCLUSIONS

We have briefly reviewed a number of approaches to developing the theory of electron waveguides; only one of these the coupled-mode theory of section 2, is really suitable for formally handling the many-electron problem but the necessity to handle very large numbers of modes in practical situations makes quantum waveguide theory a computational problem of supercomputing proportions. It is evident from the numerical studies that the practical design of good electron interferometers or directional couplers suitable for example for conservative logic switches will require extended geometry models rather than the 1-D models so far deployed.

ACKNOWLEDGMENTS

We gratefully acknowledge technical assistance from John Hague (IBM UK Ltd) in vectorising the finite-difference programs under the auspices of the IBM UK-Glasgow University-Kelvin Project on Numerically Intensive Computing. This work is supported by the Science and Engineering Research Council.

REFERENCES

Barker, J.R. (1989). Proc. San Miniato Workshop on Quantum transport in Nanostructures, ed D K Ferry, in the press.
Barker, J.R. and Finch, M. (1989) submitted for publication.
Buttiker, M. (1986). Phys.Rev.Letters 57, 1761.
Fesbach, H., and Villars, F. (1958) Rev.Mod.Phys. 30, 24.
Finch, M. (1989) PhD Thesis University of Glasgow.
Ford, C J B.,Thornton, T J., Newbury, R., Pepper, M.,Ahmed, H., Foxon, C T., Harris, J J., and Roberts, C.(1988). J.Phys.C 21 L325.
Frohne, R., and Datta, S. (1988). J.Appl.Phys. 64 4086.
Gefen, Y., Imry, Y., and Azbel, M. Ya. (1984). Phys.Rev.Letters 52 129.
Glazman et al (1988) JETP Letters 48 238.
Heinrichs, J (1986). Phys.Rev. B 35 5261.
Heinrichs, J (1988). Phys.Rev B 36 2867.
Kumar, N (1985). Phys.Rev B 31 5513.
Kirczenow, G. (1988) Phys.Rev B 38 10958.
Nixon, J., Davies, J H D, and Barker, J R., (1989) these proceedings.
Peres A, (1981). Phys.Rev. 24B:7463.
Peres A, (1983a). J.Math.Phys. 24, 1110.
Peres A, (1983b).Phys.Rev. 27B 6493.
Sporleder, F., and Unger, H-G. (1979). *Waveguide tapers transitions and couplers* (Peter Peregrinus Ltd, IEE London and New York.
Wharam, D A., Thornton, T.J., Newbury ., Pepper, M., Ahmed, H., Frost, J E F., Hasko, D.G.,Peacock, D C., Ritchie, D A., and Jones, G A C.(1988)J.Phys.C 21 L209.

HIGH MAGNETIC FIELD STUDIES OF INTRINSIC BISTABILITY, ELECTRON THERMALIZATION AND BALLISTIC EFFECTS IN RESONANT TUNNELING DEVICES

M. L. Leadbeater, E. S. Alves, L. Eaves, M. Henini,
O. H. Hughes, F. W. Sheard and G. A. Toombs

Department of Physics, University of Nottingham,
Nottingham NG7 2RD, UK

1. INTRODUCTION

Double-barrier resonant tunneling devices (RTD's) exhibit electrical properties which are dominated by the wavelike character of the conduction electrons. Here we focus on two topical problems in resonant tunneling: (1) charge buildup in the quantum well at resonance and the associated questions of electron thermalization and intrinsic bistability; and (2) ballistic motion of electrons across wide quantum wells in the presence of crossed electric and magnetic fields.

2. SPACE-CHARGE BUILDUP, INTRINSIC BISTABILITY AND ELECTRON THERMALIZATION

The occurrence of intrinsic bistability in the current-voltage characteristics of asymmetric double-barrier structures (DBS's) based on n-type GaAs/(AlGa)As has recently been reported[1,2]. Intrinsic bistability arises from an electrostatic feedback effect associated with the buildup of electronic charge in the quantum well between the barriers when resonant tunneling occurs. Although intrinsic bistability is well-founded theoretically[3,4], earlier reports of its observation proved controversial[5-7]. Our structure[1] was prepared with barriers of different thicknesses. A large buildup of charge in the well is expected at resonance when electrons are injected through the thinner (emitter) barrier and are inhibited from tunneling out due to the lower transmission coefficient of the thicker (collector) barrier. In this paper we report how studies of magneto-oscillations in the differential capacitance can be used to monitor the space-charge in the quantum well and in the accumulation layer of the emitter contact. In contrast to previous studies of magnetoquantum oscillations in symmetric DBS's[8,9], we find that the quasi-Fermi level of the electrons stored in the well lies significantly below the Fermi level of the electrons in the emitter accumulation layer. We attribute this to energy relaxation of the stored charge distribution, which will occur if the electron storage time in the quantum well is sufficiently long. Estimates of this storage time obtained experimentally and theoretically are consistent with this hypothesis. The resonant tunneling process in our asymmetric DBS is thus truly sequential[10] rather than coherent.

Our asymmetric structure consisted of the following layers, in order of growth from the n^+GaAs substrate: (i) 2 μm GaAs, doped to 2×10^{18} cm^{-3}, (ii) 50 nm GaAs, 10^{17} cm^{-3}, (iii) 50 nm GaAs, 10^{16} cm^{-3}, (iv) 3.3 nm GaAs, undoped, (v) 8.3 nm $Al_{0.4}Ga_{0.6}As$, undoped (thin barrier), (vi) 5.8 nm GaAs, undoped (well), (vii) 11.1 nm $Al_{0.4}Ga_{0.6}As$, undoped (thick barrier), (viii) 3.3 nm GaAs, undoped, (ix) 50 nm GaAs, 10^{16} cm^{-3}, (x) 50 nm GaAs, 10^{17} cm^{-3}, (xi) 0.5 μm GaAs, 2×10^{18} cm^{-3}, top contact. The layers were processed into mesas of diameter 200 μm. The conduction band profile, with the top contact biased positively, is shown in Figure 1. Owing to the low doping in the emitter contact, a quasi-bound state is formed in the accumulation layer and, at liquid helium temperatures, the associated two-dimensional electron gas (2DEG) is degenerate. In the sequential theory[10,4], resonant tunneling is regarded as two successive transitions; from the emitter into the well and then from the well into the collector contact. Energy and the transverse components of momentum are conserved in the tunneling process. Thus resonant tunneling occurs when the energies of the quasi-bound states in the accumulation layer and in the quantum well coincide, as in Figure 1. Also, conservation of kinetic energy corresponding to motion parallel to the barriers requires coincidence of the Fermi levels of the 2DEG's in the accumulation layer and quantum well. However, the states in the well are only partially occupied, since the occupancy is determined dynamically[4] by the balance between the transition rates into and out of the well. If the electron storage time is much longer than the energy relaxation time, the electron distribution in the well will thermalize and establish a quasi-Fermi level below that of the emitter contact, as in Figure 1.

We have investigated this possibility by studying magneto-oscillations in the capacitance C and tunnel current I for a magnetic field B applied perpendicular to the plane of the barrier interfaces. In this geometry the states of transverse kinetic energy of a degenerate 2DEG (in the emitter or well) are quantized into Landau levels. At a fixed applied voltage, oscillations with a definite period $\Delta(1/B)$ in $1/B$ arise from a modulation of the charge in either the accumulation layer or quantum well whenever a Landau level passes through the quasi-Fermi level. A detailed description of the oscillatory structure in I and C requires a self-consistent calculation of the electronic charge distribution and will be published elsewhere. The frequency of the oscillations, $B_f = \{\Delta(1/B)\}^{-1}$, is thus related to the Fermi energy E_F by $B_f = m^*E_F/e\hbar$, where m^* is the effective mass. For fully, as opposed to partially, occupied states $E_F = \hbar^2\pi n/m^*$, where n is the areal electron density. This gives $n = 2eB_f/h$.

The current-voltage characteristic for our asymmetric DBS at 4 K is given in Figure 2. This shows clearly the threshold for resonant tunneling at $V_{th} = 330$ mV, the turn-off at $V_p = 725$ mV where the current peaks, and the region of intrinsic bistability. When biased in the opposite direction no bistability is present[1]. The voltage dependence of the differential capacitance C and parallel conductance $G = 1/R$ are

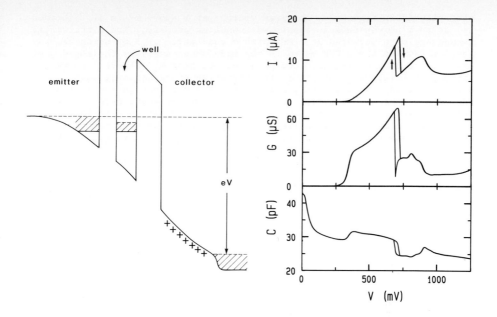

Fig. 1 (left) Conduction-band profile under applied voltage V showing bound-state levels (solid lines) in emitter and well and quasi-Fermi levels (dashed lines).

Fig. 2 (right) Voltage dependence of DC current I, AC conductance G and differential capacitance C measured at 1 MHz and 4 K.

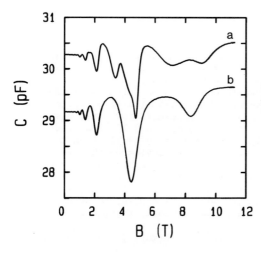

Fig. 3 Magneto-oscillations in capacitance C vs. magnetic field B for applied voltage (a) 600 mV and (b) 300 mV.

also plotted in Figure 2. The peak visible at $V \simeq 900$ mV in all 3 traces is due to a phonon-emission-assisted tunneling process. The differential parameters were measured at 1 MHz with a modulation of 3 mV using a Hewlett-Packard 4275A LCR meter in the mode which analyses the impedance of a device as a capacitor C and parallel resistor R. Figure 3 shows typical oscillatory structure in the variation of capacitance with magnetic field. Similar oscillations are also observed in the current I and conductance G. However, the capacitance measurements have higher sensitivity when the tunnel current is low. To reveal multiperiodic behaviour we have Fourier analysed the experimental capacitance traces. Figure 4 shows the Fourier spectra for different applied voltages. Each peak corresponds to a series of magneto-oscillations of period $B_f{}^{-1}$.

Below the threshold voltage V_{th}, there is a single peak in the magneto-oscillation spectrum. This peak is clearly associated with the 2DEG in the emitter accumulation layer, since the corresponding areal density n_a increases steadily with voltage V as shown in Figure 5. Moreover, the static capacitance values, given by the ratio of total accumulation charge to applied voltage, are in good agreement with the AC values shown in Figure 2.

However, between V_{th} and V_p, this magneto-oscillation frequency is independent of voltage (Figure 4) and n_a remains constant at 2×10^{11} cm^{-2}. This is the voltage range for which resonant tunneling occurs and the energies of the bound states in the accumulation layer and quantum well coincide. The voltage drop and hence electric field across the emitter barrier remain unchanged in this range, which corresponds to constant accumulation density, n_a. Also in the resonant tunneling region, a second, weaker peak appears in the magneto-oscillation spectrum (Figure 4). The frequency of this peak gives an areal density n_w, from equation (1), which increases throughout this range and approaches n_a at the voltage V_p for maximum current (Figure 5). We attribute this peak to a thermalized degenerate charge distribution stored in the quantum well. This thermalization is confirmed by comparing the lifetime τ_c of the electrons in the well with the energy relaxation rate. The lifetime τ_c is limited by tunneling through the collector barrier and is related to the current density by $J = n_w e / \tau_c$.[4] At the resonant peak ($J \simeq 0.06$ A cm^{-2}, $n_w \simeq 2 \times 10^{11}$ cm^{-2}) this gives $\tau_c \sim 0.6$ μs. The energy relaxation must be via spontaneous emission of acoustic phonons since the temperature is low (4 K) and the electron kinetic energies ($E_F \sim 7$ meV) are too small for optic-phonon processes. An estimate of the emission rate τ^{-1} from the deformation potential gives $\tau_{ph} \sim 10^{-9}$ s for a well width of 5.8 nm. This is indeed much shorter than τ_c.

When V increases above V_p a transition occurs in which charge is expelled from the well with a consequent redistribution of potential and resonant tunneling can no longer occur. This results in a step-wise increase in the accumulation layer density to $n_a \simeq 3 \times 10^{11}$ cm^{-2}, as shown in Figure 5. For $V > V_p$ only a single magneto-oscillation period is observed. The different charge states of the device on the high- and low-current parts of the hysteresis loop are also clearly shown by the different magneto-oscillation frequencies observed. We note that in the opposite bias direction, only one series of magneto-oscillations, due to the emitter accumulation layer, is observed. The corresponding sheet density increases smoothly with voltage showing that, in this case, there is little charge buildup in the well[1].

The principal features of the C(V) curve of Figure 2 can also be

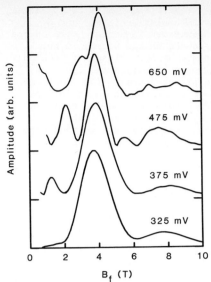

Fig. 4 *Spectrum of the magneto-oscillation frequencies B_f obtained by Fourier transforming them in $1/B$ space. The peak position B_f for each voltage gives the inverse period. The weak structure at around 8 T corresponds to the second harmonic of the series due to the charge in the accumulation layer.*

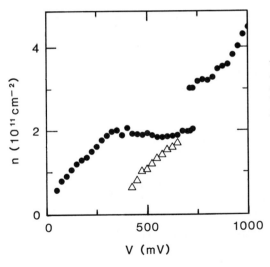

Fig. 5 *Areal density n vs. voltage V for charge in the accumulation layer n_a (circles) and well n_w (triangles). The values of n are deduced from the peaks in the Fourier spectrum.*

understood in terms of our model. We may regard the DBS as two parallel plate capacitors C_1 (emitter barrier and accumulation layer) and C_2 (collector barrier and depletion layer) connected in series. The charge on the common central plate corresponds to the charge stored in the quantum well. The steep fall in capacitance at low voltages is due to the rapid increase in depletion length in the lightly-doped (10^{16} cm^{-3}) collector layer which decreases C_2. The slower decrease for V > 50 mV is due to the less rapid rate of depletion of the layer doped to 10^{17} cm^{-3}. However, the most notable features are the sharp increase in capacitance at V_{th} and subsequent fall at V_p. When the bound states of the accumulation layer and quantum well are on resonance, the voltage drop across C_1 is essentially constant. An increase in applied voltage therefore appears almost entirely across C_2. The measured differential capacitance is thus $C \simeq C_2$ in the resonant tunneling region, whereas off-resonance $C \simeq C_1C_2/(C_1 + C_2) < C_2$.

This quasi-static argument is not obviously applicable at 1 MHz, where the measurements of differential capacitance were made. In a small-signal analysis[11] based on the sequential theory of resonant tunneling, C_1 and C_2 have parallel resistors R_1 and R_2 respectively. R_1 allows electrons tunneling from the emitter to charge the quantum well whilst R_2 allows the stored charge to leak through the collector barrier. Our previous argument is equivalent to the assumption that during resonant tunneling, R_1 becomes very small and effectively short-circuits C_1 so that the measured capacitance $C \simeq C_2$. By identifying the time constant R_2C_2 with the storage time τ_c we have $\tau_c = R_2C_2 \sim RC$ during resonance. The values shown in Figure 2 give $\tau_c \sim 0.4 \mu s$ at V_p, which is consistent with our previous estimate from the DC current and charge. An approximate theoretical estimate of τ_c can be made by calculating the attempt rate and transmission coefficient of a rectangular collector barrier. This gives $\tau_c \sim 1 \mu s$. Under bias a smaller value is expected since the average collector-barrier height is then decreased.

To conclude this section, we have observed the buildup of a thermalized charge distribution in the quantum well of a suitably asymmetric DBS. The storage time of an electron in the well was found to be $\sim 0.5 \mu s$, which is considerably shorter than the energy relaxation time. The tunneling process is thus truly sequential.

3. BALLISTIC TRANSPORT AND TUNNELING INTO HYBRID
    MAGNETO-ELECTRIC ORBITS IN WIDE QUANTUM WELLS WITH B⊥J

In this section, we consider DBS with well widths of up to 120 nm. In this case, a semi-classical picture is useful for describing the electronic transport. Figure 6(a) shows I(V) and the conductance, G = dI/dV for a double-barrier structure with a 60 nm quantum well. It comprises the following layers, in order of growth from the n$^+$ substrate: (i) 2 μm, 2 x 10^{18} cm^{-3} n$^+$GaAs buffer layer; (ii) 50 nm, 2 x 10^{16} cm^{-3} GaAs; (iii) 2.5 nm undoped GaAs; (iv) 5.6 nm undoped (AlGa)As, [Al] = 0.4; (v) 60 nm undoped GaAs quantum well; (vi) 5.6 nm undoped (AlGa)As, [Al] = 0.4; (vii) 2.5 nm undoped GaAs; (viii) 50 nm, 2 x 10^{16} cm^{-3} GaAs; (ix) 0.5 μm, 2 x 10^{18} cm^{-3} n$^+$GaAs top contact. Figure 6(b) shows similar curves for a device in which the central quantum well, layer (v), is 120 nm thick.

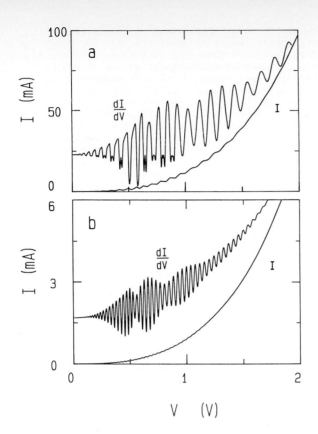

Fig. 6 Plots of I(V) and differential conductance dI/dV at 4 K and B = 0 for (a) the structure with the 60 nm well and (b) the structure with the 120 nm well.

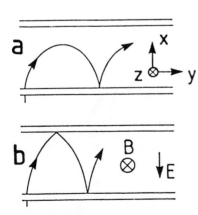

Fig. 7 Schematic diagram showing classical motion of electrons in the quantum well in crossed E and B fields; (a) skipping orbits and (b) traversing orbits.

In Figure 6, the beating of the oscillations in G at high voltage arises from the interference of electron waves scattered from the two interfaces of the collector barrier. At these voltages (> 0.5 V) the electron reaches the collector barrier at an energy in excess of potential barrier height. The standing wave resonances in the quantum well require a reflection from the collector barrier. The reflection coefficient is a minimum for the "over the barrier" resonances when an integral number of half wavelengths fit into the width of the collector barrier. The minimum in reflection thus corresponds to a minimum of the amplitude of the resonant structure in Figure 6. The existence of well-defined quantum resonances in I(V) and dI/dV means that a significant proportion of the electrons make at least two transits of the quantum well without scattering, since scattering destroys the standing wave states of the well. The time-scales for optic phonon emission and Γ-L and Γ-X scattering are ~0.1 ps. However, in the presence of the high electric fields in the well the electron transit times across the well are comparable to or less than this time scale[12].

The effect of a magnetic field applied in the plane of the barrier ($\underline{B}\|\underline{z}$, $\underline{J}\|\underline{x}$) is to modify the motion of the electrons in the well, as shown schematically in Figure 7. In classical terms, the electrons are accelerated by the large, uniform electric field ($\underline{E}\|\underline{x}$) in the well and deflected by the magnetic Lorentz force.

The electrons can tunnel through the emitter barrier into two distinct types of quantum state: (a) those associated with "skipping" orbits which interact with the emitter barrier only - these correspond to cycloidal motion under the action of crossed $\underline{E}$ and $\underline{B}$; (b) "traversing" states in which the electron wave is repeatedly reflected off both barriers[13]. The two types are also called "magneto-electric" states. The transition between the two types of state has recently been related to the quenching of the Hall effect in small structures[14].

Electrons tunnel into these states from the Fermi sea of the emitter accumulation layer 2DEG with conservation of energy and the components of wave vector, k_y and k_z, perpendicular to the tunneling direction. At low temperatures, there is a sharp cut-off in the emitter distribution function at $k_y = \pm k_F$. Here k_F is the Fermi wavevector in the emitter accumulation layer. As V or B are varied, the number of energy and momentum conserving transitions from emitter to well is sharply modulated, leading to oscillatory structure in the conductance. The resonance condition can be obtained by treating states (a) and (b) with the WKB approximation[13,15,16]. The values of $k_y = \pm k_F$ should give rise to two distinct series of oscillations from each type of magneto-electric state.

Typical magneto-oscillations arising from tunneling into magneto-electric orbits are shown in Figure 8 for the device with a 120 nm wide well. Above 6 T, tunneling occurs into type (a) skipping orbits. The two series of oscillations (0-2 T and 2-5 T) correspond to type (b) traversing states with $k_y = \pm k_F$. At high B ($\geq$ 15 T), the tunnel current is quenched, despite the large applied voltage, because there are no longer any states into which the electron can tunnel with energy and momentum conservation. The fan chart in Figure 9 shows the evolution of the type a and b orbits for the device with the 60 nm wide well. We attribute the absence of clearly defined resonances corresponding to type a_+ skipping orbits with $k_y = +k_F$ to their small tunneling probability[17]. When B $\rightarrow$ 0, the b-orbit resonances evolve into the purely electric subband resonances of the quantum well.

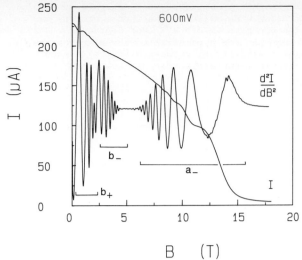

Fig. 8 Plots of $I(B)$ and d^2I/dB^2 at $V = 600$ mV, $T = 4$ K for a 100 µm diameter mesa of the structure with the 120 nm well, showing oscillatory structure due to tunneling into magneto-electric states.

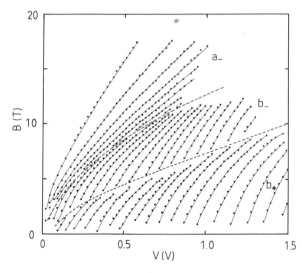

Fig. 9 Fan chart showing the positions of the maxima in the current in B-V space for the structure with the 60 nm well. The dashed lines define the regions corresponding to different types of orbit. The solid lines are guides to the eye.

271

To conclude, we have shown that in appropriately-designed double barrier structures incorporating wide wells, conduction electrons can travel ballistically at high energies (> 0.5 eV) over distances > 0.24 μm. The application of a magnetic field parallel to the plane of the barriers gives rise to magneto-oscillatory structure in I(V) due to tunneling in two types of hybrid magneto-electric orbit.

REFERENCES

1. Alves E S, Eaves L, Henini M, Hughes O H, Leadbeater M L, Sheard F W, Toombs G A, Hill G and Pate M A (1988). Electronics Lett. 24, 1190. See also Leadbeater M L, Alves E S, Eaves L, Henini M, Hughes O H, Sheard F W and Toombs G A (1988). Semicond. Sci. Technol. 3, 1060. In these papers the bias direction for occurrence of intrinsic bistability was referred to as "reverse bias".

2. Zaslavsky A, Goldman V J and Tsui D C (1988). Appl. Phys. Lett. 53, 1408.

3. Goldman V J, Tsui D C and Cunningham J E (1987). Phys. Rev. Lett. 58, 1257.

4. Sheard F W and Toombs G A (1988). Appl. Phys. Lett. 52, 1228.

5. Sollner T C L G (1987). Phys. Rev. Lett. 59, 1622.

6. Goldman V J, Tsui D C and Cunningham J E (1987). Phys. Rev. Lett. 59, 1623.

7. Toombs G A, Alves E S, Eaves L, Foster T J, Henini M, Hughes O H, Leadbeater M L, Payling C A, Sheard F W, Claxton P A, Hill G A, Pate M A and Portal J C (1988). 14th Int. Symp. on GaAs and Related Compounds, Inst. Phys. Conf. Series 91, 581.

8. Goldman V J, Tsui D C and Cunningham J E (1987). Phys. Rev. B35, 9387.

9. Payling C A, Alves E S, Eaves L, Foster T J, Henini M, Hughes O H, Simmonds P E, Sheard F W and Toombs G A (1988). Surf. Science 196, 404.

10. Luryi S (1985). Appl. Phys. Lett. 47, 490.

11. Sheard F W and Toombs G A. To be published.

12. Henini M, Leadbeater M L, Alves E S, Eaves L and Hughes O H (1989). J. Phys. Cond. Matt. 1(18), L3025.

13. Alves E S, Leadbeater M L, Eaves L, Henini M, Hughes O H, Celeste A, Portal J C, Hill G and Pate M A (1989) Superlattices and Microstructures 5 (4), 527; 4th Int. Conf. on Superlattices, Microstructures and Microdevices, Trieste, 1988

14. Beenakker C W J and van Houten H (1988). Phys. Rev. Lett. 60, 2406.

15. Snell B R, Chan K S, Sheard F W, Eaves L, Toombs G A, Maude D K, Portal J C, Bass S J, Claxton P, Hill G and Pate M A (1987). Phys. Rev. Lett. 59, 2806.

16. Eaves L, Alves E S, Foster T J, Henini M, Hughes O H, Leadbeater M L, Sheard F W, Toombs G A, Chan K, Celeste A, Portal J C, Hill G and Pate M A (1988). Physics and Technology of Submicron Structures, Springer Series in Solid State Sciences, 83, 74.

17. Sheard F W, Chan K S, Toombs G A, Eaves L and Portal J C (1988). 14th Int. Symp. GaAs and Related Compounds, Inst. Phys. Conf. Series 91, 387.

QUANTUM TRANSPORT EQUATION APPROACH TO NONEQUILIBRIUM SCREENING.*

Ben Yu-Kuang Hu
John W. Wilkins

Department of Physics
The Ohio State University
174 W. 18th Ave.
Columbus, OH 43210-1106

Sanjoy K. Sarker

Department of Physics and Astronomy
University of Alabama
Tuscaloosa, AL 35487.

1. INTRODUCTION

The very small scale on which devices are currently fabricated lead to highly nonequilibrium situations due to the very high operating electric fields within these devices. In these nonequilibrium situations, the screening due to free carriers is substantially different from screening at equilibrium. This difference manifests itself in the alteration of the carrier-impurity, carrier-phonon and carrier-carrier interactions, modifying the mobilities and the distribution functions of the carriers. Attempts have been made to characterize screening in the case of high electric fields, but these efforts have been at a formal level [1,2], or have relied on extensive computer simulations [3], or have relied on the drifted Maxwellian approximation [4-6]. Here, we report a transport equation approach to calculating nonequilibrium linear screening.

2. TRANSPORT EQUATION APPROACH

Linear screening is a consequence of the linear carrier density response $\delta n = n_1 e^{i(\mathbf{q}\cdot\mathbf{R}-\omega T)}$ to a potential $\delta U = U_1 e^{i(\mathbf{q}\cdot\mathbf{R}-\omega T)}$. The quantity that describes screening is $\chi(\mathbf{q},\omega) = n_1/U_1$, and the transport equation approach to calculating $\chi(\mathbf{q},\omega)$ is as follows. (i) Set up and solve the transport equation for the nonequilibrium situation being investigated. (ii) Linearly perturb the transport equation with an additional small sinusoidal force $iq U_1 e^{i(\mathbf{q}\cdot\mathbf{R}-\omega T)}$ to produce a distribution function response $f_1(\mathbf{p})e^{i(\mathbf{q}\cdot\mathbf{R}-\omega T)}$, and solve for $f_1(\mathbf{p})$. (iii) Integrate $f_1(\mathbf{p})$ with respect to $\mathbf{p}$ to obtain n_1. The ratio n_1/U_1 gives $\chi(\mathbf{q},\omega)$, and the dielectric constant $\epsilon(\mathbf{q},\omega) = 1 - 4\pi e^2 \chi(\mathbf{q},\omega)/q^2$.

Recently, the Boltzmann transport equation (within the relaxation time approximation) was used to study screening in a high uniform static electric field [7]. However, the Boltzmann equation, being a *classical* transport equation, cannot take into account the wave-like quantum mechanical nature of the carriers. Therefore, when the scale of the spatial variation of the external potential is short compared to the scale of the typical electron wavelength, the Boltzmann equation becomes invalid and $\chi(\mathbf{q},\omega)$ calculated from the Boltzmann equation cannot be trusted in this regime. A *quantum* transport theory is needed to obtain results in this regime.

2.1 Kadanoff-Baym Quantum Transport Equation

The quantum theory of transport of Kadanoff and Baym [8,9] is the generalization of the Boltzmann equation which takes into account spatial nonlocality of the carriers. We briefly review the formalism for this theory.

---

* Supported by the U.S. Office of Naval Research.

The Kadanoff and Baym formalism of quantum transport is based on the equations of motion for the two-time correlation functions

$$g^<(\mathbf{x}_1, t_1, \mathbf{x}_2, t_2) = i\langle \psi^\dagger(\mathbf{x}_2, t_2)\psi(\mathbf{x}_1, t_1)\rangle, \tag{1a}$$

$$g^>(\mathbf{x}_1, t_1, \mathbf{x}_2, t_2) = -i\langle \psi(\mathbf{x}_1, t_1)\psi^\dagger(\mathbf{x}_2, t_2)\rangle, \tag{1a}$$

where $\psi^\dagger(\mathbf{x}, t)$ and $\psi(\mathbf{x}, t)$ are the Heisenberg electon creation and destruction operators, and the brackets denote the thermal average. The quantum transport equation can be written in terms of the Wigner distribution function ($\hbar = 1$ in this paper),

$$f(\mathbf{p}, \mathbf{R}, T) = -i \int d\mathbf{r} e^{-i\mathbf{r}\cdot\mathbf{P}} g^<(\mathbf{R} - \tfrac{1}{2}\mathbf{r}, T, \mathbf{R} + \tfrac{1}{2}\mathbf{r}, T). \tag{2}$$

The transport equation is [8]

$$(\frac{\partial}{\partial T} + \frac{\mathbf{p}\cdot\nabla_{\mathbf{R}}}{m})f(\mathbf{p}, \mathbf{R}, T)+$$

$$i\iint d\mathbf{r} \frac{d\mathbf{p}'}{(2\pi)^3} e^{i(\mathbf{p}-\mathbf{p}')\cdot\mathbf{r}}\left[U_{\text{eff}}(\mathbf{R} + \frac{\mathbf{r}}{2}, T) - U_{\text{eff}}(\mathbf{R} - \frac{\mathbf{r}}{2}, T)\right]f(\mathbf{p}', \mathbf{R}, T)$$

$$= I[g^<, g^>, \Sigma^<, \Sigma^>], \tag{3}$$

where $I[g^<, g^>, \Sigma^<, \Sigma^>]$ is the quantum collision term, $\Sigma^<, \Sigma^>$ are the self-energies, and U_{eff} is the sum of the external potential and the Coulomb potential of the carriers.

Unfortunately, Eq. (3) is very difficult to solve in general because (a) the equation is nonlinear in $f(\mathbf{p}, \mathbf{R}, T)$ because of the dependence of U_{eff} on f, (b) I is nonlocal in time and space, and (c) $\Sigma^<$ and $\Sigma^>$ are functionals of $g^<$ and $g^>$. Therefore, progress can be made only in straightforward situations under many simplifying approximations. The situation we study is that of the application of a uniform external electric field, within the *quantum* relaxation time approximation.

2.2 Quantum Relaxation-Time Approximation

We adopt the approach given by Mermin [10], who modelled the collisions by a constant-rate relaxation of the nonequilibrium density matrix $\hat{\rho}$ to the local equilibrium density matrix $\hat{\rho}_{\text{l.eq}}$, where

$$\hat{\rho}_{\text{l.eq}} = \frac{1}{\exp(\beta[\hat{\varepsilon} - \mu - \delta\hat{\mu}]) + 1}. \tag{4}$$

Here, μ is the global chemical potential, $\hat{\varepsilon}$ is the kinetic energy operator, which is diagonal in the momentum representation, and $\delta\hat{\mu}$ is the *operator* for the change in the chemical potential, which is local and hence diagonal in the coordinate representation. The matrix elements $\delta\mu(\mathbf{q}) = \langle \mathbf{k} - \frac{\mathbf{q}}{2} \mid \delta\hat{\mu} \mid \mathbf{k} + \frac{\mathbf{q}}{2}\rangle$ are chosen so that local particle conservation is obeyed [10,11].

The equivalent relaxation-time approximation for the Wigner distribution function which will be used in the collision term in Eq. (3) is

$$I = -\frac{f(\mathbf{p}, \mathbf{R}) - f_{\text{l.eq}}(\mathbf{p}, \mathbf{R})}{\tau} \tag{5}$$

where $f_{\text{l.eq}}(\mathbf{p}, \mathbf{R})$ is given by

$$f_{\text{l.eq}}(\mathbf{p}, \mathbf{R}) = \int \frac{d\mathbf{q}}{(2\pi)^3} \langle \mathbf{p} - \frac{\mathbf{q}}{2} \mid \hat{\rho}_{\text{l.eq}} \mid \mathbf{p} + \frac{\mathbf{q}}{2}\rangle e^{i\mathbf{q}\cdot\mathbf{R}} \tag{6}$$

The matrix elements $\delta\mu(\mathbf{q})$ must be obtained self-consistently so that the particle conserving condition

$$\int d\mathbf{p} f(\mathbf{p}, \mathbf{R}) = \int d\mathbf{p} f_{\text{l.eq}}(\mathbf{p}, \mathbf{R}) \tag{7}$$

and the transport equation (3) are simultaneously satisfied.

3. LINEAR SCREENING FROM KADANOFF-BAYM EQUATION

Calculating complete nonlinear screening is far too complex to handle analytically, since the transport equation has to be solved self-consistently with arbitrary spatial density variations. Therefore we confine our attention to linear screening.

3.1 Relaxation Time Approximation in Linear Screening

If only small spatial variations in denstiy are considered (as in linear screening), the collision term simplifies considerably. Assume that the Wigner distribution function is of the form*

$$f(\mathbf{p}, \mathbf{R}, T) = f_0(\mathbf{p}) + f_1(\mathbf{p})e^{i(\mathbf{q}\cdot\mathbf{R}-\omega T)}, \tag{8}$$

with $n_0 \gg |n_1|$, where

$$n_i = \int \frac{d\mathbf{p}}{4\pi^3} f_i(\mathbf{p}), \qquad (i = 0, 1). \tag{9}$$

Then, the collision term simplifies to [11]

$$I = -\frac{f_0(\mathbf{p}) + f_1(\mathbf{p})e^{i(\mathbf{q}\cdot\mathbf{R}-\omega T)} - \left(f_{eq}(\mathbf{p}) + \delta f(\mathbf{p}, \mathbf{q})n_1 e^{i(\mathbf{q}\cdot\mathbf{R}-\omega T)}\right)}{\tau} \tag{10}$$

where

$$\delta f(\mathbf{p}, \mathbf{q}) = \frac{1}{\chi_L(\mathbf{q}, 0)} \frac{f_{eq}(\mathbf{p}+\mathbf{q}/2) - f_{eq}(\mathbf{p}-\mathbf{q}/2)}{\varepsilon(\mathbf{p}+\mathbf{q}/2) - \varepsilon(\mathbf{p}-\mathbf{q}/2)}, \tag{11}$$

and

$$\chi_L(\mathbf{q}, \omega) = \int \frac{d\mathbf{p}}{4\pi^3} \frac{f_{eq}(\mathbf{p}+\mathbf{q}/2) - f_{eq}(\mathbf{p}-\mathbf{q}/2)}{\varepsilon(\mathbf{p}+\mathbf{q}/2) - \varepsilon(\mathbf{p}-\mathbf{q}/2) - \omega} \tag{12}$$

will be recognized as the Lindhard expression for the susceptibility (f_{eq} is of course the equilibrium distribution function).

3.2 Linear Screening in High Electric Fields

As an example, we study linear screening in a uniform static electric field, within the particle-conserving relaxation time approximation described above, for carriers in a parabolic band. The procedure was described at the beginning of Section 2. The "unperturbed" quantum transport equation is identical to the Boltzmann equation within the relaxation-time approximation

$$F \frac{\partial f_0}{\partial p_z}(\mathbf{p}) = -\frac{f_0(\mathbf{p}) - f_{eq}(\mathbf{p})}{\tau}. \tag{13}$$

On addition of the small sinusoidal potential perturbation, we obtain an expression for $f_1(\mathbf{p})$ as a function of U_1. Integrating $f_1(\mathbf{p})$ with respect to $\mathbf{p}$ yields [11]

$$n_1 \left(1 - \frac{1}{\chi_L(q, 0)} \int_0^\infty \frac{dt}{\tau} \exp(-\frac{i\mathbf{F}\cdot\mathbf{q}t^2}{2m} + i\omega t - \frac{t}{\tau})\right)$$

$$\times \int \frac{d\mathbf{p}}{4\pi^3} \exp(-i\frac{t}{m}\mathbf{p}\cdot\mathbf{q})\left[\frac{f_{eq}(\mathbf{p}+\mathbf{q}/2) - f_{eq}(\mathbf{p}-\mathbf{q}/2)}{\mathbf{q}\cdot\mathbf{p}/m}\right]) =$$

$$iU_1 \int_0^\infty dt \exp(-\frac{i\mathbf{F}\cdot\mathbf{q}t^2}{2m} + i\omega t - \frac{t}{\tau}) \int \frac{d\mathbf{p}}{4\pi^3} \exp(-i\frac{t}{m}\mathbf{p}\cdot\mathbf{q})\left[f_0(\mathbf{p}+\mathbf{q}/2) - f_0(\mathbf{p}-\mathbf{q}/2)\right] \tag{14}$$

The nonequilibrium susceptibility $\chi(\mathbf{q}, \omega) = n_1/U_1$ is obtained from Eq. (14).

---

* The Wigner distribution function is always real. Therefore, in Eq. (8), we implicitly assume the real part is taken.

275

4. DISCUSSION AND SUMMARY

In the classical (Boltzmann equation) derivation of the nonequilibrium susceptibility, the only two length scales that were important were the carrier thermal mean free path $l_{th} = p_{th}\tau/m$ (where $p_{th} = (2k_B mT)^{1/2}$ is the thermal momentum) and and the carrier drift mean free path $l_d = p_d\tau/m$ (where $p_d = F\tau$ is the average drift momentum). In the regime $q \cdot \max(l_{th}, l_d) \ll 1$ and $\omega\tau \ll 1$, the carriers are not ballistic over the distance of a wavelength, and local equations apply, therefore the screening can be derived from the drift-diffusion and continuity equations [7].

The quantum mechanical nature of the carriers necessitates the introduction of a third length scale. For a nondegenerate semiconductor, this length is the thermal deBroglie wavevector, $q_{dB} = p_{th}$, which is the inverse of the average wavelength of a thermal electron. This sets the scale at which the quantum mechanical screening differs from the classical screening.

This is illustrated by the calculation of the nonequilibrium susceptibility using Eq. (14), with $f_{eq}(\mathbf{p})$ set equal to the Maxwell-Boltzmann distribution function. Fig. 1 below shows the calculated static susceptibility $\chi(\mathbf{q}, \omega = 0)$ obtained from (a) the Kadanoff-Baym quantum transport equation and (b) the Boltzmann equation, both using their respective relaxation-time approximations (for details of the calculations, see Refs. 11 and 7, respectively). For $q_{dB} \ll q$, the two approaches yield virtually

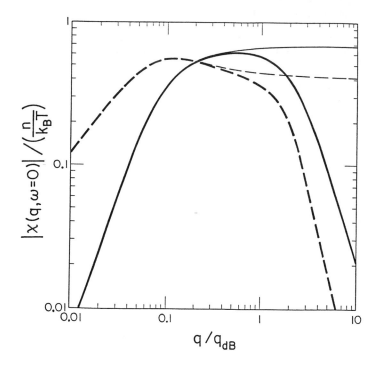

FIG. 2. *The imaginary (dashed lines) and negative real (solid lines) parts of the normalized nonequilibrium static susceptibility* $\chi(\mathbf{q}, \omega = 0)/(n/k_B T)$, *for* $\mathbf{p}_d \cdot \hat{\mathbf{q}} = 0.4 p_{th}$ *and* $q_{dB} l_{th} = 10$. *The normalization is chosen so that, at equilibrium,* $\chi(\mathbf{q} \to 0, \omega = 0)/(n/k_B T) = -1$. *The bold lines are obtained from the quantum kinetic equations, while, or comparison, the thin lines are obtained from the Boltzmann equation.*

identical results. On the other hand, for $q > q_{dB}$ the two approaches produce markedly different curves. For $q \to \infty$ the curve derived from the Boltzmann equation asymptotes to the susceptibility of a collisionless classical plasma [7]. The curve derived from the quantum kinetic equation, however, tends towards the Lindhard form, with the equilibrium distribution functions replaced by the nonequilibrium ones, i.e.,

$$\chi(\mathbf{q}, \omega) = \int \frac{d\mathbf{p}}{4\pi^3} \frac{f_0(\mathbf{p} + \mathbf{q}/2) - f_0(\mathbf{p} - \mathbf{q}/2)}{\varepsilon(\mathbf{p} + \mathbf{q}/2) - \varepsilon(\mathbf{p} - \mathbf{q}/2) - \omega - i0^+}; \qquad (q \to \infty). \tag{15}$$

To summarize, we have shown how transport equations can be used to obtain screening in nonequilibrium situations. Using the Kadanoff-Baym quantum transport equation within the relaxation-time approximation, we have studied the linear screening of parabolic-band semiconductors in a high electric field. Results for nonequilibrium screening obtained from the quantum Kadanoff-Baym and classical Boltzmann equations are virtually identical for $q \ll q_{dB}$, but differ for $q > q_{dB}$ because quantum effects due to spatial nonlocality of the carriers start to dominate.

REFERENCES

1. J. R. Barker, Solid State Comm., **32**, 1013 (1979).
2. D. Lowe, Ph.D. Thesis, Universitiy of Warwick, 1983 (unpublished).
3. P. Lugli and D. K. Ferry, Phys. Rev. Lett., **56**, 1295 (1986).
4. D. Lowe and J. R. Barker, J. Phys. C, **18**, 2507 (1985).
5. G. Berthold and P. Kocevar, J. Phys. C, **17**, 4981 (1984).
6. X. L. Lei and C. S. Ting, Phys. Rev. B, **32**, 1112 (1985).
7. B. Y.-K. Hu and J. W. Wilkins, to be published in Phys. Rev. B.
8. L. P. Kadanoff and G. Baym, *Quantum Statistical Mechanics* (Benjamin, Reading, Mass., 1962).
9. L. V. Keldysh, Zh. Eksp. Teor. Fiz., **47**, 1515 (1964) [Soviet Phys. – JEPT, **20**, 1018 (1965)].
10. N. D. Mermin, Phys. Rev. B, **1**, 2362 (1970).
11. B. Y.-K. Hu, S. K. Sarker and J. W. Wilkins, to be published in Phys. Rev. B.

CALCULATION OF BALLISTIC TRANSPORT IN TWO-DIMENSIONAL QUANTUM STRUCTURES USING THE FINITE ELEMENT METHOD [1]

Craig Lent
Srinivas Sivaprakasam
Department of Electrical and Computer Engineering
University of Notre Dame
Notre Dame, Indiana 46556

David J. Kirkner
Department of Civil Engineering
University of Notre Dame
Notre Dame, Indiana 46556

I. INTRODUCTION

Current fabrication technology permits the construction of ultra-small semiconductor structures in one dimension, usually the direction of crystal growth. This capability has spawned a wealth of experiments and theory describing transport in the ballistic regime. This development has been aided greatly by the fact that ballistic transport can be understood, qualitatively at least, by simply solving the one-particle Schrödinger equation. The one-dimensional form of the Schrödinger equation is fairly easily solved so that transmission coefficients and currents can be calculated.

Fabrication technology is becoming increasingly sophisticated and is now beginning to create structures quantized in two and three spatial dimensions. The leap to two dimensions makes the solution of the Schrödinger equation considerably more challenging. Analytic textbook solutions become inadequate for guiding intuition and design. Unbound states which carry current require particularly careful analysis in two dimensions. It is important in improving our understanding of ballistic transport in two-dimensional electron wave-guide devices that sufficiently powerful and flexible numerical methods be developed.

We have used the Finite Element Method (FEM) to solve the single-particle Schrödinger equation for two-dimensional potentials. While calculations of bound

[1]Supported by the Air Force Office of Scientific Research under grant no. AFOSR-88-0096 and by an IBM Faculty Development Award.

279

state wavefunctions have been done previously [3], this represents the first method to yield wavefunctions for states which carry current. We present solutions for the transmission coefficients of double-cavity electron wave-guide structures. The FEM provides a very flexible, elegant way of handling boundary conditions for very complex structures. Ultimately self-consistent solutions, at least in the Hartree approximation, are required. The method presented here lends itself well to such an extension because it yields the wavefunctions directly. As demonstrated in the double-barrier resonant tunneling problem, single-particle solutions can nevertheless reveal most of the important transport features.

II. THEORY

There are several ways in which two-dimensional electron waveguide structures might be fabricated. A metal pattern deposited on an AlGaAs-GaAs heterojunction can be used to create channels in the two-dimensional electron gas formed at the heterojunction interface. This technique has been used by Bernstein and Ferry in making very fine grid structures on a FET gate [1]. The technique might be refined by using a quantum well instead of the heterojunction potential to confine carriers in the plane. Another technique which may prove useful involves etching and regrowth of lithographically defined patterns in quantum well layers. We do not concern ourselves here with the exact method used, but look instead for the basic transport features such structures would exhibit.

We consider a system in which electrons are confined in the xy-plane by some potential which is such that only the ground state z-eigenfunction is ever occupied. The potential in the xy-plane, defined by some lithographic means, is assumed to take the form of rectangular waveguides which act as input and output leads, connected to a device region. The geometry for the double-cavity structure is shown in Figure 1. In this case the device region is simply the two rectangular cavities and the short channel which connects them. For simplicity

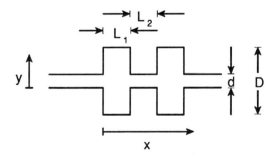

Figure 1: Geometry for the double resonant cavity. The two cavities are here assumed to have the same width D and length L_1.

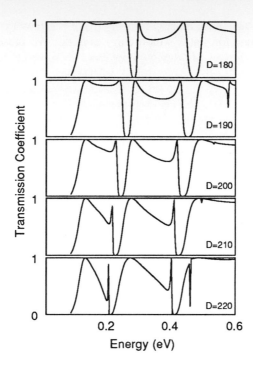

Figure 2: The transmission coefficient versus energy for the double cavity shown in Figure 1 with $d = 100$ Angstroms, $L_1 = 100$ Angstroms, $L_2 = 50$ Angstroms, and various values of D.

we assume the rectangular waveguides of width d have infinite potential walls so the wavefunctions in the leads consist of plane waves in the x-direction and sine functions in the y-direction. We assume an incoming plane wave from the left, which produces a reflected wave and a transmitted plane wave. The wavefunction in the left ($x < 0$) contact is then,

$$\psi^L(x, y) = A_m e^{ik_m x} sin\left(\frac{m\pi y}{d}\right) \tag{1}$$

$$+ \sum_{n=1}^{N} b_n e^{-ik_n x} sin\left(\frac{n\pi y}{d}\right) + \sum_{n=N+1}^{\infty} b_n e^{k_n x} sin\left(\frac{n\pi y}{d}\right).$$

In the right contact the wavefunction is

$$\psi^R(x, y) = \sum_{n=1}^{N} a_n e^{ik_n x} sin\left(\frac{n\pi y}{d}\right) + \sum_{n=N+1}^{\infty} a_n e^{-k_n x} sin\left(\frac{n\pi y}{d}\right), \tag{2}$$

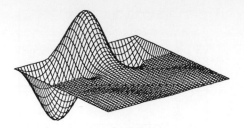

Figure 3: The real part of the wavefunction for an energy of 0.23 eV, for the double cavity problem. The geometry is the same as for Figure 2 with $D = 200\text{Å}$.

where

$$k_n = \sqrt{|E - \frac{\hbar^2}{2m^*}\left(\frac{n\pi}{d}\right)|}$$

The incoming wave has amplitude A_m and represents an excitation of the m-th transverse mode of the input waveguide. N is the number of traveling modes possible in the waveguide leads. The modes with $n > N$ are evanescent modes with complex wavenumber and carry no current.

We solve the two-dimensional effective-mass Schrödinger equation in the device region using the Finite Element Method (FEM). The details of the method are presented elsewhere [4]. The FEM enables us to straightforwardly include the condition that the wavefunction and its normal derivative match the analytical form of equations (1) and (2) as an additional set of constraint equations. The region is discretized into small elements on which the wavefunction is approximated by bilinear shape functions. This discretization yields a set of algebraic equations for the values of the wavefunction at the nodal points. For simplicity we assume infinite potential barriers at boundary walls. The boundary conditions are the that the wavefunction be zero on the cavity walls and that the wavefunction and its first derivative match those for the analytical expressions given in equations (1) and (2) above at the interface between the leads and the device region. The incoming and outgoing current are calculated directly from the wavefunction and the transmission coefficient is obtained from their ratio.

III. RESULTS

The transmission coefficient as a function of energy is shown for the double-

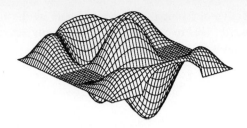

Figure 4: The real part of the wavefunction for an energy of 0.27 eV, for the double cavity problem. The geometry is the same as for Figure 2 with $D = 200$Å.

cavity structure in Figure 2. It has been calculated for several values of D, the cavity width. The input and output channels are 100 Å wide and each cavity is 100 Å in length with a 50 Å long channel connecting them. The FEM discretization is accomplished on square 5 Å elements. Results were sufficiently converged at this mesh size that further refinements had no significant effect on the results.

The two main minima, at energies slightly above 0.2 eV and 0.4 eV, are due to excitations of bound states of a single cavity. The real part of the wavefunction near such a minimum is shown in Figure 3. The plateau regions of high transmission between these minima owe their shape to the coupling between the two cavities. The maxima at the high-energy and low-energy ends of these plateaus are due to states which excite both wells. A sample wavefunction is shown in Figure 4.

It is interesting to note that the shape of the transmission coefficient vs. energy curves is quite different for cavity structures than for constrictions. In a constricted geometry, one observes sharp peaks, corresponding to the modes allowed in the narrow region, and broad valleys. In the cavity resonance case examined here, the transmission dips sharply at resonances and is generally high between them.

The energy dependence of the transmission coefficients shown in Figure 2 suggests possible applications of such cavities in waveguide devices. For some implementations of lithographically defined structures, such as the metal-on-heterojunction technique, the cavity dimensions can be changed dynamically by applying voltages to the metal overlayers and shrinking or enlarging depleted regions. This effectively would sweep the transmission characteristic through a family of curves such as those shown in the figure. This presents the possibility of transistor action in such devices [2]. In addition, as the cavity width increases, a large region of negative differential resistance occurs as the high-transmission plateaus slope downward. This region, while not especially steep, has the advantages of being broad and tunable (through changes in D).

References

[1] G. Bernstein and D.K. Ferry, J. Vac. Sci. Technol. **B5** 964 (1987).

[2] Fernando Sols, M. Macucci, U. Ravaioli, and Karl Hess, Appl. Phys. Lett. **54**, 350 (1989).

[3] Steven E. Laux and Frank Stern, Appl. Phys. Lett. **49**, 91 (1986).

[4] Craig S. Lent, Srinivas Sivaprakasam, and David J. Kirkner, submitted to the International Journal of Numerical Methods in Engineering.

RESONANT TUNNELING IN DOUBLE BARRIER DIODES

Karim Diff and George Neofotistos

Department of Physics and Center for Advanced Computational Science
Temple University, Philadelphia, PA 19122

Hong Guo

Department of Physics
Mc Gill University, Montreal, Quebec Canada H3A 2T8

James D. Gunton

Department of Physics
Lehigh University, Bethlehem, PA 18105

Abstract

We present results of a study of resonant tunneling in double-barrier GaAs-AlGaAs diodes. We solve the time-dependent Schrödinger equation for Gaussian wavepackets to study the time characteristics of the device. In addition to the time constant of the exponential decay, we study the build-up time, which is the time needed for the probability of finding the electron inside the well to reach its peak. The difference between the effective mass of the electrons in the well and in the barriers has a strong effect on the decay time constant and the magnitude of the trapped wavepacket inside the well, but does not modify the build-up time. The latter is only affected by the size of the wavepackets and, for a given packet size, by the geometry of the structure. We also present results for the current-voltage (I-V) characteristics using wavepackets to represent the wavefunction of the tunneling electrons instead of plane-waves. We obtain results for the peak current density and the peak-to-valley ratio that are in reasonably good agreement with the experimental values given the approximations made.

1. Introduction

With the currently available semiconductor fabrication technology, microelectronic devices have entered the submicron regime and structures with active regions of less than few hundred Angstroms are now easily fabricated. The most studied of these structures is undoubtedly the resonant tunneling double-barrier structure first conceived by Tsu and Esaki [1]. Recently, the interest in this structure was renewed after Sollner et al. reported results for structures that operated at very high frequencies and displayed large peak-to-valley ratios in the negative-differential-regime of their I-V characteristics [2]. However, despite the efforts focused on this structure, a significant difference remains between the theoretical predictions and the experimental results [3],[4],[5]. In this paper we present a study of the double-barrier structures described in reference 2 using wavepackets to represent the wavefunction of the electron. This approach is more appropriate than the conventional plane-waves treatment to study the time characteristics of the device, and allows for a more intuitive picture of the scattering process to get the current-voltage (I-V) characteristic.

2. Time characteristics

We have studied the propagation of wavepackets across double-barrier diodes by solving the one-dimensional time-dependent Schrödinger equation written in the effective-mass approximation for the envelope function

$$i\hbar\frac{\partial\Psi}{\partial t} = -\frac{\hbar^2}{2}\frac{\partial}{\partial x}\left[\frac{1}{m^*(x)}\frac{\partial\Psi}{\partial x}\right] + V(x)\Psi(x,t) \tag{1}$$

where $V(x)$ is the potential profile of the double-barrier. In this work we neglect all effects due to scattering, band bending, and space-charge accumulations. The transmission through quantum tunneling is enhanced when the kinetic energy of a particle incident on the structure is equal to the energy of the quasi-bound state inside the well. The value of the resonant energy is obtained by computing the transmission coefficient across the device using the time-independent Schrödinger equation. The form used for the kinetic energy operator preserves the hermiticity of the Hamiltonian in the case of position-dependent masses [6], and its validity has recently been confirmed experimentally [7]. To study the time characteristics of the resonant tunneling we compute the probability of finding the tunneling electron inside the well $P_{well}(t) = \int_{well}dx\,|\Psi(x,t)|^2$, as a function of time. The result for the case of a structure with 50 A barriers and well width and a barrier height of 0.23 eV is shown in the solid curve of figure 1. The decay time constant τ corresponds to the inverse width of the resonant level obtained from the full width at half maximum of the transmission coefficient, namely $\tau=\frac{\hbar}{\Gamma}$. Most discussions of the time characteritics of double-barrier devices focus on this quantity. However, it has been pointed out that the build-up time could be as important for this purpose [8]. This is defined as the time needed to reach the maximum of $P_{well}(t)$. We have previously presented results for various well widths [9]. These results are only an upper bound since the time is measured from the beginning of the simulation when the wavepacket is outside the barriers. As it reaches the structure, the propagating wavepacket oscillates very strongly and a precise determination of the moment at which it enters the double barrier is problematic. However, our results indicate that the build-up time is not exponentially long as was previously concluded [3]. The value of the decay constant is fixed by the geometry of the structure but also depends on the choice of the effective mass inside the barriers as can be seen from figure 1. The solid curve corresponds to the case of different effective masses while the dotted curve is obtained using the same mass throughout the device. We have assumed that the energy in the direction parallel to the GaAs-AlGaAs interface is zero. The difference in the results for the time scales clearly indicate that the neglect of effective mass differences is a simplification that is not justified for quantitative esti-mates. In figure 1, solid curve and dashed curve, we show results obtained for different choices of the wavepacket width σ. The decay time constant is the same for both choices since it does not depend on the shape of the packet, but the build-up time and the peak value of the probability are different. This is due to the fact that a wide wavepacket in real space has a narrow spread in momentum space and on resonance this leads to a stronger build-up inside the well because fewer of the Fourier component are off-resonance [8]. In principle this would mean that a maximal effect can be obtained when the wavepacket spread in real-space is so large that its associated energy spread is narrower than the resonant level width. However this would involve wavepackets wider than the entire device which is of the order of 5000 A. A more realistic value for σ is obtained from the mean free-path which is typically around 500 A for GaAs. This order of magnitude implies that in most cases the energy spread of the electrons will be such that a less than perfect resonance is achieved. This is to be contrasted with calculations that assume that the wavefunction of the electron can be represented by a plane-wave and thus on resonance the transmission coefficient is equal to one.

3. I-V Characteristics

Using Gaussian wavepackets to represent the wavefunction of tunneling electrons, we have calculated the I-V characteristics for the double-barrier structure described above. Using the formalism of scattering theory we write the wavefunction of an electron, moving from left to right, after tunneling as a superposition of eigenstates in the corresponding region. If the wavepacket is far from the double-barrier, the contribution of the discrete quasi-bound state can be neglected since it is exponentially damped. We assume that the potential drop due to the external electric field is linear and is limited to the barriers and well regions. For an electron with average initial momentum k we can write the transmitted wavefunction as

$$\Psi^{tr}{}_k = \int dk' \, T(k',V) \, a(k'-k)\phi_{k'} e^{-i\varepsilon_{k'}\frac{t}{\hbar}} \tag{2}$$

where $T(k',V)$ is the transmission coefficient in the presence of the potential drop V, $a(k'-k)$ is the Fourier transform of $\Psi_k(x,t=0)$, $\phi_{k'}$ is the eigenfunction of the Hamiltonian in the region on the right of the double-barrier and $\varepsilon_{k'}$ is the corresponding energy. These quantities are obtained from the time-independent Schrodinger equation for which we have used the Airy functions as solutions in the barriers. The current associated with this wavepacket is

$$j_k = e <\Psi^{tr}{}_k | n_k \frac{\hbar}{im}\frac{\partial}{\partial x} | \Psi^{tr}{}_k>, \tag{3}$$

where n_k is the electron density $<\Psi^{tr}{}_k | \Psi^{tr}{}_k>$. In the expression for n_k the integration boundaries are in principle limited to the half-space on the right of the double-barrier, but since the transmitted wavepacket, far to the right, does not overlap with the scattering region and the reflected packet, we can extend the integration over the entire space. The final result is then

$$j_k = \frac{e\hbar\sigma}{\pi^2}\int dk'k' \, |a(k'-k)|^2 |T(k',V)|^2 \tag{4}$$

Then the total current density is obtained from the usual formula

$$J = 2e \int \frac{d^3k}{(2\pi)^3} \, j_k \, [\, f(\varepsilon_k)-f(\varepsilon_k+eV) \,], \tag{5}$$

where $f(\varepsilon_k)$ is the Fermi-Dirac distribution. In the usual treatment of this problem, the electrons are represented by plane-waves for which $j_k = \frac{\hbar k}{m^*} |T(k,V)|^2$. In figure 2 we show results obtained for various wavepacket widths and for the case of plane-waves. A constant effective mass throughout the device has been assumed. We notice that the peak current density J_{max} and the peak-to-valley ratio (P/V) increase with the packet width and that the case of plane-waves represents the limit of infinite width. As mentionned before, the plane-wave treatment leads to a maximum transmission that is not realistic for wavepackets of finite width. This could account for the great discrepancy between the experimental results and the results obtained with plane-waves. For the temperature considered here (T=25K), the experimental results are $J_{max} = 0.2 \ 10^4 A/cm^2$, and P/V=6. The corresponding results with plane-waves are $J_{max} = 2.9 \ 10^4 A/cm^2$ and P/V=51.8. With wavepackets we obtain values that vary from $J_{max} = 0.9 \ 10^4 A/cm^2$ to $J_{max}=1.2 \ 10^4 A/cm^2$ and from P/V=1.2 to P/V=41. However, this raises some questions about the importance of the form of the incident states used for the calculation of the I-V characteristics. In our model, each electron wavefunction (equation 2) represents a linear superposition of eigenstates of the corresponding Hamiltonian, and this reflects the fact that in general for a device of finite spatial dimensions the electron cannot be adequately described by a plane-wave. These wavefunctions are distributed according to the Fermi-Dirac statistics (equation 5) and it should be shown whether such a distribution of wavepackets is equivalent to a distribution of plane-waves when thermodynamic quantities such as the current density are computed. This important question will be adressed in a future publication. In the calculation presented here a single width σ has been assumed for simplicity. Preliminary results for the more realistic case of a distribution of widths show little deviation from the results presented here and a complete study of this aspect of the problem will be presented elsewhere. Taking into account the simplicity of our model, and although our results for the peak current density are larger than the experimental ones we obtain a relatively good agreement with the experimental data. The major limitation of our results is the value of the external bias for which J_{max} is obtained. However it has been shown previously that space-charge effects can increase this value as well as reduce the P/V ratio [10]. Results that incorporate this effect in our formalism will be presented in a future publication.

4. Acknowledgements

This work was supported by U.S. Office of Naval Research under the grant N00014-83-K--0382.

References

1. Tsu, R., and Esaki, L., Appl. Phys. Lett. **22**, 562, (1973)

2. Sollner, T.C.L.G.,Goodhue, W.D., Tannenwald, P.E., Parker, C.D., Peck, D.D., Appl. Phys. Lett. **43**, 588, (1983).

3. Ricco, B., Azbel, M.Ya., Phys. Rev. **B29**, 1970, (1984).

4. Capasso, F., Mohammed, K., Cho, A.Y., J. Quant. Elect. **QE-22**, 1853, (1986).

5. Barker, J. R., Physica **134B**, 22, (1985).

6. Morrow, R.A., and Brownstein, K.R., Phys. Rev. **B30**, 678, (1984).

7. Galbraith, I. and Duggan, G., Phys. Rev. **B38**, 10057, (1988).

8. Jauho, A.P., Proceedings of Advanced Summer School on Microelectronics, Kivenlahti, Espoo, Finland, June 1987, Eds. T. Stubb and R. Paanen.

9. Guo, H., Diff, K. Neofotistos, G. and Gunton, J.D., Appl. Phys. Lett. **53**, 131, (1988).

10. Cahay, M., McLennan, M., Datta, S., and Lundstrom, M.S., Appl. Phys. Lett. **50**, 612, (1987).

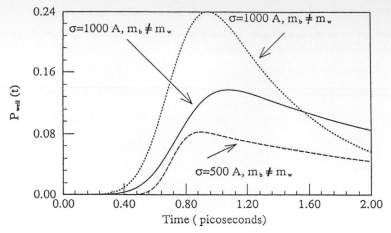

Fig.1 P_{well} (t) for various combinations of the wavepacket width
σ and the effective masses m_b (barriers) and m_{well} (well).
The barriers height is 0.23 eV and their width is 50 A.
The width of the well is 50 A. m_b =0.09 m_o . m_{well} =0.067 m_o

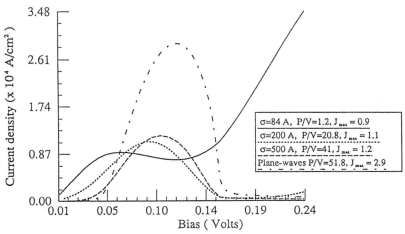

Fig.2 I-V curves for double-barrier diodes. A single effective mass
$m^* = 0.067$ m_o was used for the barriers and the well. The temperature
is T=25 K and the doping is 10^{18} cm^{-3}

A SIMPLIFIED METHOD FOR QUANTUM SIZE EFFECT ANALYSIS IN SUBMICRON DEVICES INCLUDING FERMI-DIRAC STATISTICS

Wayne W. Lui

AT&T Bell Laboratories
Allentown, Pennsylvania

Jeffrey Frey

Department of Electrical Engineering
University of Maryland

I.   Introduction

One way to model the electron states resulting from the quantum confinement in modern submicron devices has been by solving self-consistently the Poisson Equation, which governs the electrostatic potential, and the Schrœdinger Equation, which governs the electron distribution [1]. The solving of the latter equation, however, involves solving an eigenvalue problem. Although 1-D calculations of this sort have become common practice, it is nevertheless CPU-time consuming. Two-dimensional self-consistent numerical solutions to the Poisson and Schrœdinger equations, it was reported [2], requires in the order of two CPU-hours on a scalar machine for a single self-consistent solution; with 95-98% of the total calculation time devoted to the solution of the eigenvalue problem. This is perhaps the reason why the self-consistent solution of the Poisson and the Schrœdinger equations method has been limited to 1-D cases.

Similar to a previous work [3], we shall describe a simplified method to obtain electron distribution of a two-dimension electron gas (2DEG) without involving the Schrœdinger Equation. In this approach, the diagonal elements of the density matrix (or equivalently the carrier distribution) are calculated directly. As a result, the eigenvalue problem is avoided, thereby greatly reducing the CPU computation time. Initial 1-D calculation results suggest more than an order of magnitude CPU time savings over the conventional Schrœdinger Equation approach. Using the method developed here, quantum size effect analysis will not have to be limited to only 1-D cases.

NANOSTRUCTURE PHYSICS
AND FABRICATION

II. Model and Analysis

The outline of our methodology is as follows:

(1) Express the electron distribution in terms of a quantum density operator ρ.

(2) Then, express this operator ρ as the solution of a partial differential equation with respect to β ($= 1/k_B T$).

(3) Assuming that the Taylor expansion of ρ in h (Planck's constant) exists, linearize the PDE in (2) and express it up to $O(\hbar^2)$ terms.

(4) Using the Wigner function formalism, transform the resulting equation and simplify it. An equation governing the spatial distribution of electrons is retrieved.

(5) Solve this simplified equation numerically to obtain the electron distribution.

Step 1:

The distribution of electrons in a 2DEG, at equilibrium, is given by [1]:

$$n = \sum_{\text{all states}} \frac{m}{\pi \hbar^2 \beta} \langle i \mid \log_e \left[1 + e^{\beta(E_F - H)} \right] \mid i \rangle = \sum_{\text{all states}} \frac{m}{\pi \hbar^2 \beta} \langle i \mid \rho \mid i \rangle$$

(1)

with

$$H = -\frac{\hbar^2}{2m} \nabla^2 + V(r) = \text{Hamiltonian operator}$$

and m = electron mass; E_F = Fermi-level; $V(r)$ = potential function.

The energy distribution, on the other hand, is found to be:

(i) If $\beta(E_F - E_i) \gg 1$, then $u_i \approx \frac{1}{2}(E_F - E_i) n_i$

(ii) If $\beta(E_i - E_F) \gg 1$, then $u_i \approx \frac{1}{\beta} n_i$

(iii) If $E_i \approx E_F$, then $u_i \approx \frac{\pi^2}{12 \beta \log_e 2} n_i \approx \frac{1.2}{\beta} n_i$

At equilibrium, most electrons reside at states where the eigenenergies lie below the Fermi-level. Furthermore, at low temperatures where quantum size effect is more pronounced, most electrons fall under category (i). It is not a bad approximation, therefore, to write

$$u \approx \sum_{\text{all states}} \frac{m}{\pi \hbar^2 \beta} \langle i \mid \frac{1}{2}(E_F - H) \rho \mid i \rangle$$

292

Step 2:

Through elementary manipulations, the operator ρ is found to obey the following partial differential equation:

$$\left(\frac{\partial \rho}{\partial \beta}\right)^2 + \frac{\partial^2 \rho}{\partial \beta^2} = (E_F - H)\frac{\partial \rho}{\partial \beta}$$

(2)

Step 3:

Assuming that ρ can be expanded in a Taylor Series in $\hbar^2$ such that

$$\rho = \rho_0 + \hbar^2 \rho_1 + O(\hbar^4)$$

(3)

Equation (2) can be rewritten as

$$\left[2\frac{\partial \rho_0}{\partial \beta} - E_F\right]\frac{\partial \rho}{\partial \beta} + \frac{\partial}{\partial \beta}(H\rho) + \frac{\partial^2 \rho}{\partial \beta^2} - \left(\frac{\partial \rho_0}{\partial \beta}\right)^2 = O(\hbar^4)$$

(4)

with

$$\rho_0 = \rho_0 = \log_e\left[1 + e^{\beta(E_F - V)}\right]$$

Step 4:

Let the Wigner equivalent [4] of the quantum operator ρ be written as $\rho_w(r,p)$. To retrieve the equation governing the electron distribution, we employ a property of Wigner equivalents, where

$$\int dp\,\rho_w(r, p) = \langle r \mid \rho \mid r \rangle = \text{diagonal matrix elements of } \rho$$

Furthermore, it can be shown, from discussions in Step 1, that for a 2DEG,

$$\int dp\,(H\rho)_w = \left(-\frac{\hbar^2}{12m}\nabla^2 + \frac{2}{3}V + \frac{1}{3}E_F\right)\rho + O(\hbar^4) \quad \text{for } V(r) \le E_F$$

(5a)

and

$$\int dp\,(H\rho)_w = \left(-\frac{\hbar^2}{8m}\nabla^2 + V\right)\rho + O(\hbar^4) \quad \text{for } V(r) > E_F$$

(5b)

As a result, the equation governing the electron distribution $\rho(r,\beta)$ is obtained:

$$\left[-\frac{\hbar^2}{12m}\nabla^2 + (V - E_F)\left\{\tanh\left[\frac{\beta}{2}(V - E_F)\right] - \frac{1}{3}\right\} + \frac{\partial}{\partial \beta}\right]\frac{\partial \rho}{\partial \beta} = \left[\frac{E_F - V}{1 + \exp\left[\beta(V - E_F)\right]}\right]^2 + O(\hbar^4)$$

(6a)

for $V(r) < E_F$; and

$$\left[-\frac{\hbar^2}{8m}\nabla^2 + (V - E_F)\tanh\left[\frac{\beta}{2}(V - E_F)\right] + \frac{\partial}{\partial \beta}\right]\frac{\partial \rho}{\partial \beta} = \left[\frac{E_F - V}{1 + \exp\left[\beta(V - E_F)\right]}\right]^2 + O(\hbar^4)$$

(6b)

for $V(r) \ge E_F$.

The initial conditions for these equations are:

$$\rho(r, \beta=0) = \log_e 2 \quad \text{and} \quad \left[\frac{\partial \rho}{\partial \beta}\right]_{\beta=0} = \frac{1}{2}(E_F - V)$$

(7)

If the $O(\hbar^4)$ terms can be considered negligible, the electron distribution $\rho(r,\beta)$ can be obtained by solving the above set of initial-value equations numerically. (Notice also that unlike the Schrœdinger Equation, equation (6) does not require boundary conditions.)

Step 5:

The Crank-Nicholson finite-difference scheme [5] is used to solve equation (6) numerically. At each β-step a matrix equation in the form $\mathbf{Ax} = \mathbf{b}$, where $\mathbf{A}$ is a sparse matrix, need to be solved. (In particular, for a 1D potential well, the matrix $\mathbf{A}$ is a tridiagonal matrix.) It is found that the total CPU time required for each solution is roughly linearly dependent on the number of grid-points used to discretize the solution space. It should also be mentioned that the discretization error for this finite-difference scheme is in the order of $O[(\Delta\beta)^2 + (\Delta r)^2]$. This numerical method has been found to be very efficient in solving these simplified equations.

III.   Conclusions

A simplified method for quantum size effect analysis including Fermi-Dirac statistics is obtained. The electron distribution, using this method, can be obtained without solving any eigenvalue problem. When equation (6) is solved self-consistently with the Poisson Equation, this means a tremendous CPU-time savings compared with the conventional Poisson/Schrœdinger Equation method. 2D and 3D device simulations including quantum size effect are thus feasible using this method. Initial numerical experiments indicate that this method is robust and efficient. Considering other uncertainties in material parameters, the method developed is also sufficiently accurate for most engineering level analyses.

Acknowledgements

The authors would like to thank Reinhard Erwe for helpful discussions, and Jim Prendergast and Peter Lloyd for support and encouragements.

References

1. Stern, F. and Sarma, S.D. , Physical Review B, 30, 840 (1984).
2. Laux, S. et al, IEDM Proceedings, 567 (1986).
3. Lui, W.W. and Frey, J., Journal of Applied Physics, 64, 6790 (1988).
4. Imre, K. et al, J. Math. Phys., 8, 1097 (1967), and references therein.
5. Carnahan et al, Applied Numerical Methods, Ch.7, John Wiley&Sons, New York (1969).

Title: Electron Distribution in a Narrow Potential Well
Legend: Solid - Results using the simplified method;
Legend: Dotted - Solutions from Schroedinger Equation;
Legend: Broken - Potential Function (right axis)
Other Data: E_F = 2V; Temperature = 77K; m^* = 0.19 m_e

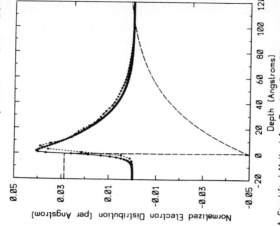

Title: Electron Distribution in a Narrow Potential Well
Legend: Solid - Results using the simplified method;
Legend: Dotted - Solutions from Schroedinger Equation;
Legend: Broken - Potential Function (right axis)
Other Data: E_F = 1V; Temperature = 77K; m^* = 0.19 m_e

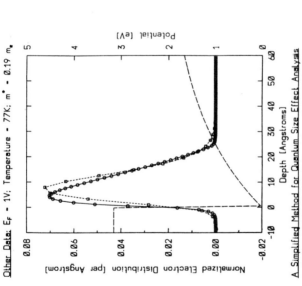

A Simplified Method for Quantum Size Effect Analysis in Submicron Devices Including Fermi-Dirac Statistics
by Wayne Lui (Bell Labs, AT&T) and Jeffrey Frey (University of Maryland)
International Symposium on Nanostructure Physics and Fabrication 1989, Texas

A MONTE CARLO STUDY OF THE INFLUENCE OF TRAPS ON HIGH FIELD ELECTRONIC TRANSPORT IN a-SiO$_2$ [1]

R. L. Kamocsai and W. Porod
Department of Electrical and Computer Engineering
University of Notre Dame
Notre Dame, IN 46556

We study the influence of trapping and detrapping processes on high field electronic transport in a-SiO$_2$. Detrapping provides an energy-loss mechanism for energetic conduction electrons and is modelled as a pair collision event. A dynamic equilibrium between these processes leads to stable hot electronic distributions.

I. INTRODUCTION

Silicon dioxide is of vital importance for silicon MOS technology because of its use as an insulator. As device dimensions shrink, gate insulators have to withstand huge electric fields which are, even under normal operating conditions, close to electrical breakdown. Despite its enormous technological importance, electronic transport in SiO$_2$ under such high field conditions is not well understood.

In recent years, the understanding of high field transport in the pre-breakdown regime has been advanced by experimental findings of stable electronic distributions with average energies of several electron volts [1]. These findings were puzzling, since until then it was assumed that the dominant mechanism for energy loss was due to the emission of LO phonons, based on the strong electron - phonon coupling in a polar material such as SiO$_2$ [2-4]. According to this picture, one would expect electronic distributions with average energies on the order of a polar - optical phonon energy, which is at variance with these experiments.

This behavior can be explained by introducing additional scattering mechanisms for high - energy electrons. One explanation [5] considered the inclusion of nonpolar Umklapp processes, whereas in an alternate explanation, the effects of additional higher conduction bands were taken into account [6].

Another puzzling observation in SiO$_2$ under high electric field conditions is the appearance of a positive charge [7-9]. This phenomenon reveals itself as a shift in the flat band voltage of MOS capacitors exposed to high field stress [10]. For

---

[1]This work was supported by the SDIO/IST and managed by the Office of Naval Research.

low fields, a negative shift is observed corresponding to the trapping of electrons. At high fields, positive charge appears which is frequently attributed to the high - field injection of holes [11].

In this paper, we concentrate on the role of trapping and detrapping processes for high - field electronic transport in SiO_2. In particular, we develop a microscopic model for pair collisions involving electrons in trap states. The influence of traps on electronic transport is twofold. (1) Electrons in trap levels can exchange energy with electrons in the conduction band. (2) The occupation of traps contributes to the space charge. We believe that the inclusion of these two effetcs will allow a unified understanding of the above mentioned puzzling features of high field transport in SiO_2.

This paper is organized as follows. In Section II, we discuss a formulation of pair collision rates which allows us to study band - band and trap - band events. The results of our simulation of electronic transport are presented in Section III. Finally, we conclude in Section IV.

II. PAIR COLLISION RATES

We use Kane's theory [12] for the calculation of pair collision rates. Consider the process in which an electron with momentum k_4 in band n_4 interacts with an energetic electron with momentum k_1 in band n_1. The momenta of the two particles residing in bands n_2 and n_3 after the collision are labeled by k_2 and k_3, respectively. The rate, $\omega(E)$, for this event, averaged over all initial states with the same energy, is given by:

$$\omega(E) = \frac{2\pi}{\hbar} \left(\frac{V}{(2\pi)^3} \right)^2 \sum_{n_1,n_2,n_3,n_4} \frac{\int dk_1 dk_2 dk_3 \delta(E_1 - E)}{\sum_{n_1} \int dk_1 \delta(E_1 - E)} |M|^2 \, \delta_E.$$

Here, δ_E represents energy conservation and the matrix-element, M, involves the wavefunction of the initial and final states, as well as the interaction potential, which is the Coulomb repulsion between the charge carriers.

The calculation of the pair collision rate in this form presents a formidable problem. The rate involves the summation over all the final states which leads to a nine - fold integral over a rapidly oscillating function. Furthermore, the detailed wavefunctions have to be known, which represents a problem, in principle, for a material like amorphous SiO_2.

Kane demonstrated that this calculation of the pair collision rates can be simplified significantly. If one relaxes the constraint of the conservation of momentum (random - k approximation), then the summation over the final states is straight forward. In that case each state contributes with a weight proportional to its density of states, denoted by ρ, to the ionization rate,

$$\omega(E) = \omega_0 \int \int \rho_2(E_2)\rho_3(E_3)\rho_4(E_2 + E_3 - E)dE_2 dE_3.$$

The matrix-element, assumed to be independent of energy, is included in the pre-factor ω_0. This random - k approximation [12] leads to surprisingly good results in silicon when compared to the full calculation, which includes the detailed wave-function dependence of the matrix-elements and the conservation of momentum.

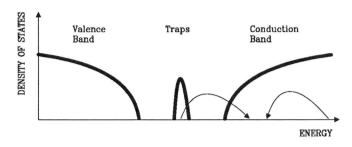

Figure 1: Schematic picture of a pair collision leading to a detrapping event.

III. RESULTS

The above form of the pair collision rate is of particular interest to us since it does not require knowledge of the wavefunctions, but only knowledge of the density of states. Although it was originally derived for the calculation of band - band impact ionization rates, the random - k approximation provides us with a method of including the influence of traps on pair collision events. Each trap is assumed to contribute via its contribution to the density of states. We may distinguish between two principal processes. (1) An electron transfers from the valence band into an empty trap state. The energy required for this process is obtained from an energetic electron in the conduction band, thereby representing an effective energy loss mechanism for conduction band electrons. This process thus helps in stabilizing the electronic distribution in the pre - breakdown regime. (2) An electron transfers from an occupied trap state into the conduction band. The energy required for this process is provided by an energetic electron in the conduction band losing some of its energy. This detrapping event, schematically shown in figure 1, creates new electrons in the conduction band and may thus initiate an electron avalanche in the breakdown regime.

We have developed a Monte Carlo computer simulation of electronic transport in SiO_2 which includes trapping and detrapping processes. Our simulation includes all the standard scattering events [5,6] known to be important for high - field transport in SiO_2. These include (i) acoustic deformation potential scattering with a value of 15 eV used for the deformation potential, (ii) optical phonon scattering with energies of 0.063 eV and 0.143 eV, and (iii) intervalley scattering, representing the density of states contributed by higher conduction bands. Also included in our simulation is a detrapping rate. This has been calculated using Kane's random - k approximation [12] and assumes a constant trap density of

$n_T = 10^{19} cm.^{-3}$ The trap density of states is centered at 4 eV below the bottom of the conduction band and is 1 eV wide. Electrons in the conduction band can also become trapped. We model this process with a constant scattering rate of 10^{12} Hz.

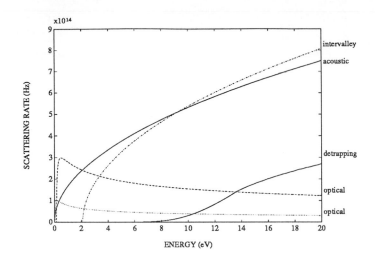

Figure 2: Scattering rates used in the Monte Carlo simulation.

As shown in figure 2, the detrapping rate is strongly dependent on the energy. The average detrapping rate therefore depends sensitively on the electronic energy distribution. Figure 3 shows the transients of this average detrapping rate averaged over an ensemble of 2000 electrons. The averaging is performed each femtosecond. Notice that the rates level off, indicating a stable electronic energy distribution. Since the threshold for the detrapping event is the energy necessary to promote an electron from the trap to the conduction band, this process is only important if the average electronic energy is sufficiently high. In this paper, this threshold energy (the distance between the top of the trap band and the bottom of the conduction band) is 3.5 eV.

Figure 3 shows that electric fields on the order of 4 MV/cm are required to reach this energy threshold. This figure also allows us to understand the shift in the flat band voltage of MOS capacitors as a function of the applied field stress. If the electric field leads to a detrapping rate which is lower than the trapping rate, negative charge is predicted. On the other hand, if the field is strong enough and leads to distributions with energies above the threshold, detrapping events will dominate. The remaining ionized trap states contribute to the space charge and are believed to account for the observed positive shift of the flat band voltage.

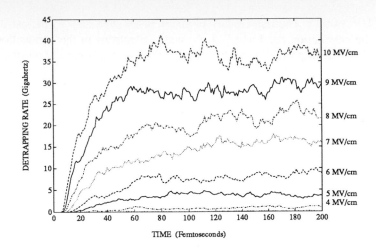

Figure 3: Transient behavior of the ensemble - averaged detrapping rates for a sequence of electric fields.

IV. CONCLUSION

We include the effects of trap states on high field electronic transport in SiO_2. Stable electronic distributions correspond to a dynamical equilibrium between trapping and detrapping events. When trapping events dominate, negative space charge results. For high electric fields, detrapping events become dominant due to the heating of the electronic distribution and positive space charge is generated.

REFERENCES

[1] D. J. DiMaria, T. N. Theis, J. R. Kirtley, F. L. Pesavento, D. W. Wong, and S. D. Broerson, *Journal of Applied Physics* **57**, 1214 (1985).
[2] H. Fröhlich, *Proceedings of the Royal Society* **A160**, 230 (1937).
[3] K. K. Thornber and R. P. Feynman, *Physical Review* **B1**, 4099 (1970).
[4] D. K. Ferry, *Applied Physics Letters* **27**, 689 (1975).
[5] M. V. Fischetti, *Physical Review Letters* **53** 1755 (1984)
[6] W. Porod and D. K. Ferry, *Physical Review Letters* **54**, 1189 (1985).
[7] E. Harari, *Journal of Applied Physics* **49**, 2478 (1978).
[8] Y. Nissan - Cohen, J. Shappir, and D. Frohman - Bentchkowsky, *Journal of Applied Physics* **60**, 2024 (1986).
[9] E. Avni and J. Shappir, *Journal of Applied Physics* **64**, 734 (1988).
[10] I. C. Chen, S. Holland, and C. Hu, *IEEE Elect. Dev. Letters* **7**, 164 (1986).
[11] M. V. Fischetti, *Physical Review* **B31**, 2099 (1985).
[12] E. O. Kane, *Physical Review* **159**, 624 (1967).

TUNNELING TIMES AND THE BÜTTIKER-LANDAUER MODEL[1]

J. A. Støvneng[2]
E. H. Hauge[2]

Department of Physics
The Ohio State University
Columbus, Ohio 43210

1. INTRODUCTION

For vertical transport in semiconductors, in which tunneling processes are crucial, an important question is how long time the tunneling process takes. Different candidates for general expressions for tunneling times have been suggested. In Sec. 2, we briefly discuss these proposals in view of an exact requirement involving the dwell time.

A related question is this: What is the characteristic frequency of the inelastic process when tunneling particles couple to, say, phonons, which typically represent the heatbath in this context? And what is the precise connection between the characteristic frequency and the duration of the tunneling process? Sec. 3 addresses these questions. We generalize Büttiker and Landauer's model of an oscillating, rectangular barrier [1] to barriers of arbitrary shape. Through examples we show that no direct, general relation exists between the characteristic frequency of an oscillating barrier and the duration of the tunneling process.

Our conclusions are collected in Sec. 4.

[1] Supported in part by the Office of Naval Research.
[2] Permanent address: Institutt for fysikk, Universitetet i Trondheim, NTH, N-7034 Trondheim-NTH, Norway.

2. DWELL TIME vs TRANSMISSION AND REFLECTION TIMES.

Assume that the stationary, one-dimensional scattering problem of Fig. 1 has been solved exactly at any energy, $E = \hbar^2 k^2 / 2m$, with a wave function of the form

$$\psi(x;k) = \begin{cases} e^{ikx} + \sqrt{R}e^{i\beta}e^{-ikx} ; & x < b \\ \chi(x;k) ; & b < x < a \\ \sqrt{T}e^{i\alpha}e^{ikx} ; & a < x \end{cases} \qquad (2.1)$$

$R(k)$ and $T(k) = 1 - R(k)$ are the reflection and transmission probabilities, respectively, and $\beta(k)$ and $\alpha(k)$ are the corresponding phase shifts. Assume, for convenience, that $b < 0$ and $a > 0$.

The *dwell time*, τ_D, is a measure of the time, averaged over all scattering channels, spent by a particle in a region of space [2]. In our one-dimensional case, the dwell time in the barrier region is defined as [3]

$$\tau_D(b, a; k) = \frac{1}{v(k)} \int_b^a dx |\psi(x;k)|^2. \qquad (2.2)$$

The incoming particle flux is $v(k) = \hbar k / m$. If it is meaningful to speak of separate tunneling times for particles ending up in different scattering channels, the dwell time must be a weighted sum of these times:

$$\tau_D(b, a; k) = R(k)\tau_R(b, a; k) + T(k)\tau_T(b, a; k), \qquad (2.3)$$

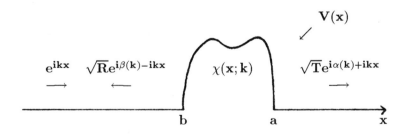

Fig. 1. *The one-dimensional tunneling problem. Indicated on the figure are the incoming, transmitted and reflected waves, and the wavefunction under the barrier.*

304

where τ_R and τ_T are the reflection and transmission times, respectively. The two outcomes, transmission and reflection, are mutually exclusive events. Which of them is realized, one can check from the final state, i.e., without disturbing the tunneling process [4].

The *phase times* are derived by letting a wave packet impinge on the barrier [5-10]. A simple-minded derivation based on a stationary phase approximation gives $\Delta\tau_T^\phi(b,a;k) = [a - b + d\alpha/dk]/v(k)$ and $\Delta\tau_R^\phi(b,a;k) = [-2b + d\beta/dk]/v(k)$. For reasons that will become clear, we shall call these the *extrapolated* phase times. They cannot [11] be identified with the duration of the tunneling process since they obey the identity [8]

$$\tau_D(b,a;k) = R(k)\tau_R^\phi(b,a;k) + T(k)\tau_T^\phi(b,a;k) + \frac{\sqrt{R}}{kv}\sin(\beta - 2kb) \quad (2.4)$$

which contradicts (2.3). However, more careful derivations [7-10] of the phase times reveal that they are *asymptotic* characteristics of *completed* scattering events, involving wave packets narrow in k-space (width σ), hence necessarily wide in x-space. Replacing the interval (b,a) with (x_1, x_2), such that $|x_1|, x_2 \gg \sigma^{-1}$, and averaging (2.4) over the initial packet, leaves the interference term negligible. Thus

$$\overline{\tau}_D(x_1, x_2; k) = \overline{R}(k)\overline{\tau}_R^\phi(x_1, x_2; k) + \overline{T}(k)\overline{\tau}_T^\phi(x_1, x_2; k) + O(\sigma). \quad (2.5)$$

The discrepancy between (2.4) and (2.3) results from a formal extrapolation of the asymptotic phase times back to the barrier region, assuming free particle motion everywhere outside the barrier [8]. As pointed out by Leavens and Aers [11], this is not correct in front of the barrier, where the incoming wave interferes with the reflected part and gives rise to an extra time delay. This self-interference delay can be shown to be represented by the last term in (2.4).

A different approach was suggested by Baz' [12]: Cover the region of interest with an infinitesimal, homogeneous magnetic field, $\mathbf{B} = B_0\hat{z}$, and take advantage of the constant Larmor precession of the spin of the particles in the field. With the incident beam polarized in the x-direction, the *local Larmor times* become

$$\tau_{yT/R}^L(b,a;k) = \lim_{\omega_L \to 0} \langle s_y\rangle_{T/R}/(-\tfrac{1}{2}\hbar\omega_L). \quad (2.6)$$

Here, ω_L is the Larmor frequency, and $\langle s_y\rangle_{T/R}$ is the mean value of the y-component of the spin of transmitted/reflected particles. The local Larmor times obey Eq. (2.3), but one simple example shows that they are not always reliable: Let the field region, (x_1, x_2), be entirely to the right of the barrier. Then $\tau_{yT}^L(x_1, x_2; k) = (x_2 - x_1)/v + \sqrt{R}[\sin(2\alpha - \beta + 2kx_1) - \sin(2\alpha - \beta + 2kx_2)]/2kv$, whereas the dwell time gives $\tau_D(x_1, x_2; k) = T(x_2 - x_1)/v$.

Clearly, if reliable, τ_{yT}^L should coincide with the dwell time, $\tau_D^c = \tau_D/T = (x_2 - x_1)/v$, *conditional* on the particles having tunneled through.

However, the Larmor clock shows the phase times when used in an asymptotic sense, i.e., with wavepackets narrow in k-space, and the magnetic field covering a region large compared to the size of the wavepackets [9].

Effectively, the magnetic field represents a decrease or increase of the barrier height, depending on whether the incident particle is spin-up or spin-down with respect to the field [3]. As a result, the spin of the transmitted and reflected particles acquire nonzero z-components, $\langle s_z \rangle_T$ and $\langle s_z \rangle_R$. This lead Büttiker to define Larmor "times", τ_{zT}^L and τ_{zR}^L, analogous to (2.6). However, the basis for an interpretation of these objects as times intrinsically characteristic of the tunneling process, is not clear. Moreover, they do not obey (2.3), but rather conservation of angular momentum [3], $R\tau_{zR}^L + T\tau_{zT}^L = 0$.

A third set of Larmor "times", the so-called *Büttiker-Landauer times*, are defined as [3]

$$(\tau_{T/R}^{BL})^2 = (\tau_{yT/R}^L)^2 + (\tau_{zT/R}^L)^2. \tag{2.7}$$

Clearly, τ^{BL} does not meet the requirement (2.3). In our opinion, it can, at best, be viewed as a hybrid quantity reflecting a combination of time and energy aspects of the tunneling process.

There is a close connection between the local Larmor times and a *complex time*, τ^Ω, derived by Sokolovski and Baskin [13] from a generalization of the classical time concept into the quantum domain:

$$\mathrm{Re}\tau_{T/R}^\Omega = \tau_{yT/R}^L \quad ; \quad \mathrm{Im}\tau_{T/R}^\Omega = \tau_{zT/R}^L. \tag{2.8}$$

Sokolovski and Baskin proved that τ^Ω satisfies (2.3). However, we take the common sense attitude that the duration of a process should be a *real* time. In general, τ^Ω is complex. In our view, τ^Ω can only be given physical meaning in connection with clearly defined model situations. Eq. (2.8) represents one such example, and we shall give another below.

3. THE BÜTTIKER-LANDAUER MODEL. CHARACTERISTIC FREQUENCY vs TUNNELING TIME.

Büttiker and Landauer [1] investigated the inelastic process of tunneling through an oscillating barrier, $V = V_0 + V_1 \cos \omega t$. Particles interacting with the barrier can emit or absorb modulation quanta, $\hbar \omega$, leading to sidebands in the transmitted and reflected beam. We have studied a generalized version of this model, when the rectangular barrier is replaced by one of arbitrary shape, $V(x) = V_0(x) + V_1(x)\cos\omega t$. The time dependent Schrödinger equation for this problem must have the general form

$$\Psi(x,t) = \sum_n \phi_n(x)e^{-\frac{i}{\hbar}(E+n\hbar\omega)t}. \tag{3.1}$$

A formal solution for ϕ_n, can be obtained in the low-frequency limit, $\omega \to 0$. For the transmitted sideband amplitudes one finds

$$A_{\pm n}^{(0)} = \int_{-\frac{d}{2}}^{\frac{d}{2}} dx_1 \cdots \int_{-\frac{d}{2}}^{\frac{d}{2}} dx_n \frac{V_1(x_1)\cdots V_1(x_n)}{2^n n!} \frac{\delta^n A_0[V_0(x)]}{\delta V(x_1)\cdots \delta V(x_n)} \tag{3.2}$$
$$+ O(V_1(x)^{n+2}),$$

where A_0 of the unperturbed problem is considered as a functional of $V_0(x)$, and $\delta/\delta V(x)$ denotes a functional derivative. The result for reflected sidebands is analogous, with B replacing A. For results beyond the $\omega \to 0$ limit one must rely on a perturbation expansion in $V_1(x)$, which was already introduced in Ref. 1 for rectangular barriers.

For $V_1(x) = V_1 = const.$, but for general $V_0(x)$, one can show that

$$\frac{A_{\pm 1}^{(0)}}{A_0} = -i\frac{V_1}{2\hbar}\tau_T^\Omega \quad ; \quad \frac{B_{\pm 1}^{(0)}}{B_0} = -i\frac{V_1}{2\hbar}\tau_R^\Omega. \tag{3.3}$$

Thus, the complex "times", τ_T^Ω and τ_R^Ω, have a direct physical interpretation as relative sideband amplitudes for the oscillating barrier, in the $\omega \to 0$ limit.

Now we turn to the question of whether there is a connection between the characteristic frequency of oscillating barriers and the duration of the tunneling process. For the important case of opaque, rectangular barriers, $\exp(-2\kappa d) \ll 1$ with $\kappa^2 = [2m(V_0 - E)]/\hbar^2$, the relative intensities of the transmitted sidebands were found in Ref. 1 to be

$$I_{\pm 1}^T(\omega) = \left|\frac{A_{\pm 1}}{A_0}\right|^2 = \left(\frac{V_1}{2\hbar\omega}\right)^2 \left(e^{\pm\omega\tau_T^{BL}} - 1\right)^2, \tag{3.4}$$

where $\tau_T^{BL} = md/\hbar\kappa$ (and $\hbar\omega$ has been assumed to be small compared to E and $V_0 - E$). Clearly, the characteristic frequency is $\omega_c = 1/\tau_T^{BL}$ in this case. It determines the low frequency limit through

$$I_{\pm 1}^T(0) = \left(\frac{V_1\tau_T^{BL}}{2\hbar}\right)^2 \tag{3.5}$$

and also the sideband asymmetry $(I_{+1}^T - I_{-1}^T)/(I_{+1}^T + I_{-1}^T) = \tanh\omega\tau_T^{BL}$. For general barriers, Büttiker and Landauer take (3.5) as the *definition* of the traversal time, and it is easy to show that τ_T^{BL} as defined by (2.7) and (3.5) is one and the same quantity.

For opaque rectangular barriers, one has for the lowest *reflected* sidebands:

$$I_{\pm 1}^R(\omega) = \left(\frac{V_1 \tau_R^{BL}}{2\hbar}\right)^2 \left(1 \pm \frac{1}{2}\omega\tau_\kappa + O(\omega^2)\right), \tag{3.6}$$

where $\tau_R^{BL} = (2mk)/(\hbar\kappa(\kappa^2+k^2))$ and $\tau_\kappa = (2m)/(\hbar\kappa^2)$. Although (3.6) does not provide a characteristic frequency (this requires the full ω-dependence of $I_{\pm 1}^R$), nothing suggests that $\omega_c = 1/\tau_R^{BL}$ in this case. The corresponding calculation for transparent barriers, $\kappa d \ll 1$, yields

$$I_{\pm 1}^T(\omega) = \left(\frac{V_1}{2\hbar}\frac{md}{\hbar k}\right)^2 \frac{1 + \frac{2}{3}\eta^2 + \frac{1}{36}\eta^4}{1 + \frac{1}{4}\eta^2}(1 \mp \omega\tau_k\theta(\eta) + O(\omega^2)),$$

$$\theta(\eta) = \frac{1 + \frac{2}{3}\eta^2 + \frac{1}{90}\eta^4 - \frac{13}{360}\eta^6 - \frac{7}{4320}\eta^8}{1 + \frac{11}{12}\eta^2 + \frac{7}{36}\eta^4 + \frac{1}{144}\eta^6}, \tag{3.7}$$

with a similarly complicated expression for $I_{\pm 1}^R(\omega)$. Here $\eta = kd$ and $\tau_k = \hbar/E$. Again, there are no simple relations between τ_T^{BL} (or τ_R^{BL}), as found from the $\omega \to 0$ limit, and the characteristic times associated with the sideband asymmetry. Note, furthermore, that τ_T^{BL} as given by (3.7) tends to zero with d as $d \to 0$, whereas τ_R^{BL} tends to $\hbar/V_0$ in this limit. Clearly, the latter result is in conflict with an interpretation of τ_R^{BL} as the duration of a tunneling process.

We have also investigated the oscillating δ-function barrier, for which the entire frequency dependence of the sideband intensities can be calculated explicitly. Starting from the oscillating rectangular barrier, we let $d \to 0$, $V_0 \to \infty$ and $V_1 \to \infty$ so that $V_0 d \equiv \hbar c_0$ and $V_1 d \equiv \hbar c_1$ are finite quantities, measuring the strength of the static and the oscillating part of the δ-barrier, respectively. To leading order in c_1, we find

$$I_{\pm 1}^T(\omega) = \left|\frac{A_{\pm 1}}{A_0}\right|^2 = \frac{c_1^2}{4(v^2 + c_0^2)}\left(1 \pm \frac{\hbar\omega}{E + \frac{1}{2}mc_0^2}\right)^{-1}, \tag{3.8}$$

$$I_{\pm 1}^R(\omega) = \left|\frac{B_{\pm 1}}{B_0}\right|^2 = \left(\frac{v}{c_0}\right)^2 I_{\pm 1}^T(\omega). \tag{3.9}$$

The time, τ_T^{BL}, defined from $I_{\pm 1}^T(0)$, vanishes with d, as one would expect for the duration of a tunneling process. However, τ_R^{BL}, defined from $I_{\pm 1}^R(0)$, remains nonzero. The transmission and reflection sideband asymmetry follows from (3.8) and (3.9) as $(I_{+1} - I_{-1})/(I_{+1} + I_{-1}) = -(\hbar\omega)/(E + \frac{1}{2}mc_0^2)$. Clearly, the characteristic frequency is finite. Thus, the example of the oscillating δ-function barrier shows unambiguously that *no direct, general relation exists between the characteristic frequency and the duration of tunneling*. In general, one must expect the characteristic frequency to represent a combination of the energy sensitivity of the tunneling process, the energy dependence of the density of states, *and* the duration of the tunneling process. A simple connection between sideband asymmetry and tunneling times cannot be made.

4. CONCLUSIONS

We have sketched a picture which relates various proposals for the tunneling time. Furthermore, we have discussed the connection between tunneling times and the characteristic frequency of a modulated system. More detailed accounts of our work will be published elsewhere. Our main points are the following:

1. All asymptotic treatments of tunneling, in which completed scattering events are considered, give results that converge on the classic phase times as the width, σ, in k-space of the wave packets tends to zero. But, since phase times include self-interference effects, the extrapolated phase times can in general not be identified with the duration of tunneling through the barrier.

2. The dwell time is complementary to the phase times in that it offers an exact local statement on tunneling times. However, it can not distinguish between different scattering channels.

3. The claim of the local Larmor times, τ_y^L, is that they correctly distribute the dwell time over scattering channels. We have shown, by a counterexample, that this claim can, at best, have restricted validity.

4. Since τ_T^{BL} and τ_R^{BL} do not satisfy (2.3), they cannot be accepted as generally valid expressions for tunneling times.

5. The example of the oscillating δ-function barrier shows that there exists no direct, general relation between the characteristic frequency of a modulated process and the intrinsic tunneling time.

6. None of the proposed expressions for the duration of a tunneling process (which distinguish between transmitted and reflected particles) satisfy all necessary conditions on such expressions. They can therefore not be generally valid. Whether generally valid expressions exist for the duration of tunneling processes for particles with given energy is not clear.

7. This does not mean that all proposed expressions are uninteresting or irrelevant. As an example, the characteristic frequency in Büttiker and Landauer's oscillating barrier clearly sets the important time scale for the

coupling between the tunneling process and external degrees of freedom. That this time scale has not been convincingly shown to be identical with the duration of the tunneling process itself, does not detract from its significance.

ACKNOWLEDGEMENTS

The authors are grateful to John Wilkins and his colleagues at the Ohio State University for their hospitality during the academic year 1988-89.

REFERENCES

1. M. Büttiker and R. Landauer, *Phys. Rev. Lett.* **49**, 1739 (1982); *Phys. Scripta* **32**, 429 (1985); *IBM J. Res. Dev.* **30**, 451 (1986).
2. F. T. Smith, *Phys. Rev.* **118**, 349 (1960).
3. M. Büttiker, *Phys. Rev.* **27**, 6178 (1983).
4. R. P. Feynman and A. R. Hibbs, *Quantum Mechanics and Path Integrals* (McGraw-Hill, 1965).
5. E. P. Wigner, *Phys. Rev.* **98**, 145 (1955).
6. T. E. Hartmann, *J. Appl. Phys.* **33**, 3427 (1962).
7. J. R. Barker in *The Physics and Fabrication of Microstructures and Microdevices*, edited by M. J. Kelly and C. Weisbuch (Springer, New York, 1986).
8. E. H. Hauge, J. P. Falck and T. A. Fjeldly, *Phys. Rev.* B **36**, 4203 (1987).
9. J. P. Falck and E. H. Hauge, *Phys. Rev.* B **38**, 3287 (1988).
10. W. Jaworski and D. M. Wardlaw, *Phys. Rev.* A **37**, 2843 (1988); **38**, 5404 (1988).
11. C. R. Leavens and G. C. Aers, *Phys. Rev.* B **39**, 1202 (1989).
12. A. J. Baz', *Yad. Fiz.* **4**, 252 (1966); **5**, 229 (1966) [*Sov. J. Nucl. Phys.* **4**, 182 (1967); **5**, 161 (1967)].
13. D. Sokolovski and L. M. Baskin, *Phys. Rev.* A **36**, 4604 (1987).

VARIANCE FLUCTUATIONS AND SAMPLE STATISTICS FOR FLUCTUATING VARIABLE-RANGE HOPPING CONDUCTION[1]

R.A. Serota
J. Yu

Department of Physics
University of Cincinnati
Cincinnati, OH 45221

The discovery of the reproducible structure of resistance fluctuations in narrow-channel Si-MOSFETs has triggered the emergence of the field of mesoscopic physics[1]. It was quickly realized that for small gate voltages the fluctuations are intrinsic to variable-range Mott hopping conduction[2]. The approach developed in Ref. 2 was subsequently put on firm analytical footing[3] allowing for a clear explanation of the size of the fluctuations. Another consequence of the analytical treatment of Ref. 3 was the prediction of a very slow decay of the size of the fluctuations with the length of the wire. The approach of Refs. 2 and 3 was based on the percolation approximation for a Miller-Abrahams resistor network[4]; since in a short wire only a few hops are needed to traverse the sample, the largest resistor of the percolating path will control its resistance for a range of values of the chemical potential (regulated by the gate voltage), until a switch to another percolating path occurs. As it happens, the largest resistor in the new path will in turn control the resistance of that path for a range of values of the chemical potential. Since each hop is exponentially activated, such a switch may lead to orders of magnitude fluctuations of the resistance[2].

In Ref. 3 it was shown that one should distinguish between two length-scales in a percolating path: the "long hop", controlling the average resistance, and the "average hop", controlling the size of the resistance fluctuations. According to Mott[5], each hop should be viewed as a compromise between the activation energy and the overlap factor which limits search for nearby energies at faraway sites. In higher dimensions, it never occurs that a site nearby in energy in a percolating path is too far away; an electron can always "jump around" such an obstacle. In one dimension[3] or in

---

[1]Supported by the Army Research Grant Number DAAL03-88-K-0139.

a very narrow[6] two-dimensional wire, on the other hand, an electron cannot hop around such an obstacle because of topological restrictions. The existence of two length and resistance scales in the problem is thus due to this "bottle-neck effect" with the largest resistor in the path defining the average resistance and the average resistor defining the fluctuation in the value of the largest resistor when the path is switched. Put in analytical form, the above argument yields for the dimensionless average logresistance and its root-mean-square fluctuation[3]:

$$\langle \ln \rho \rangle \approx (T_0/T)^{1/2} [\ln(2\alpha L)]^{1/2};$$
(1)

$$\Delta \equiv \left(\left\langle (\ln \rho - \langle \ln \rho \rangle)^2 \right\rangle \right)^{1/2} \approx (T_0/T)^{1/2} [\ln(2\alpha L)]^{-1/2}.$$
(2)

Here L is the length of the wire, $\ell = \alpha^{-1}$ is the localization length, T_0 is defined as $k_B T_0 = \alpha /N(0)$, and $N(0)$ is the density of states given by $N(0) = (a w)^{-1}$, where a is the distance between nearest impurity sites and w is the impurity band width. Two remarks should be made regarding these results. First, we notice an extremely slow decay of the resistance fluctuations with the length of the wire. Second, when the fluctuations do finally disappear, a crossover to the activated regime with $\langle \ln \rho \rangle \approx (T_0/T)$ takes place in agreement with the predictions of Ref. 7.

In Ref. 3, the predictions of Eqs. (1) and (2) have been checked via a numerical simulation along the lines of Ref. 2. An unexpected result of our simulation was the large fluctuation of the root-mean-square fluctuation Δ of the logresistance from wire to wire, that is from one ensemble of impurities to the other. In fact, it is the median of Δ over the large number of ensembles that satisfies the Eq. (2). Here, we explain this fact in terms of sampling statistics with the understanding that the well-established facts about the latter apply only to samples of *randomly* chosen data points. Along the way, we determine the form of the distribution function of the logresistance and show that it is slightly skewed with respect to the normal distribution.

Our numerical solution of the resistor network is identical to those of Refs. 2 and 3. A hundred chains of 1000 uniformly spaced localized sites with the localization length $\ell = \alpha^{-1} = 50$ were considered, with the energies of the sites chosen at random from the band of width $\pm w/2 = \pm 0.5$. For $k_B = 1$, we find $T_0 = 0.02$. We have set $T = 0.001$. For a given chemical potential, we considered hopping between two reservoirs located at the left and the right ends of the chain. Details of our procedure are given in Ref. 3. For each ensemble, we swept the values of the chemical potential from -0.4 to 0.4 with a step of 0.02, 400 values overall. In Fig. 1, we show a typical one ensemble structure of resistance fluctuations. In Fig. 2 we plot variations of the mean and the variance of the logresistance from ensemble to ensemble. Note the large fluctuations of the variance[3].

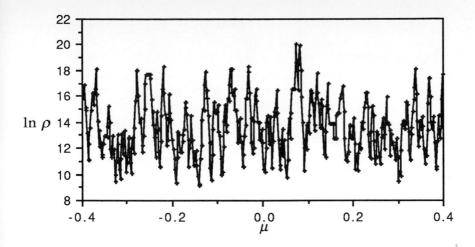

Fig. 1. Logresistance as a function of the chemical potential for one ensemble.

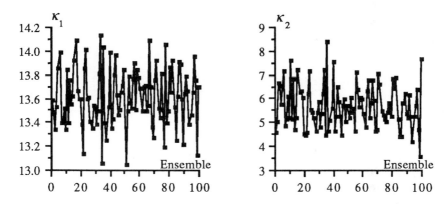

Fig. 2. Variations of the mean and variance from ensemble to ensemble.

Next, we investigate the convergence towards a limiting resistance distribution function by gradually accumulating resistances from all ensembles. After absorbing the resistances of a new ensemble, we compute the cumulants of the resistance distribution function. In Fig. 3, we depict the results of such an accumulation procedure. It is clear that the mean and variance saturate faster than the higher-order cumulants. We use their saturation values as fitting parameters for the resistance distribution function which is a subject of determination. In Fig. 4, we plot the limiting resistance distribution function

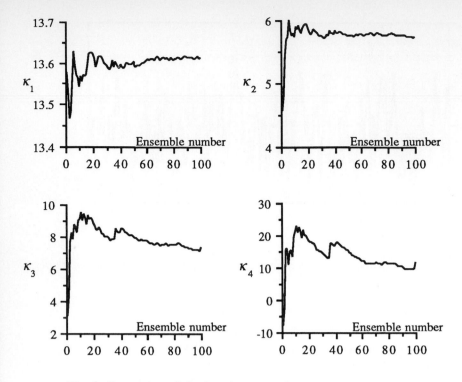

Fig. 3. *Saturation of the first four cumulants.*

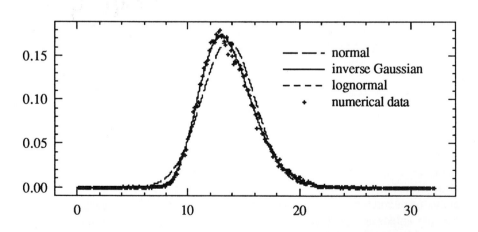

Fig. 4. *Distribution function of logresistance.*

against the normal, lognormal, and inverse Gaussian distributions with the same mean and variance. Apparently, the latter two provide a better fit than the normal distribution and are hardly distinguishable here. Indeed, when the mean and variances satisfy the condition $\overline{\ln \rho} \gg \overline{\Delta}$, which is the case here, one finds for both distributions[8]

$$f(\ln \rho) \approx \left(\frac{1}{2\pi \overline{\Delta}^2 \left(1 + \dfrac{\ln \rho - \overline{\ln \rho}}{\overline{\ln \rho}} \right)^{\alpha}} \right)^{1/2} \exp\left(- \frac{(\ln \rho)^2}{2\overline{\Delta}^2 \left(1 + \dfrac{\ln \rho - \overline{\ln \rho}}{\overline{\ln \rho}} \right)} \right), \qquad (3)$$

where α is of order unity. The fact that $\ln \rho$ is a positively defined quantity[3] leads to a slight skewness of the distribution towards larger resistances.

To explain large fluctuations of the variance Δ^2 from ensemble to ensemble, we note that for a random sample the variance of the mean and variance are given by[8]

$$\text{var} m_1 = \frac{1}{n} \mu_2, \qquad\qquad \text{var} m_2 = \frac{1}{n}\left[\mu_4 - \mu_2^2 \right], \qquad (4)$$

where n is the sampling size and μ_4 and μ_2 are the fourth and the second moments of the distribution. Both above equations imply that $n \approx 93$ rather than $n = 400$ for the number of the values of the chemical potential for each ensemble. Obviously, not all 400 resistances are independent; a glance at Fig. 1, confirmed by numerical analysis, proves that there are on the average two resistance points on each line leading to a peak or valley. It means that these points correspond to the same largest resistor in the same path. Moreover, even when a switch occurs, it is not always to an entirely different pair of sites; one of the new pair of sites often belongs to the pair of sites of the previous largest resistor. It points to the correlations between the resistances of each ensemble obtained in sweeping the chemical potential.

REFERENCES

1. S. Washburn and R.A. Webb, (1986). Adv. Phys. 35, 412; M.A. Kastner, R.F. Kwasnick, J.C. Licini, and D.J. Bishop, (1987). Phys. Rev. B36, 8015.
2. P.A. Lee, (1984). Phys. Rev. Lett. 53, 2042.
3. R.A. Serota, R.K. Kalia, and P.A. Lee, (1986). Phys. Rev. B33, 8441.
4. A. Miller and E. Abrahams, (1960). Phys. Rev. 120, 745; V. Ambegaokar, B.I. Halperin, and J.S. Langer, (1971). Phys. Rev. B4, 2612.
5. N.F. Mott and E.A. Davis, (1971). In "Electronic Processes in Non-Crystalline Materials", Clarendon Press, Oxford.
6. R.A. Serota, (1988). Solid State Commun. 67, 1031.
7. J. Kurkijarvi, (1973). Phys. Rev. B8, 922.
8. M.G. Kendall and A. Stuart, (1958). In "The Advanced Theory of Statistics," Hafner, New York.

Chapter 6

Quantum Wires and Ballistic Point Contacts

WHEN IS THE HALL RESISTANCE QUANTIZED?

M. Büttiker

IBM Research Division,
IBM Thomas J. Watson Research Center
P. O. Box 218, Yorktown Heights, New York 10598

I. INTRODUCTION

In this paper, we apply to the quantum Hall effect (1) some elements of a theory which expresses electric conduction in terms of global transport coefficients (2). This approach relates transmission probabilities directly to global longitudinal resistances and global Hall resistances. It is an extension and modification of earlier work (3) with the key distinction that current and voltage probes are treated on the same physical principles. As a consequence this approach (2) explicitly reflects the fundamental symmetries of electrical conduction (4-6). This approach (2,6) has recently been used to address a wide range of problems: voltage fluctuations in small metallic conductors (7), transport in ballistic conductors (8), low field anomalies of the Hall effect (9-14), the effect of phase randomizing events on the series addition of resistances including resonant tunneling (15), and electron focusing (5). The application to the quantum Hall effect of this approach (2,6) was noted independently by Beenakker and van Houten (10), Peeters (11) and the author (16). Remarkably the generalized resistance expressions found in (2,6) are valid over the entire range of fields. Furthermore, in contrast to other recent work (17,18, 19) on the quantum Hall effect, the discussion based on (2) does not invoke any a priori assumptions about what is measured at a voltage contact. This approach permits the study of the Hall effect in highly non-uniform samples and highlights the role of current and voltage contacts. Under special circumstances the theory (16) predicts simultaneously *quantized* Hall resistances and *quantized* longitudinal resistances at values which are not given by the number of bulk Landau levels below the Fermi energy. Such anomalous plateaus have been observed in experiments by Washburn et al. (20) and Haug et al. (21). The approach predicts that contacts are important, whenever there exists a non-equilibrium population of current carrying states (16). A striking demonstration of this effect has been given by van Wees et al. (22). Experiments by Komiyama et al. (23) also require a proper treatment of the contacts and point to extremely long equilibration lenghts. Van Wees et al. (24) have demonstrated the suppression of the Shubnikov-de Haas oscillations if current is injected or detected with a point contact. Very small samples, as observed by Chang et al. (25), exhibit fluctuations both in the longitudinal resistances and the Hall resistances (26). The longitudinal resistance can even be negative (25,27). The approach advanced here invokes transport along edge states (28,29). A direct test of transport along edge states is provided by the analysis of the Aharonov-Bohm effect (30). Motion along edge states leads to the prediction (16,18) that the Aharonov-Bohm effect is *suppressed* in the quantum Hall regime for ring structures which are wide compared to a cyclotron radius, and this has indeed been observed by Ford et al., Timp et al., van Loodsrecht et al., and van Wees et al. (31). Aharononv-Bohm like oscillations require backscattering either in the bulk (18) or at the contacts (19). The suppression of the Aharonov-Bohm effect is important, since some of the pioneering papers do invoke the sensitivity of a two-dimensional electron gas to an Aharonov-Bohm flux (28,32) to explain the quantum Hall effect. Below, we address the question posed in the title of this paper by discussing a few representative simple examples. We are concerned with situations where electron motion can be completely described by edge states with localized interactions between them. More complex situations at arbitrary fields can be analyzed computationally as is nicely demonstrated by the work of Ravenhall et al. (12) and Kirczenow(13). Finally, we address some older experiments, which by measuring voltages at interior contacts, have led to the conclusion that current transport cannot be along edge states. We point to a differing interpretation of these experiments.

II. GLOBAL RESISTANCES

Fig. 1a shows an electric conductor consisting only of resistive elements and connected to a number of terminals at chemical potentials μ_i, i = 1,2,3,4... The terminals are sources and sinks of carriers and energy and are assumed to have a density of states so large that the current density can always be assumed to be zero. The terminals are at equilibrium. The currents incident from the terminals on the conductor are related to the voltages $V_i = \mu_i/e$ by the conductances (33),

$$I_i = \sum_{j \neq i} G_{ij} V_j \tag{1}$$

The G_{ij} in Eq. (1) are global conductances; they describe conduction from one terminal to the other. A quantum transport theory is obtained by spacing the terminals so closely that electron motion from one terminal to the other is phase-coherent. The conductor scatters carriers only elastically. Inelastic, phase-randomizing and energy dissipating events occur only in the reservoirs. We can then follow (3) and view the conductor as a target which permits transmission and reflection of carriers. Ref. 2 finds for the currents I_i incident on the conductor,

$$I_i = \frac{e}{h} [(M_i - R_{ii}) \mu_i - \sum_{j \neq i} T_{ij} \mu_j], \tag{2}$$

where M_i are the number of channels in reservoir i and R_{ii} and T_{ij} are the total probabilities for reflection at terminal i and the total probability for transmission from probe j to probe i. Comparison of Eq. (2) and Eq. (1) taking into account that each row (and colum) of the matrix defined by Eq. (2) adds to zero, yields $G_{ij} = (e^2/h) T_{ij}$. In a configuration where reservoirs m and n are used as a source and sink and contacts k and l are voltage probes, the resistance is $\mathcal{R}_{mn,kl} = (\mu_k - \mu_l)/eI$. Here $I = I_m = -I_n$ is the current impressed on the sample. At the voltage contacts, there is zero net current flow, $I_k = I_l = 0$. These conditions on the currents determine the resistance (2,6)

$$\mathcal{R}_{mn,kl} = (h/e^2)(T_{km} T_{ln} - T_{kn} T_{lm})/D. \tag{3}$$

D is a subdeterminant of rank three of the matrix formed by the coefficients in Eq. (2), which multiply the chemical potentials. All subdetrminants of rank three of this matrix are equal and independent of the indices m,n,k, and l. Microreversibility, implies $T_{ij}(B) = T_{ji}(-B)$, $R_{ii}(B) = R_{ii}(-B)$. Using this in Eq. (3) gives rise to the reciprocity of four-terminal resistances, $\mathcal{R}_{kl,mn}(B) = \mathcal{R}_{mn,kl}(-B)$. We cite here only two recent experimental demonstrations of this symmetry (4,5), which is a manifestation of the Onsager-Casimir symmetry of the transport coefficients of Eqs. (1) and (2), and refer the reader to (6) for additional references. Ref. 6 discusses conductors with more than four terminals and shows that a mapping exists such that a generalization of Eq. (3) applies. If the terminals cannot be assumed to be closely spaced, fictitious contacts can be used to bring incoherence into the conduction process (14,15) or alternatively inelastic scattering has to be incorporated in the calculation of the global conductances from the outset (7). A derivation of Eqs. (1-3) using linear response formalism has been given by Baranger and Stone, and Stone and Szafer (34).

III. THE QUANTUM HALL EFFECT

Let us next consider how the quantum Hall effect is established. For the scope of this paper, it will be sufficient to assume that electron motion occurs in a potential eU(x,y) which varies slowly compared to the cyclotron radius. The states at the Fermi energy E_F are then determined by the solution of the equation

$$E_F = \hbar\omega_c(n + 1/2) + eU(x,y) \qquad (4)$$

where n is a positive integer and x,y is a path in two-dimensional space. We emphasize, that it is the equilibrium potential eU which matters, i.e. the potential in the absence of a net current flow ($\mu_1 = \mu_2 = \mu_3 = \mu_4$). Eq. (4) admits two types of solutions: There are open paths which necessarily originate and terminate at a contact. In addition, there are closed paths. We refer to the open paths as "edge" states since typically the open states form close to the boundary of the sample. In Fig. 1 the open paths are indicated by faint solid lines. At zero temperature and in the absence of tunneling between open states and the closed paths it is only the open paths (edge states) which contribute to electric conduction. It is the connection of the contacts via edge states (open paths) which determines the measured resistance. Each edge state provides a path along which carriers can traverse the conductor without backscattering (16). If

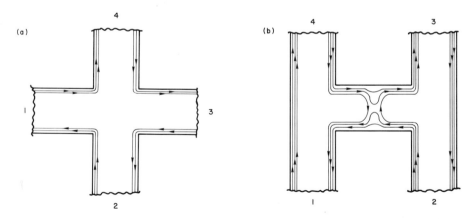

Fig. 1. (a) Conductor with Hall probes. (b) Conductor with barrier reflecting K edge states.

each of the incident edge states carries a unit current (is full), all outgoing edge states carry also a unit current (are full). Given N edge states, the total transmission probabilities in the conductor of Fig. 1a are $T_{41} = N$, $T_{34} = N$, $T_{23} = N$, and $T_{12} = N$. All other T_{ij} are zero. The total reflection probabilities in the absence of internal reflection are $R_{ii} = M_i - N$. The Hall resistance $\mathscr{R}_{13,42}$ is determined by $T_{41}T_{23} - T_{43}T_{21}$, which is equal to N^2. Evaluation of the subdeterminant yields $D = N^3$. All Hall resistances of the conductor of Fig. 1a are quantized and yield $\pm h/e^2N$. The "longitudinal" resistances (for example $\mathscr{R}_{12,43}$) are zero. Note, that the resistance is determined by states at the Fermi energy only.

IV. ANOMALOUSLY QUANTIZED FOUR-TERMINAL RESISTANCES

There are many arguments which can be put forth to explain the quantum Hall effect. To test such arguments, they should be applied to make predictions for situations which are not ideal (35). The discussion given above can easily be applied to the case where various current and voltage probes are interconnected with the result that the two-terminal resistance is $\mathscr{R}_{12,12} = (h/e^2)(p/q)$ with p and q integers as found experimentally by Fang and Stiles (35). Below we analyze situations where the connection of edge states to various contacts is changed in a well controlled fashion. Fig. 1b shows a conductor where a gate creates a barrier to carrier flow. For a certain range of barrier height K edge states are reflected. Application of Eq. (3) predicts Hall resistances (16)

$$\mathscr{R}_{13,42} = \left(\frac{h}{e^2} \right) \frac{1}{(N - K)} \tag{5}$$

$$\mathscr{R}_{42,13}(B) = \mathscr{R}_{13,42}(-B) = - \left(\frac{h}{e^2} \right) \frac{N - 2K}{N(N - K)} \tag{6}$$

and *quantized* longitudinal resistances (16) which are symmetric in the field,

$$\mathscr{R}_{12,43}(B) = \mathscr{R}_{12,43}(-B) = \mathscr{R}_{43,12}(B) = \left(\frac{h}{e^2} \right) \frac{K}{N(N - K)} . \tag{7}$$

All other four-terminal resistance measurements on the conductor of Fig. 1b are zero. The plateaus predicted by Eqs. (5-7) have been observed in strikingly clear experiments by Washburn et al. (20) and Haug et al. (21). Interestingly, van Houten et al. (36) found Eq. (7) to be a good approximation to the low field four-terminal magneto-resistance of a constriction, if there is some equilibration of the carriers between the constriction and the probes. In contrast, the derivation of Eqs. (5-7) assumes equilibration only in the reservoirs.

So far we have assumed that carriers which reach a contact from the interior of the sample can escape into the reservoir with probability 1. This is called a contact without internal reflection (16). Correspondingly, if carriers approaching a contact have a probability of less than 1 to escape into the reservoir, we have a contact with *internal* reflection. A current source contact with internal reflection populates edge states in a non-equilibrium fashion, similar to the barrier discussed above. If contacts with no internal reflection and contacts with internal reflection *alternate* along the perimeter of the sample all Hall resistances are still quantized (proportional to $1/N$) and all longitudinal resistances are zero (16). But if two contacts with internal reflection are adjacent there is at least one Hall measurement which depends on the detailed scattering properties of the contacts. A clear demonstration of this has come with the work of van Wees et al. (22). They consider two contacts spaced closely compared to an equilibration length. The width of the contacts can be varied and thus a barrier is created at the contacts which permits only a limited number of edge states to transmit. Fig. 2a shows a particular situation, where N-K edge states transmit at contact 1 and N-L edge states transmit at contact 2. The Hall resistance $\mathcal{R}_{13,42}$ is determined by the number N of bulk edge states, since carrier flow is from contact 1 to contact 4 and contact 4 provides equilibration. But if carrier flow is from contact 2 to contact 4 the Hall resistance is

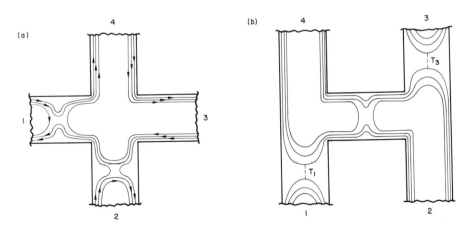

Fig. 2. (a) Conductor with two contacts with internal reflection. (b) Conductor with barrier and two weakly coupled contacts.

$$\mathscr{R}_{24,13} = \left(\frac{h}{e^2}\right)\frac{1}{(N-M)} \tag{8}$$

where $M=\min(K,L)$. It is the contact which provides less reflection, i. e. which exhibits the larger conductance which determines the outcome of the measurement (5,22).

Let us briefly return to the conductor of Fig. 1b. Clearly, the anomalous quantization found in the conductor of Fig. 1b also hinges on the properties of the contacts (at least as long as the contacts are within an equilibration length of the barrier). To show this, consider the conductor in Fig. 2b where two of the contacts are separated by barriers from the main conducting channel. If this barrier forms a smooth saddle, the contacts interact only with the outermost edge state. Let the probability for transmission from this contact to the outermost edge be $T_1 < 1$ at contact 1 and $T_3 < 1$ at contact 3. The Hall resistances are anti-symmetric in the field and given by $(h/e^2)1/(N-K)$ independent of T_1 and T_2. *All longitudinal resistances are zero (!)* in contrast to the example discussed above. That this is so is seen by inspection of Fig. 2b. It is only the outermost edge state which is measured and this state penetrates the barrier. These simple examples show the significance of the properties of contacts. This of course is only true as long as the contacts are close enough to the barrier to sense the differing population of the edge states. If the contacts are further than an equilibration length away from the barrier then the outcome of the measurement is independent of the properties of the contacts and given by Eqs. (5-7). Komiyama et al. (23) have performed experiments on conductors with contacts which exhibit internal reflection. (The contacts in these experiments are probably much more complex then the simple examples discussed here). These experiments are interesting in many ways: they demonstrate that an equilibration of the population of edge states does not occur even over distances of *several hundred* μm. Furthermore, they drastically demonstrate the highly non-local nature of conductance in the quantum Hall regime. The absence of longitudinal resistance in Fig. 2b is closely related to a recent experiment by van Wees et al. (24) which demonstrates the suppression of the Shubnikov-de Haas oscillations by selectively populating or detecting edge states with a point contact.

V. INTERIOR CONTACT

The approach advanced here which emphasizes the transport along edge states, which is supported by the experiments described above, has been objected to in the past on seemingly clear experimental evidence (37-40). Cage (40) writes "There is considerable confusion about this question, due mainly to a series of theoretical papers dealing with current-carrying edge states..." and goes on to defend a classical current distribution pattern based on the notion of local electric conductivities σ_{xx} and σ_{xy}. A number of experiments have been performed with contacts in the interior of the conductor, as shown schematically in Fig. 3a. Even for fields at which the Hall resistance is quantized, the voltage measured with reference to an interior contact exhibits large fluctuations as the field is increased through the plateau. It is argued that since carrier motion in high magnetic fields is along equipotential lines, the measurement of a voltage which differs from that of the current source or current sink in the interior of the sample indicates current flow in the bulk of the sample. Below we point out that the voltage measured at an interior contact can exhibit large swings without an appreciable change in the current flow pattern.

The conductor of Fig. 3a has three reservoirs and it is, therefore, sufficient to consider Eq. (2) for $i=1$, 2, and 3. In the presence of zero net current flow into reservoir 3 the chemical potential is (14)

$$\mu_3 = \frac{T_{31}\mu_1 + T_{32}\mu_2}{T_{31} + T_{32}}.$$

(9)

The measured chemical potential is determined by the probabilities of carriers entering contacts 1 and 2 to reach contact 3. The bottleneck for this processes is the transmission from the edge states to the contact. Motion along the edge

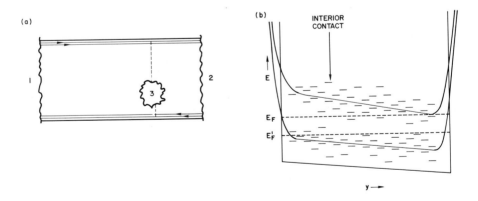

Fig. 3. (a) Conductor with interior contact. (b) Typical energy spectrum of the conductor. The arrow indicates the position of the contact.

states occurs with probability 1. At low temperatures transmission form the form the edge states to the contact can occur via quantum tunneling and at higher temperatures via Mott hopping through impurity states. We can thus assume that the transmission probabilities T_{31} and T_{32} are extremely small. The current exchanged between the edge states and the measurement reservoir is negligible compared to the current flowing along the edges. Despite the minuscule current which flows from the upper edge to the lower edge the measured chemical potential exhibits very large fluctuations. Indeed if $T_{32} << T_{31} << N$, the measurement yields a potential $\mu_3 \sim \mu_1$ and if $T_{31} << T_{32} << N$, the chemical potential is $\mu_3 \sim \mu_2$. Large swings in the chemical potential of interior contacts are indeed experimentally observed and in macroscopic conductors are likely due to small but large scale non-uniformities of the sample which for some fields make a Mott hopping path to one edge more likely than to the other. Fig. 3b shows the spectrum of the conductor in the vicinity of the interior contact. The interior contact is taken to be closer to the lower edge and in Fig. 3b its position is indicated by an arrow. If the Fermi energy is slightly above the bulk Landau level N (onset of the N Hall resistance plateau) carrier flow to the upper edge is more likely since there are more impurity states available. The measured chemical potential is slightly less but close to μ_1. As the magnetic field is increased (the Fermi energy lowered) the density of impurity states between the contact and the lower edge increases and this path becomes favorable. Hence the potential at the interior contact approaches that of the lower edge and is closer to μ_2. In large samples it is only the average impurity distribution which counts. In smaller samples we can expect the sample specific impurity distribution to become important and to produce very irregular chemical potential fluctuations (7). If additional exterior contacts are added to the conductor in Fig. 3b, our model gives, in complete agreement with experiment, that a two terminal measurement, using an interior contact and an arbitrary exterior contact (without internal reflection or arbitrary contacts but spaced further apart than an equilibration length) does not depend on the exterior contact. That is because motion along the edge occurs without resistance. If open edge states were to reach the interior contact the resistance would be of the order of h/e^2 and not as experimentally observed in the mega ohm range.

Transport along edge states combined with a formula which expresses resistances in terms of global transport coefficients (2,6) leads to an appealing and simple picture of the quantum Hall effect and is applicable to a variety of highly non-uniform sample geometries. Recent experiments by Chang and Cunningham (41) indicate that an extension of this approach to the fractional quantum Hall effect should also be possible.

REFERENCES

1.   K. von Klitzing, G. Dorda and M. Pepper, Phys. Rev. Lett. **45**, 494 (1980).
2.   M. Büttiker, Phys. Rev. Lett. **57**, 1761 (1986).
3.   J. Frenkel, Phys. Rev. **36**, 1604 (1930); R. Landauer, IBM J. Res. Dev. **1**, 223 (1957); Z. Phys. **B21**, 247 (1975); ibid. **68**, 217 (1987); H. L. Engquist and P. W. Anderson, Phys. Rev. **B24**, 1151 (1981); M. Büttiker, Y. Imry, R. Landauer, S. Pinhas, Phys. Rev. **B31**, 6207 (1985).
4.   A. D. Benoit, S. Washburn, C. P. Umbach, R. B. Laibowitz, and R. A. Webb, Phys. Rev. Lett. **57**, 1765 (1986).
5.   H. van Houten, C. W. J. Beenakker, J. G. Williamson, M. E. I. Broekaaart, and P. H. M. van Loosdrecht, B. J. van Wees and J. E. Moij, C. T. Foxon and J. J. Harris, (unpublished); C. W. J. Beenakker, H. van Houten, and B. J. van Wees, Supperlattices and Microstructures **5**, 127 (1989).
6.   M. Büttiker, IBM J. Res. Develop. **32**, 317 (1988).
7.   M. Büttiker, Phys. Rev. B **35**, 4123 (1987); H. U. Baranger, A. D. Stone and D. P. DiVincenzo, Phys. Rev. B **37**, 6521 (1988); S. Maekawa, Y. Isawa, and H. Ebisawa, J. Phys. Soc. Jpn. **56**, 25 (1987); C. L. Kane, P. A. Lee, and D. P. DiVincenzo, Phys. Rev. B **38**, 2995 (1988); S. Herschfield and V. Ambegoakar, Phys. Rev. **B38**, 7909 (1988). S. Datta and M. J. McLennan, (unpublished).
8.   G. Timp, A. M. Chang, P. Mankiewhich, R. Behringer, J. E. Cunningham, T. Y. Chang, and R. E. Howard, Phys. Rev. Lett. **59**, 732 (1989); Y. Takagaki, K. Gamo, S. Namba, S. Ishida, S. Takaoka, K. Murase, K. Ishibashi and Y. Aoyagi, Solid State Communic. **68**, 1051 (1988); Y. Avishai and Y. B. Band, (preprint).
9.   M. L. Roukes, A. Scherer, S. J. Allen, Jr., H. G. Craighead, R. M. Ruthen, E. D. Beebe, and J. P. Harbison, Phys. Rev. Lett. **59**, 3011 (1987); A. M. Chang and T. Y. Chang, (unpublished); C.J. B. Ford, S. Washburn, M. Büttiker, C. M. Knoedler and J. M. Hong, (unpublished).
10.   C. W. J. Beenakker and H. van Houten, Phys. Rev. Lett. **60**, 2406 (1988).
11.   F. M. Peeters, Phys. Rev. Lett. **61**, 589 (1988).
12.   D. G. Ravenhall, H. W. Wyld and R. L. Schult, Phys. Rev. Lett. **62**, 1780 (1989).
13.   G. Kirczenow, Phys. Rev. Lett. **62**, 1920 (1989).
14.   H. Baranger and A. D. Stone, (unpublished).
15.   M. Büttiker, Phys. Rev. **B33**, 3020 (1986); IBM J. Res. Develop. **32**, 63 (1988); D. Sokolovski, Phys. Lett. **A123**, 381 (1988); C.W. J. Beenakker and H. van Houten, (unpublished).
16.   M. Büttiker, Phys. Rev. **B38**, 9375 (1988).
17.   P. Streda, J. Kucera and A. H. MacDonald, Phys. Rev. Lett. **59**, 1973 (1987); see also Ref. 21.
18.   J. K. Jain, Phys. Rev. Lett. **60**, 2074 (1988); J. K. Jain and S. A. Kievelson, Phys. Rev. Lett. **60**, 1542 (1988).
19.   U. Sivan, Y. Imry, and C. Hartzstein, Phys. Rev. **B39**, 1242 (1989).
20.   S. Washburn, A. B. Fowler, H. Schmid, and D. Kern, Phys. Rev. Lett. **61**, 2801 (1988).
21.   R. J. Haug, A. H. MacDonald, P. Streda, and K. von Klitzing, Phys. Rev. Lett. **61**, 2797 (1988); R. J. Haug, J. Kucera, P. Streda and K. von Klitzing, (unpublished).
22.   B. J. van Wees, E. M. M. Willems and C. J. P. M. Harmans, C. W. J. Bennakker, H. van Houten and J. G. Williamson, C. T. Foxon and J. J. Harris, Phys. Rev. Lett. **62**, 1181 (1989).
23.   S. Komiyama, H. Hirai, S. Sasa and S. Hiyamizu, (unpublished); S. Komiyama, H. Hirai, S. Sasa and F. Fuji, (unpublished).

24. B. J. van Wees, E. M. M. Willems, L. P. Kouwenhoven, C. J. P. M. Harmans, J. G. Williamson, C. T. Foxon, and J. J. Harris, (unpublished).

25. A. M. Chang G. Timp, J. E. Cunningham, P. M. Mankiewich, R. E. Behringer and R. E. Howard, Soli State Commun. **76** , 769 (1988).

26. M. Büttiker, Phys. Rev. Lett. **62**,229 (1989).

27. M. Büttiker, Phys. Rev. **B38**, 12724 (1988).

28. B. I. Halperin, Phys. Rev. **B25**, 2185 (1982).;

29. A. H. MacDonald and P. Streda, Phys. Rev. **B29**, 1616 (1984).

30. M. Büttiker, Y. Imry and R. Landauer, Phys. Lett. **A96**, 365 (1983); Y. Gefen, Y. Imry, M. Ya. Azbel, Phys. Rev. Lett. **52**, 129 (1984); M. Büttiker, Y. Imry and M. Ya. Azbel, Phys. Rev. **A30**, 1982 (1984); U. Sivan and Y. Imry, Phys. Rev. Lett. **61**, 1001 (1988).

31. C. J. B. Ford, T. J. Thornton, R. Newbury, M. Pepper, H. Ahmed, D. C. Peacock, D. A. Ritchie, J. E. F. Frost, and G. A. C. Jones, Appl. Phys. Lett. **54**, 21 (1989); G. Timp, P. Mankiewich, P. de Vegvar, R. Behringer, J. E. Cunningham, R. E. Howard, H. U. Baranger, and J. Jain, Phys. Rev. **B39**, 6227 (1989); P. H. M. van Loosdrecht, C. W. J. Beenakker, H. van Houten, J. G. Williamson, B. J. van Wees, J. E. Mooij, C. T. Foxon, and J. J. Harris, Phys. Rev. **B38**, 162 (1988); B. J. van Wees, L. P. Kouwenhoven, C. J. P. M. Harmans, J. G. Williamson, C. E. Timmering, M. E. I. Broekaart, C. T. Foxon, and J. J. Harris, (unpublished).

32. R. B. Laughlin, Phys. Rev. **B23**, 5632 (1981).

33. G. F. C. Searle, *The Electrician* , **66**, 999 (1911).

34. A. Baranger and A. D. Stone, (unpublished). A. D. Stone and A. Szafer, IBM J. Res. Develop. **32**, 384 (1988).

35. F. F. Fang and P. J. Stiles, Phys. Rev. **B29**, 3749 (1984); D. A. Syphers and P. J. Stiles, Phys. Rev. **B32**, 6620 (1985).

36. H. van Houten, C. W. J. Beenakker, and P. H. M. van Loosdrecht, T. J. Thornton, H. Ahmed, and M. Pepper, C. T. Foxon, and J. J. Harris, Phys. Rev. **B37**, 8534 (1988).

37. G. Ebert, K. von Klitzing and G. Weimann, J. Phys. **C18**, L257 (1985).

38. E. K. Sichel, H. H. Sample and J. P. Salerno, Phys. Rev. **B32**, 6975 (1985).

39. H. Z. Zheng, D. C. Tsui and A. M. Chang, Phys. Rev. **B32**, 5506 (1985).

40. M. E. Cage, in *The Quantum Hall Effect* , edited by R. E. Prange, and S. M. Girvin, Springer Verlag, New York, 1987. page 56.

41. A. M. Chang and J. E. Cunnigham, (unpublished).

WHEN ISN'T THE CONDUCTANCE
OF AN ELECTRON WAVEGUIDE QUANTIZED?

G. Timp, R. Behringer, S. Sampere, J.E. Cunningham, and R.E. Howard

AT&T Bell Laboratories, Holmdel, New Jersey 07733

Presumably, a one-dimensional, ballistic wire has no resistance. However, in a conventional resistance measurement, leads contact each end of the one-dimensional conductor and a finite resistance is measured. The resistance is due to the redistribution of carriers between the electronic states in the region of the contact or lead, and in the one-dimensional conductor. A two terminal measurement of a one-dimensional constriction in a two-dimensional electron gas reveals a conductance of approximately $2e^2N/h$ which is quantized in steps of $2e^2/h$ as the width of the conductor narrows, corresponding to the successive depopulation of the N one-dimensional subbands in the constriction. We have examined the two, three and four terminal conductances of a one-dimensional ballistic constriction in a two-dimensional electron gas to determine under what conditions the conductance is quantized. The two terminal conductance is quantized to $2e^2/h$ with an accuracy of about 1% as the width of the constriction is varied, but only if the constriction has a submicron length.. A three terminal measurement of the conductance of two ballistic constrictions in series and in close proximity to one another is also quantized, but only for low values of N. The implications of these observations on four terminal measurements are discussed.

1. *INTRODUCTION*

Van Wees et al. [1] and Wharam et al. [2] recently discovered that the two terminal conductance of a ballistic constriction in a high mobility 2D electron gas, i.e. an electron waveguide, is quantized in steps of approximately $2e^2/h$ to about 10% accuracy as the width of the constriction is varied, corresponding to the depopulation of successive spin-degenerate one-dimensional (1D) subbands. Although reminiscent of the quantized Hall effect, [3] the two terminal conductance of an electron waveguide is quantized even in the *absence of a magnetic field.* This remarkable result was not anticipated theoretically because it was widely held that the nature of the contact, used to measure the conductance of the ballistic constriction, was not ideal. Imry [4] suggested that the conductance of a constriction measured between two semi-infinite contacts could be quantized provided that: (1) the contacts are adiabatically tapered to match to the constriction; and (2) that the constriction is both short enough to be ballistic and long enough to prevent evanescent modes from carrying appreciable current. However, it was believed that the scattering from the contacts, and the discrepency between the measured potential and the actual potential in the vicinity of the contact to the ballistic constriction would result in a two terminal conductance with fluctuations e^2/h in amplitude obscuring any quantization. It has subsequently been shown [5] [6] [7] [8] [9] [10] theoretically that the quantization persists to about 0.1% of $2e^2/h$ even if these features are incorporated into the model, but the steps are not as sharp, in qualitative agreement with the observations by van Wees et al. [1] So, despite the apparent nonadiabatic nature of the contact to the 1D constriction, and the scattering at the contact, the two terminal conductance of an electron waveguide is quantized.

When isn't the conductance quantized, and how robust is the quantization? In what follows we show that the absence of scattering in the constriction represents a strigent condition for the observation of a quantized conductance. Impurity scattering[6] on a submicron scale and scattering from a lead in close proximity to the constriction [11] may cause the quantization of the conductance to deteriorate. We report our observations of the two terminal conductance of a single ballistic constriction as a function of length and mobility, and of the two and three terminal conductances of two ballistic constrictions in series. The ballistic constrictions are made using a split-gate geometry which laterally constrains the 2D electron gas in a GaAs/AlGaAs heterostructure to the region in the gap between the split gates. We find that the conductance is quantized to about 1% accuracy in two terminal measurements of constrictions with a lithographic gap 300nm wide and a length of 200nm in heterostructures where the 2D electron mobility $\mu \approx 120 m^2/Vs$, but we find that the conductance is not well quantized for constrictions which are lithographically longer than about 600nm. We attribute the absence of quantization to impurity scattering within the 600nm constriction. We show that a second ballistic constriction in series with and in close proximity (300nm) to the first can also be detrimental to the quantization. The latter observation, the absence of quantization due to scattering from a second ballistic constriction in series, differs from a report by Wharam et al. [12] of quantized resistances for constrictions about $1\mu m$ apart. Moreover, we demonstrate, using two constrictions in series and a three terminal measurement scheme, that the lowest energy 1D subbands can be well quantized and do not scatter effectively around a bend in a ballistic constriction. The latter finding is crucial for understanding the four terminal resistances associated with a bend in the ballistic constriction [13] and the Hall resistance [14] because it clearly delineates experimentally which subbands couple to the leads in a multiterminal measurement.

2. TWO TERMINAL CONDUCTANCE

The ballistic constrictions we examined were obtained using a split Schottky gate fabricated on a high mobility AlGaAs/GaAs heterostructure. Figure 1

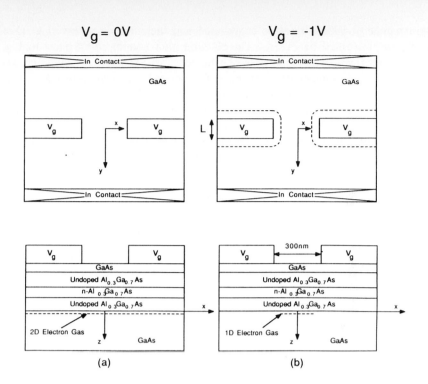

Figure 1. Figures 1a and b show a top view (top) and cross-section (bottom) of the split-gate geometry on a heterostructure for gate voltages of $V_g=0$ and -1V respectively. Electrons from the n-doped AlGaAs layer are transferred to the lower AlGaAs/GaAs heterojunction and localized within 10nm of that interface. If $V_g \leq -1V$, the 2D electron gas at the AlGaAs/GaAs interface around the split gate is depleted forming a constriction (indicated by the dashed line in the top view) in the gap between the gate electrodes.

illustrates schematically one of the configurations of gate electrodes on the top surface of an AlGaAs/GaAs heterostructure that we used. We fabricated devices in two different heterostructures; one consisting of a 7nm GaAs cap, a 25nm AlGaAs buffer layer, a Si-doped monolayer and a 70nm AlGaAs spacer layer on top of a $1\mu m$ GaAs buffer layer grown on a semi-insulating GaAs substrate; and a second consisting of a 6nm GaAs cap, a 24nm AlGaAs buffer layer, a Si-doped monolayer and a 42nm AlGaAs spacer layer on top of a $1\mu m$ GaAs buffer layer also grown on a semi-insulating substrate. Using photolithography and a wet etch, a Hall bar geometry comprised of either $3\mu m$ wide or $100\mu m$ wide wires is transferred onto the heterostructures. On top of the Hall bar geometry, a (Schottky) split-gate with a gap of 300nm between the gate electrodes is fabricated using electron beam lithography. Electron beam lithography is used to prepare a mask for lift-off. A gold or gold-palladium alloy film approximately 100nm thick is evaporated and the mask is removed to give the split-gate illustrated by Fig. 1. The electron micrographs in the inset to Fig. 2a, Fig. 3a and Fig. 5 are typical of the devices we examined. Using the $100\mu m$ Hall bridge we evaluated the mobility and two-dimensional (2D) carrier density at the lower AlGaAs/GaAs interface from the quantized 2D Hall effect and longitudinal magnetoresistance at 280mK. The first heterostructure had a mobility $\mu=85m^2/Vs$ and a 2D carrier density of $n=4.2\times10^{15}m^{-2}$ at the lower AlGaAs/GaAs interface, while the second heterostructure showed $\mu=120m^2/Vs$ with $n=2.7\times10^{15}m^{-2}$.

Figure 2a shows the two terminal resistance, $R_{12,12}$, as a function of the applied gate voltage that we found in a device like that shown in the inset. The device was fabricated on the lower mobility heterostructure. Following the convention of Buttiker [15] $R_{lm,jk}(G_{lm,jk})$ denotes a resistance (conductance) measurement in which positive current flows into lead l and out lead m with a positive voltage measured between leads j and k respectively. The convention for numbering the leads is given in the inset to Fig. 2a. The series resistance found at $V_g=0V$ was subtracted from the resistance measured as a function of gate voltage. The gate length (the length L along the y-axis in Fig. 1b) for the device of Fig. 2a is about 200nm. By applying a voltage more negative than -0.5V to the split Schottky gates, the 2D electron gas at the AlGaAs/GaAs interface immediately beneath the gate electrodes is depleted and so the 2D electron gas is laterally constrained (along the x-axis in Fig. 1b) within the gap in the electrodes. As the gate voltage decreases further, the constriction narrows; the carrier density within the constriction decreases; and plateaus are observed in the resistance. The two terminal conductance obtained by inverting the resistance versus gate voltage is also shown in Fig. 2a. The *average* conductance (minus the series resistance) is approximately $2e^2N/h$ and is evidently quantized in steps of $2e^2/h$ with about 10% accuracy as the gate voltage narrows the constriction. Eleven steps (N=11) are shown in Fig. 2a, but up to fifteen steps have been clearly resolved in data obtained from similar devices. For voltages more negative than about -2.95V, the 300nm gap pinches off.

Figure 2b shows the two terminal resistance as a function of the applied gate voltage (minus the two terminal resistance found at $V_g=0.0V$) that we found at 280mK in a device fabricated on the higher mobility heterostructure. Six steps are shown in Fig. 2b. The quantization of conductance is accurate to about 1%. Figure 3 compares the conductance obtained in a two terminal measurement at zero magnetic field with the four terminal conductance measured using the Hall bridge on which the gates are fabricated with all of the gate electrodes grounded. The accuracy of the quantization to $2e^2/h$ of the two terminal conductance as a function of gate voltage does not approach the accuracy or precision that can be achieved in a routine measurement of the quantum Hall effect. The inset to Fig. 3 shows that deviation from perfect quantization that we find as a function of the step N. The deviation can qualitatively be represented as an additional resistance in series with the ballistic constriction; a resistance we have no justification for subtracting.

The mobility of the 2D electron gas used to fabricate the device of Fig. 2a is approximately $85m^2/Vs$ (at T=280mK) which, with $n=4.6\times10^{15}m^{-2}$, corresponds to a low temperature mean free path of about $L_e=\hbar k_F\mu/e=6.8\mu m$. A similar

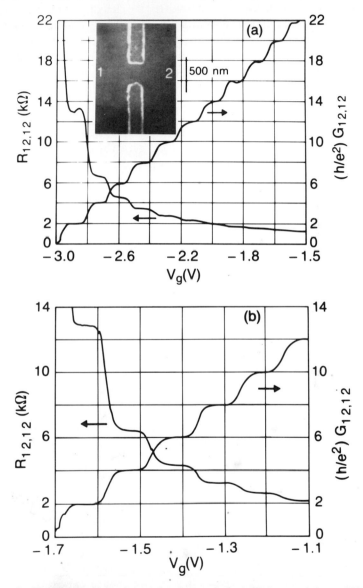

Figure 2. The two terminal resistance (conductance) of a ballistic constriction at 280mK as a function of the gate voltage (or width of the constriction) where L=200nm.

estimate can be made for the device of Fig. 2b and we find that $L_e=7.3\mu m$. Since the 300nm gap in each device pinches off for large negative gate voltages, the depletion around the gate must be less than 150nm under conditions where the wire conducts, and so we estimate the actual length of the constriction to be less than 500nm; much less than the mean free path deduced from the mobility of the 2D electron gas. If the constriction is ballistic, we expect the quantization of the conductance plateaus to improve as the length increases because the current carried by evanescent modes diminishes. [8] In addition, resonances in the conductance may occur when the condition $k_yL=n\pi$ is satisfied, where k_y is the wavevector along y.[5][7] However, we find that as the length of the constriction increases, the quantization deteriorates. Figure 4a shows the resistance and conductance determined as above for a constriction with gate electrodes of length L= 600nm, spaced with a 300nm gap. The length of the constriction is estimated to be less than 900nm for this geometry which is still a factor of eight less than the mean free path estimated from the 2D mobility and carrier density, yet the quantization is destroyed. Figure 4b shows the resistance and conductance obtained in a similar device fabricated in the second heterostructure with

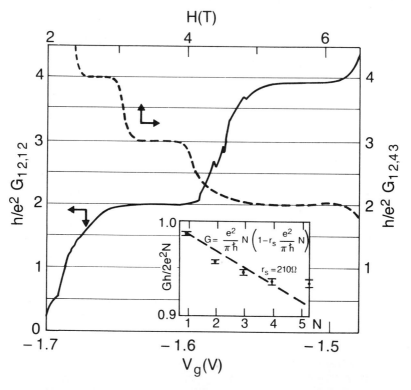

Figure 3. The two terminal conductance $G_{12,12}$ of a ballistic constriction as a function of the applied gate voltage at zero magnetic field, and the four terminal magnetoconductance, $G_{12,43}$, measured in the same device but with $V_g=0V$.

$\mu=120m^2/Vs$. While the quantization of the resistance is still apparent in the higher mobility device; the accuracy of the quantization is less than 20% for the first three steps with no quantization observed for higher N.

There is a direct correspondence between the number of populated 1D subbands and the measured conductance provided that there is no impurity scattering in the ballistic constriction. If there is no impurity scattering and if the current is equally distributed between subbands, then the total current is $I=N(2e/h)\Delta\mu$ where $\Delta\mu$ is the chemical potential difference between two infinitely wide contacts to the 1D ballistic wire, and N is the number of 1D subbands in the constriction. Since the voltage difference between the wide reservoirs is $V=\Delta\mu/e$, the two terminal resistance of an N subband wire is given by:

$$G=(2e^2/h)N. \tag{1}$$

The ideal, two terminal resistance is a quantized function of the number of 1D subbands.

However, if there are impurites within the wire, carriers can be scattered from one subband to another and reflected from the wire. Under these circumstances only a fraction of the current incident in subband j (Eq. 2.2) is transmitted, i.e. $I_j=(2e/h)\sum_{i=1}^{N}t_{ij}\Delta\mu$, where t_{ij} is the probability intensity for transmission from subband j into subband i and we ignore the energy dependence of the t_{ij} which arises from the difference in chemical potential between the two contacts. Summing the contributions from each of the subbands gives the total current, and the resistance becomes:

$$G=(2e^2/h)\sum_{i,j=1}^{N}t_{ij}. \tag{2}$$

This is the Landauer formula for the two terminal conductance. If (1) k_BT is much smaller than the energy interval between subbands; (2) the length of the perfect wire is much longer than the decay length for evanescent modes; and (3) there is no scattering in the narrow wire (i.e. $t_{ij}=\delta_{ij}$); the conductance given by Eq. (2) reduces to that of Eq. (1).

The correspondence between Eq. (1) and the data of Fig. 2 is suprising because to obtain Eq. (1) it is assumed that the current is distributed evenly among the 1D subbands and and it is assumed that the contacts to the constriction are of infinite width. The reflections at the contacts to the constriction and the finite width of the contacts are completely ignored in the simple analysis which gave Eq. (1). We do not expect the $3\mu m$ or the $100\mu m$ wide wires, which we routinely use to contact the constriction, to be adequately represented by an infinitely wide reservoir. In particular, in addition to the quantized contact resistance between the constriction and the 2D electron gas, we expect[4] a contact resistance between the point at which the voltage is actually measured and the local potential potential at the contact to the constriction. The observation of a well quantized two terminal conductance implies: (1) that the latter contact resistance is not significant, at least for contacts as narrow as $3\mu m$; (2) that the wire is ballistic; and (3) that the energy separation between subbands is much larger than 280mK.

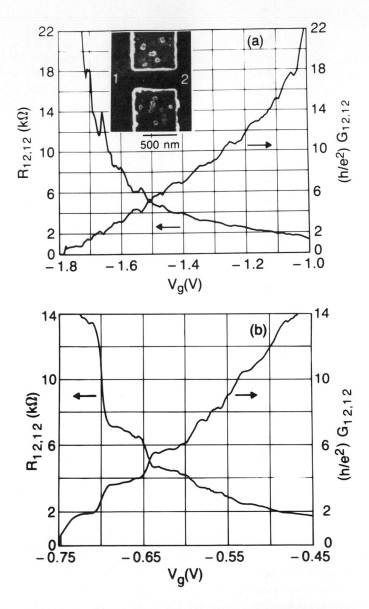

Figure 4. The two terminal resistance (conductance) of a two ballistic constrictions at 280mK as a function of the gate voltage (or width of the constriction) where L=600nm.

Impurity scattering within the ballistic constriction may be responsible for the absence of quantization in the longer constrictions. The mean free path estimated from the mobility of the wide wire is not necessarily an appropriate estimate for the elastic scattering length in the ballistic constriction especially for gate voltages near pinch-off. Parenthetically, the estimate of L_e obtained from electron focussing may not be relevant, either. Electron focusing data obtained by using ballistic constrictions as both a monochrometer and analyzer do not provide an indication of L_e within the constriction, but measure L_e within the 2D electron gas where the focussing occurs. [16] The 2D electron gas effectively screens potential fluctuations, however. As the constriction in the 2D electron gas narrows and the carrier density is reduced, the Fermi wavevector k_F becomes shorter and the screening of the Coulombic impurities in the doped layer by the electron gas becomes less effective. Consequently, the elastic scattering length may be reduced to approximately $1 \mu m$ or less because the 1D electron gas ineffectively screens potential fluctuations. [17] As shown illustrated by Fig. 4b, we expect the quantization of the low index subbands to be less fragile than the high index subbands because of the limited possibilities for backscattering when only a few subbands are occupied.

3. THREE TERMINAL CONDUCTANCE

Two terminal measurements of a ballistic constriction do not directly measure the potential of the conductor because of the contact resistance. Three and four terminal measurements do not generally measure the local potential of the conductor either, even if the leads are made from the same material and have the same dimensions as the conductor and there is no net current in the voltage leads. Although the net current in the voltage leads vanishes, a carrier may propagate coherently from the conductor into a lead and back into the conductor. The lead then becomes part of the measured resistance and must be included in the computation of the resistance.

Following Buttiker,[15] we can generalize the computation of the resistance to a multi-terminal measurement. The current in lead i is given by:

$$I_i = -(2e/h) \sum_{j=1}^{N} T_{ij} \mu_j \tag{3}$$

where T_{ij} is the trace over the subbands, i.e. $T_{ij} = \sum_{m,n} t_{ij,mn}$ with $t_{ij,mn}$ representing the probability for a carrier in subband n and lead j to be transmitted to subband m in lead i, and $T_{ii} = \sum_{m,n} r_{ii,mn} - N_i$ with N_i the number of subbands in lead i and $r_{ii,mn}$ the probability for a carrier in subband n and lead i to be reflected into subband i and lead m. The potential, V_j, associated with a reservoir of chemical potential μ_j is given by $V_j = \mu_j/e$.

From Eq. (3) we can obtain a formula for the three terminal or four terminal resistance by imposing a current, I, between two leads, requiring the net current in the other leads to vanish, and solving for the voltage difference. For purposes of illustration we consider the mathematically simpler problem associated with the determination of the three terminal resistance first. If we impose a current between leads 1 and 2 so that a positive current enters through lead 1 and exits through lead 2, then we can calculate the potential differences $\mu_1-\mu_3$ and $\mu_3-\mu_2$, and so find the conductances:

$$G_{12,13}=eI/(\mu_1-\mu_3)=\frac{e^2}{h}\frac{D}{T_{32}} \tag{4a}$$

$$G_{12,32}=eI/(\mu_3-\mu_2)=\frac{e^2}{h}\frac{D}{T_{31}} \tag{4b}$$

with $D=T_{21}T_{31}+T_{21}T_{32}+T_{31}T_{23}$, whereas the two terminal conductance is

$$G_{12,12}=\frac{e^2}{h}\frac{D}{(T_{31}+T_{32})} \tag{5}$$

Thus, the two terminal conductance in the presence of one additional lead generally differs from Eq. (2). The additional lead scatters carriers according to the probablities T_{13} and T_{32}.

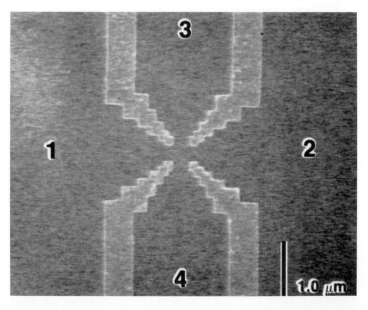

Figure 5. An electron micrograph of two series ballistic constrictions in close proximity fabricated on the higher mobility heterostructure. Associated with each port numbered 1 through 4 there is a contact so that multi-terminal resistance measurements can be performed on a junction in which the geometry can be manipulated.

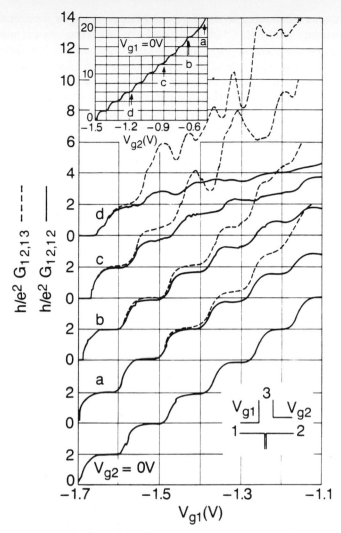

Figure 6. The three terminal conductance $G_{12,13}$ and the two terminal conductance $G_{12,12}$ as a function of the gate voltage applied to the forward constriction. For the biases shown port 4 is pinched off. The inset shows the quantization of the two terminal conductance $G_{12,12}$ for the second constriction under circumstances where there is no forward constriction.

To examine the effect of a lead on the conductance of ballistic constriction, we investigated two ballistic constrictions in series and in close proximity to one another. Figure 5 is an electron micrograph which illustrates the disposition of the split gate electrodes in a typical device fabricated in the higher mobility heterostructure. The convention for numbering the leads is given in Fig. 5. Associated with each port of the device there are contacts to the 2D electron gas. Figure 6 shows the conductances $G_{12,12}$ and $G_{12,13}$ obtained in such a device. The conductance versus V_{g1}, found if the second series constriction (the pair of electrodes on the right in Fig. 5) is grounded, is illustrated in the lower portion of Fig. 6 for reference. The data corresponding to traces (a) through (d) in Fig. 6 were obtained with the two lower gates in the micrograph of Fig. 5 biased so that there was no conduction through the port 4. The traces (a), (b), (c) and (d) were obtained by biasing V_{g2} at voltages of: -0.55V, -0.67V, -0.9V, and -1.15V respectively while varying V_{g1}. In the inset to Fig. 6, the two terminal conductance $G_{12,12}$ through the second constriction as a function of gate voltage is show for the condition $V_{g1}=0$, and the voltages corresponding to the traces (a)-(d) are indicated. The shift in the threshold for conduction as a function of V_{g1} observed for different values of V_{g2} is only about 50mV, less than the voltage difference between different conductance plateaus.

The solid lines in traces (a), (b), (c) and (d) of Fig. 6 indicate that the conductance $G_{12,12}$ is determined by the value of the narrowest constriction. In particular, $G_{12,12}$ is suppressed to values below $(11)2e^2/h$, $(9)2e^2/h$, $(6)2e^2/h$, and $(3)2e^2/h$ in traces (a)-(d) at a negative voltage of about -1.1V on V_{g1} corresponding respectively to N=11, 9, 6, and 3 1D subbands occupied in the second constriction (see the inset to Fig. 6). The conductance found for lower index N remains well quantized, however, even when the second constriction is narrow (see trace (d)). The dashed lines in Fig. 6 depict the results of the three terminal measurement $G_{12,13}$. We notice that the three terminal conductance is generally larger than $G_{12,12}$ and deviates substantially from the quantized values except for low N. For low N, $G_{12,13} \approx G_{12,12} \approx 2e^2 N/h$. It follows from Eq. (4a) and Eq. (5) that $T_{31} \ll 1$ for low N. Thus, the lead does not scatter or couple to all modes in the same way. Unlike the high index subbands, the low lying 1D subbands do not turn the corner, but are focussed in the forward direction. Because $T_{31} \ll 1$, the low N modes do not scatter into or couple to lead 3.

4. FOUR TERMINAL RESISTANCE

The low probability for the low N modes to propagate around a bend has important implications for four terminal magnetoresistance measurements of a ballistic constriction. In particular, the four terminal magnetoresistance associated with a bend in a constriction is negative and the Hall effect is suppressed from the conventional 2D value. Following Buttiker[15] we can express the four terminal resistance in terms of transmission probabilities by using Eq. (3). If we impose a current, I, between leads m and n, we can then solve Eq. (3) for the voltage difference, V_k-V_l, between leads k and l. The four terminal resistance, $R_{mn,kl}=(V_k-V_l)/I$, is given by:

$$R_{mn,kl}=(h/2e^2)(T_{km}T_{ln}-T_{kn}T_{lm})/D \qquad (6)$$

where D is always positive and depends on the lead geometry but is independent of permutations in the indices mn,kl. If the gate electrodes of Fig. 5 are biased similarly, the four port junction is approximately symmetric. If $T_{31}, T_{24} \ll 1$ and $T_{21}, T_{34} \approx 1$ for low N, we expect that $R_{14,32} = (T_{31} T_{24} - T_{34} T_{21})/D < 0$ and $R_{12,43} = (T_{41} T_{32} - T_{42} T_{31})/D \approx 0$ in the absence of a magnetic field.

Baranger and Stone[18] have shown that the four terminal formula for the resistance (Eq. (6)) developed by Buttiker applies for an arbitrary magnetic field.

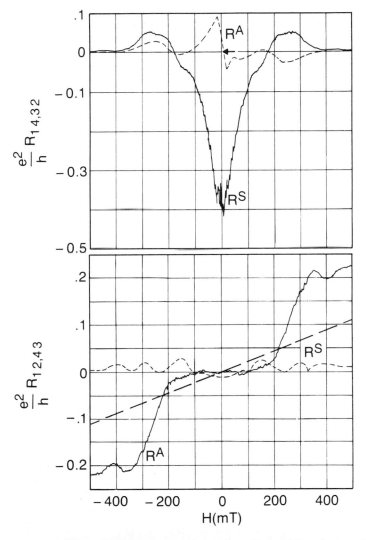

Figure 7. The magnetoresistances $R_{14,32}$ (top) and $R_{12,43}$ bottom. The resistances have been decomposed into the respective symmetric and antisymmetric components where $R^S_{mn,kl} = \frac{1}{2}[R_{mn,kl}(H) + R_{kl,mn}(H)]$ and $R^A_{mn,kl} = \frac{1}{2}[R_{mn,kl}(H) - R_{kl,mn}(H)]$.

However, we anticipate that the transmission coefficients will change for magnetic fields larger than $W \approx \sqrt{hc/eH}$, where W is the width of the constriction, because beyond this field range the conduction occurs via edge states and backscattering is suppressed.[19] Consequently, we expect that $R_{14,32}(H)<0$ and $R_{12,43}(H) \approx 0$ only for $-hc/eW^2 < H < hc/eW^2$.

Figure 7 shows the magnetoresistances $R_{14,32}$ and $R_{12,43}$ observed when the two series ballistic constrictions of Fig. 5 are biased symmetrically so that N=2 in each of the constrictions which comprise the junction. Both the symmetric and antisymmetric components of the resistances are shown as a function of magnetic field. We estimate that the width of the constrictions is approximately 100nm using a hard wall confinement potential, the carrier density deduced from the high field Hall effect, and the number of 1D subbands deduced from the two terminal resistances. We find that the resistance $R_{14,32}$ is grossly negative for -200mT < H < 200mT, [20] in correspondence with our expectations. The arrow in the top portion of the figure indicates the positive value of the resistance obtained using the same measurement scheme with $(V_g=0V)$. The resistance of a bend is large and negative because the probability for a carrier in a low index mode to propagate around a corner is small. Similarly, we find that $R_{12,43}$ is suppressed for -200mT < H < 200mT below the resistance expected for a classical Hall effect. The dashed line in the lower portion of Fig. 7 shows the slope extrapolated from the high field Hall effect which is characteristic of a 2D electron density of $1.1 \times 10^{15} m^{-2}$. The Hall effect is suppressed for low fields because the Hall voltage associated with the low index modes is poorly coupled to the leads used for measuring it. Thus, both the negative resistance associated with a bend in the constriction, and the suppression of the Hall effect in a ballistic junction are related phenomenon associated with the inability of the low index modes to propagate around a corner.

REFERENCES

1. B.J. van Wees, H. van Houten, C.W.J. Beenakker, J.G. Williamson, L.P. Kouwenhoven, D. van der Marel, and C.T. Foxon, Phys. Rev. Lett. 60, 848 (1988).

2. D.A. Wharam, T.J. Thornton, R. Newbury, M. Pepper, H. Ahmed, J.E.F. Frost, D.G. Hasko, D.C. Peacock, D.A. Ritchie, and G.A.C. Jones, J. Phys. C 21, L209 (1988).

3. Marvin E. Cage, *The Quantum Hall Effect*, edited by R.E. Prange, and S.M. Girvin, Springer-Verlag, New York (1987) p. 37-68.

4. Y. Imry, *Physics of Mesoscopic Systems*," Directions in Condensed Matter Physics, G. Grinstein and G. Mazenko, eds., World Scientific Press, Singapore, 1986 p.101.

5. A. Szafer and A.D. Stone, Phys. Rev. Lett. 62 300 (1989).

6. E.G. Haanappel and D. van der Marel, Phys. Rev. B 39 (1989).

7. G. Kirczenow, Solid State Commun. 68, 715 (1988).

8. L.I. Glazman, G.B. Lesovick, D.E. Khmel'nitskii and R.I. Shekhter, Pis'ma Zh. Eksp. Teor. Fiz 48, 218 (1988) [JETP Lett. 48 238 (1988)].

9. S. He and S. Das Sarma, "Quantum Conduction in Narrow Constrictions," preprint.

10. Y. Avishai and Y.B. Band, "Ballistic Electronic Conductance of an Orifice and of a Slit," preprint.

11. R. Landauer, "Which version of the formula for conductance as a function of transmission probabilities is correct?", unpublished.

12. D.A. Wharam, M. Pepper, H. Ahmed, J.E.F. Frost, D.G. Hasko, D.C. Peacock, D.A. Ritchie and G.A.C. Jones, J. Phys. C 21, L887 (1988).

13. G. Timp, H.U. Baranger, P. deVegvar, J.E. Cunningham, R.E. Howard, R. Behringer, and P.M. Mankiewich, Phys. Rev. Lett. 60, 2081 (1988).

14. M.L. Roukes, A. Scherer, S.J. Allen Jr., H.G. Craighead, R.M. Ruthen, E.D. Beebe and J.P. Harbison, Phys. Rev Lett. 59, 3011 (1987).

15. M. Buttiker, Phys. Rev. Lett. 57 1761 (1986) and M. Buttiker, IBM J. Res. Develop. 32, 317 (1988).

16. See: H. van Houten, C.W.J. Beenakker, J.G. Williamson, M.E.I. Broekaart, P.H.M. van Loosdrecht, B.J. van Wees, J.E. Mooij, C.T. Foxon and J.J. Harris, " Coherent electron focusing with quantum point contacts in a two dimensional electron gas," unpublished.

17. J.H. Davies and J.A. Nixon, "Fluctuations from random charges under a short gate in a narrow-channel MODFET," Phys. Rev. B, 39 3423 (1988).

18. H.U. Baranger and A.D. Stone, submitted to Phys. Rev. B.

19. G. Timp, P.M. Mankiewich, P. deVegvar, R. Behringer, J.E. Cunningham, R.E. Howard, H.U. Baranger, and J.K. Jain, Phys. Rev B 33 (1989).

20. Y. Takagaki, K. Gamao, S. Namba, S. Ishida, S. Takaoka, K. Murase, K. Ishibashi and Y. Aoyagi, Solid State Commun. 68, 1051 (1989).

ELECTRON BEAMS AND WAVEGUIDE MODES: ASPECTS OF QUANTUM BALLISTIC TRANSPORT

H. van Houten

Philips Laboratories Briarcliff, NY 10510, USA

C.W.J. Beenakker

Philips Research Laboratories, 5600 JA Eindhoven, The Netherlands

Quantum ballistic transport in a two-dimensional electron gas can ideally be explored by means of quantum point contacts. A horn-like shape or a potential barrier in the point contacts gives rise to collimation of the injected electron beam, and thereby has a dramatic effect on magneto-transport effects. The theoretical concepts underlying this new field are discussed, and a geometry for their investigation is proposed. This geometry has potential applications as a novel hot-electron transistor.

1. INTRODUCTION

The new field of quantum ballistic transport (1) in a two-dimensional electron gas (2DEG) can be approached from two complementary points of view, similar to geometrical and wave optics. One is the *trajectory picture*, in which the ballistic motion of electrons along classical trajectories is taken as a starting point. Quantum effects can subsequently be incorporated in a semi-classical approximation. The other is the *mode picture*, which focuses instead on the quantum states of electrons in a confined geometry. These states are the propagating waveguide modes of the transport problem.

Classical ballistic transport in metals is very similar to geometrical optics. It has been studied extensively after the pioneering work on point contacts by Sharvin (2). The 2DEG in a GaAs-AlGaAs heterostructure with its large Fermi wavelength (40 nm) and long transport mean free path (10 μm) offers the possibility to extend these studies to the quantum ballistic transport regime. New effects have been found, suggesting an analogy to wave optics. An example is the conductance quantization of quantum point contacts (3..5), which can be treated in terms of transmission through an electron waveguide. A description of transport as a quantum mechanical transmission problem was proposed in 1957, in a seminal paper by Landauer (6). An extension of his approach to multi-terminal measurements in the presence of a magnetic field has been developed by Büttiker (7). Analogies such as the one with optics are useful, because they can suggest new experiments. In this paper we explore in particular an analogy with Knudsen effusion (8) of a rarefied molecular gas, which is a text-book example of purely classical ballistic transport. This is followed by a consideration of transport in a geometry with point contacts in series, which mimics the physics of molecular beams.

2. QUANTUM POINT CONTACTS

2.1 Knudsen Flow and Sharvin Point Contacts

Consider the three-dimensional flow of a rarefied molecular gas through a narrow orifice (see Fig. 1a). The physics of this Knudsen (8) effusion problem is determined by the condition that the mean free path l is much larger than the linear dimension W of the orifice. The current of molecules J between two reservoirs, between which a gas density difference δn is externally maintained, is given by

$$J = \frac{1}{4}\delta n \ \bar{v} \ W^2 \tag{1}$$

for a square orifice of side W. Note that δn does not have to be small compared to the average density. Here $\bar{v} = (8kT/\pi m)^{1/2}$ is the mean thermal speed of the molecules of mass m. This purely classical result follows readily on integrating the Maxwell-Boltzmann equilibrium distribution function over velocity magnitude and direction (8).

The ballistic electron flow through a Sharvin point contact, connecting two metallic half spaces (2) is illustrated in Fig. 1b. At low temperatures, and for a small voltage difference across the contact, the electrons contributing to the transport are those which move with the Fermi velocity v_F. In the ballistic limit $l \gg W$, the electron flux through a 2D constriction expressed in terms of a density difference is

$$J = \frac{1}{2}\delta n <v_F> W, \tag{2}$$

where the brackets denote an angular average over the half space

$$<v_F> = \frac{1}{\pi} \int_{-\pi/2}^{\pi/2} v_F \cos \phi \ d\phi \ , \tag{3}$$

$$\rightarrow \ J = \frac{1}{\pi}\delta n \ v_F W \ ; \quad \delta n \ll n \ . \tag{4}$$

In the analogous 3D case $J = \delta n \ v_F W^2/4$, which differs from Eq. (1) because of the different statistics for the two problems (Maxwell-

348

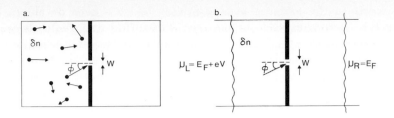

Fig. 1 (a) Knudsen effusion through a narrow orifice connecting two reservoirs with a gas density difference δn. (b) Ballistic transport through a Sharvin point contact.

Boltzmann versus Fermi-Dirac). Accordingly, Eq. (4) only applies if the density difference is small (see sec. 4). An electron density difference cannot be maintained, because of the cost in electro-static energy. Screening charges will reduce δn without changing the electro-chemical potential difference $\delta\mu = eV$, which is the thermodynamic driving force for the current. The two quantities are related by $\delta n = D(E_F)\delta\mu$, with $D(E_F) = m/\pi\hbar^2$ the 2D density of states at the Fermi level, including spin degeneracy. One thus finds (3) for the conductance $G \equiv e^2 J/\delta\mu$,

$$ G = \frac{2e^2}{h}\frac{k_F W}{\pi} \ . \tag{5} $$

The 3D Sharvin point contact conductance [1] is $G = (2e^2/h)(k_F^2 W^2/4\pi)$.

Note that the derivation given here does not have to be modified for a channel of finite length connecting the two half spaces, provided the scattering of the particles from the channel boundaries is purely specular (corresponding to conservation of momentum in the channel direction). As it happens, this is a much more realistic assumption for ballistic transport in a constricted 2DEG than in the case of free molecular flow.

2.2 A Refractive Index Step And Transport Over a Barrier

The result Eq. (5) is readily extended to the case where, in addition to the geometrical constriction, a local potential barrier limits the current flow. Such a barrier is indeed present in typical 2DEG point contacts, because of a reduction in local carrier density (3). Although the electro-

---

[1] In the 3D case the point contact resistance for *small* mean free path ($l \ll W$) is known as the Maxwell spreading resistance which is of order ρ/W, the resistance of a cube of size W. The backscattering associated with the reduced finite mean free path for hot electrons allows one to perform point contact spectroscopy (see *e.g.* Ref. 9). In 2D the analogous spreading resistance does not dominate the sample resistance, since it is only of order ρ, the resistance of a square.

static potential is likely to have a smooth shape, we discuss for simplicity an abrupt potential step in the constriction (see Fig. 2a). Only electrons with sufficient longitudinal momentum will be transmitted, while the transverse momentum is conserved. For a potential barrier of height E_o this leads to a reduction of the *cone of acceptance* (10) of the constriction from its original value of π to $2\alpha_{max}^{barrier} = 2\arccos(E_o/E_F)^{1/2}$. The conductance is still given by Eq. (5), but with the reduced Fermi wave vector $k_{F,min} = (1/\hbar)\,[2m(E_F - E_o)]^{1/2}$. A reduction in carrier concentration and a reduction in width thus both reduce the point contact conductance. Note that a potential barrier in the constriction is analogous to a refractive index step in optics.

2.3 The Conical Reflector And Adiabatic Transport

The analogy of ballistic transport with optics suggests a purely geometrical effect which also limits the cone of acceptance of the point contact. The device we have in mind is the conical reflector (see Fig. 2b), although horns of somewhat different shape obey essentially the same physics (11,10). For such weakly flared constrictions (which may also contain a slowly varying potential barrier), the quantity $S = k_F W \sin\alpha$ is a constant of the motion, or *adiabatic invariant* (S is proportional to the action for motion transverse to the channel (12)). In the semi-classical approximation, the invariance of S implies that trajectories entering the point contact within a cone of opening $2\alpha_{max}^{horn} = 2\arcsin$ $[(W_{min}/W_{max})\,(1 - E_o/E_F)^{1/2}]$ are transmitted, the others being reflected. The conductance is still given by Eq. (5), but with $k_{F,min} W_{min}$ replacing $k_F W$. Both the flaring of the constriction and the presence of a potential barrier tend to *collimate* the injected electron beam at the exit of the constriction. It would be of interest to investigate to what extent diffraction reduces this classical effect, which may be important in geometries with point contacts in series (see below) (10).

2.4 Quantum Mechanical Aspects And Finite Temperature Averaging

So far quantum mechanical aspects other than the degeneracy of the electron gas have been ignored. The conductance quantization of the point contact can be explained semi-classically (3,13), but is most naturally treated in terms of the mode picture, using the multi-channel generalization of the Landauer formula for two-terminal conductances (14)

$$G = \frac{2e^2}{h}\ Tr\ tt^\dagger\ .$$

(6)

where t denotes the transmission matrix, and the trace has to be taken

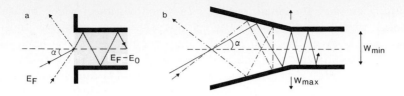

Fig. 2 (a) Ballistic transport through an abrupt constriction containing a potential barrier of height E_o. (b) Ballistic transport through a constriction flared in the form of a conical reflector.

over all populated 1D subbands, or modes, occupied in the 2DEG leads attached to the constriction. If width and potential barrier change sufficiently slowly, the constriction is adiabatic (see before), and the problem becomes remarkably simple as noted by Glazman *et al.* (15). The quantum number n labeling 1D subbands or modes is an adiabatic invariant (in fact, πn corresponds to the classical invariant $S = k_F W \sin \alpha$ mentioned above). Consequently, there is no off-diagonal mode coupling, the N lowest index modes or subbands are perfectly transmitted, and the higher index modes in the 2DEG are perfectly reflected. From this argument and Eq. (6) the conductance quantization follows directly

$$G = \frac{2e^2}{h} N \quad . \tag{7}$$

Eq. (5) is recovered in the limit of large quantum numbers, since N is equal to the largest integer smaller than $k_F W/\pi$. We note that contact resistances of order h/e^2 were first proposed by Imry (16).

For an abrupt constriction the physics is more complicated. Recent numerical and analytical work has shown, surprisingly, that even in that case clear plateaux in the conductance as a function of width are obtained (17..20). A mean field approximation in good agreement with the numerical results has been formulated (18), which assumes strong off-diagonal mode coupling. This is consistent with the trajectory picture: Since $\sin \alpha = n\pi/k_F W$ is invariant for motion on a straight trajectory, it follows that n increases appreciably when the width W increases abruptly. One then has highly non-adiabatic, but perfectly ballistic transport [2].

The question arises which model is most realistic. The interference effects observed at low temperatures in the absence of a magnetic field (21) are in conflict with perfect adiabatic transport, and thus would favor the second theoretical model (which predicts transmission resonances

---

[2] *cf.* our discussion (10) of the interesting series resistance experiment by Wharam *et al.* (22).

(17..19)). However, the single point contact conductance, which measures the total transmission probability, is not the most sensitive tool to discriminate between the two models. Experiments with two point contacts, acting as injector and collector, are much better suited to this purpose. This includes electron focusing experiments (4) and measurements on point contacts in series (22). Coherent electron focusing experiments in a weak magnetic field have unequivocally demonstrated that for extremely narrow point contacts ($W < \lambda_F$), where diffraction is important, a range of magnetic edgestates (see below) is coherently excited (4). This is a clear example of non-adiabatic transport. On the other hand, an interesting phenomenon has recently been found, which demonstrates that adiabatic transport can indeed occur. This is the anomalous quantum Hall effect, which is observed in the electron focusing geometry (23,4), for somewhat wider point contacts in a strong magnetic field. It was found that quantum Hall plateaux in that geometry are due to *selective* excitation and detection of Landau levels by the quantum point contacts. The observation of adiabatic transport between two adjacent point contacts indicates that the potential landscape was sufficiently smooth on the scale of the magnetic length $l_m = (\hbar/eB)^{1/2}$, which in a high magnetic field takes over the role of λ_F.

We pause to discuss the influence of a finite temperature. In ballistic transport, its most important effect is the smearing of the Fermi-Dirac distribution. Two regimes can be distinguished. If $kT \ll \delta E$, the energy difference between successive 1D subbands at the Fermi level, the major effect is to reduce the interference structure (21). In the opposite limit, the 1D subband structure is no longer important, and the semi-classical result (5) (and its 3D counterpart) should be adequate. A finite temperature also affects the coherence length related to inelastic scattering, which is the dominant effect in the diffusive transport regime (1). So far, no evidence has been found which shows that this is equally important for quantum ballistic transport. [Of course, at elevated temperatures (above 10 K) inelastic scattering reduces the transport mean free path, and thus induces a gradual transition to diffusive transport.] This suggests an optimum temperature for well defined plateaux in the point contact conductance, as was indeed observed experimentally (about 0.6 K) (3,5).

2.5 Skipping Orbits And Traversing Trajectories

In this section we give a brief summary of magnetic field effects (see also (24)). It is sufficient to consider electrons at the Fermi level since only they contribute to the non-equilibrium current. A magnetic field affects the nature of the trajectories of these electrons. In the bulk of the 2DEG the electrons move in cyclotron orbits with a radius $l_{cycl} = mv_F/eB$, and at the 2DEG boundary in skipping orbits with the same radius of curvature. In the narrow constriction skipping orbits confined to a single boundary, and traversing trajectories (25,24) interacting with both boundaries coexist if $2l_{cycl} > W$. The two-terminal resistance of the entire sample is essentially a bottle neck problem. In the absence of a magnetic field the dominant bottle neck is the point contact. However, once $2l_{cycl} < W$, an even narrower bottle neck is constituted by the contact between the alloyed ohmic contact, and the wide 2DEG regions. The reason is that the cyclotron orbits do not contribute to the current in ballistic transport because of their circular motion, so that the current is carried by electrons in skipping orbits localized within a distance $2l_{cycl}$ from the boundary. The role of the point contact width W is thus gradually taken over by that of the cyclotron radius. This classical argument is substantiated by a treatment in terms of depopulation of subbands (3,24), which gives for strong fields $2l_{cycl} < W$

$$ G = \frac{2e^2}{h} N_L \quad , \tag{8}$$

with $N_L \approx E_F/\hbar\omega_c$ the number of occupied edge states, corresponding to an equal number of bulk Landau levels below the Fermi level. A formula describing the point contact conductance in weak and strong fields has been given elsewhere (3,4). Note that if a potential barrier is present in the constriction $N_L \approx (E_F - E_o)/\hbar\omega_c$, so that in that case the conductance is limited by the point contact, even in a strong magnetic field.

As mentioned before, adiabatic transport is at least approximately realized in a strong magnetic field. The energy of the edge states can be separated in a part labeled by the quantum number $n = 1,2,...,N_L$, corresponding to the quantized circular motion, and a part corresponding to the motion of the guiding center of the skipping orbit along the 2DEG boundary: $E_G = E_F - (n - \frac{1}{2})\hbar\omega_c$. Because of energy conservation, the adiabatic motion of the guiding center is thus along equipotentials. This picture replaces that of the classical trajectories, appropriate in weak magnetic fields [3] .

3. ELECTRON BEAMS AND MULTI-TERMINAL MEASUREMENTS

So far we have concentrated on the two-terminal conductance of a single point contact. An important distinction between two- and four-terminal measurements of the conductance has to be made if a magnetic field is applied perpendicular to the 2DEG. A novel negative

---

[3] Tunneling between right- and left-moving edgestates localized at opposite boundaries can occur at the potential steps at entrance and exit of the constriction. This mechanism was invoked in Ref. 26 to explain an Aharonov-Bohm effect in the point contact resistance.

magnetoresistance effect was found in four-terminal measurements, and it was explained in terms of the reduction of backscattering by the point contact on increasing the magnetic field. This effect follows from the nature of the electron trajectories, discussed earlier (27). In geometries with more point contacts an even richer magnetotransport behavior can be expected (10), as is evident from electron focusing experiments (4). In this section we illustrate the crucial role of the measurement set-up, and the injection of ballistic electrons in a geometry with point contacts in series (see Fig. 3). The physics is similar to that of collimated molecular beams (28). The 2DEG regions between the point contacts can be equipped with ohmic contacts, and we assume that near these contacts a local equilibrium has been established as a result of inelastic scattering. The following discussion is based on the Landauer-Büttiker formalism, which treats transport as a transmission problem (this can be done equally well for classical as for quantum mechanical transport problems). The ohmic contacts (and a region of 2DEG around them) act as *reservoirs* (6,7) with well-defined electro-chemical potentials. For a two-terminal measurement current and voltage probes [4] coincide, and the obtained conductance reflects the total transmission probability. We refer the chemical potentials to a zero ground level. The terminals at ground level act as sinks for electrons which are not *directly* transmitted through the point contacts (like the pumps in a molecular beam apparatus). Another terminal, connected to a 2DEG region behind a point contact acting as injector, is maintained at a chemical potential μ_i. The resulting electron beam is detected by measuring the potential of a terminal attached to a 2DEG region behind a point contact acting as collector. Alternatively, one can measure the collector current. The transmission through the individual point contacts can also be measured separately, thereby allowing a characterization of their properties.

The symmetry properties of multi-terminal measurements in the presence of an external magnetic field have been discussed by Büttiker (7). A four-terminal measurement can be interpreted as a generalized longitudinal resistance measurement if the current flow does not intersect an imaginary line between two voltage probes, and as a generalized Hall resistance measurement otherwise. Three-terminal measurements correspond to one or the other, depending on the sign of the magnetic field. Experimentally, this has been demonstrated in an electron focusing experiment (4). A theoretical description of transport in multi-terminal geometries can be based on the Büttiker formula

---

[4] A current probe is defined as an ohmic contact with an externally maintained electrochemical potential, while the electro-chemical potential on a voltage probe follows from the condition that no *net* current flows in the probe.

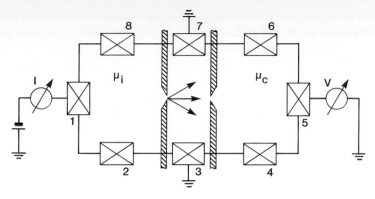

Fig. 3 A geometry with two point contacts in series. Electrons are injected in the 2DEG by the first point contact, and collected by the second. The ohmic contacts 3 and 7 act as sinks for electrons which are not directly transmitted. This structure has also potential use as a novel hot-electron transistor (see sec. 4).

$$\frac{h}{2e} I_\alpha = (N_\alpha - R_\alpha)\mu_\alpha - \sum_{\beta \neq \alpha} T_{\beta \to \alpha}\, \mu_\beta \quad , \tag{9}$$

which relates the current in lead α (with N_α quantum channels) to the chemical potentials μ_β of each of the reservoirs, via transmission and reflection probabilities $T_{\beta \to \alpha}$ (from reservoir β to α) and R_α (from reservoir α back to the same reservoir). These equations may be solved using symmetry relations, and a model (classical or quantum mechanical) for the various transmission and reflection probabilities. This approach has recently been applied to a wide range of phenomena: the quenching of the Hall effect (25), the quantum Hall effect (29,27,23,4,30), propagation around a bend (31), point contacts in series (10), and classical and coherent electron focusing (4). A detailed discussion of these phenomena is beyond the scope of this paper. As an illustration, we apply Eq. (9) to the geometry of Fig. 3.

First, we consider the injection of a total current I. According to Eq. (9), we have

$$\frac{h}{2e} I = \frac{h}{2e^2} G_i \mu_i - T_{c \to i}\, \mu_c \quad , \tag{10}$$

Here we used the definition of the conductance of the injector $G_i = (2e^2/h)(N_i - R_i)$. [Eq. (7) is recovered if $N_i - R_i = N$, the number of subbands occupied in the point contact itself]. The equation for the collector gives us

$$0 = \frac{h}{2e^2} G_c \mu_c - T_{i \to c}\, \mu_i \quad . \tag{11}$$

Only the *direct* transmission probability $T_{i \to c}$ enters here, because all other reservoirs are connected to ground. We thus find for the collector voltage (normalized by the current)

$$\frac{V_c}{I} = \frac{2e^2}{h} \frac{T_{i \to c}}{G_c G_i - \delta} \quad , \tag{12}$$

with $\delta = (2e^2/h)^2 T_{i \to c} T_{c \to i}$. Note the explicit injector-collector reciprocity (7,4). This simple but powerful result has also served as a starting point for a treatment of low field electron focusing (where $T_{i \to c}$ is an oscillating function of B for one field direction, and 0 for the other), and of the anomalous quantum Hall effect in the same geometry (where $T_{i \to c}$ follows from the assumption of adiabatic transport) (23,4). Quite generally $\delta \ll G_i G_c$, as is certainly true for electron focusing, where $T_{c \to i} \approx 0$, but also for narrow point contacts in series because $T_{i \to c} \ll N_i - R_i$. If this is the case, the collector voltage is linearly proportional to the transmission probability $T_{i \to c}$. The geometry of Fig. 3 permits a straightforward determination of the *direct* transmission probability, in contrast to the series resistance experiment of Wharam (22), where the analysis is complicated due to the contribution of indirectly transmitted electrons (10).

The direct transmission probability for the geometry of Fig. 3 can be estimated classically, along the lines of Ref. 10. In the absence of a magnetic field this is quite simple: for abrupt identical point contacts with N quantum channels, separated by a distance $L \gg W$, one finds from the geometry $T_{i \to c} = N(W/2L)$. We note that the transmission probability can be enhanced by flaring of the constrictions, and the introduction of a potential barrier in the point contacts, because of their collimating effects (10). Finally, in the opposite limit of adiabatic transport through identical point contacts with N quantum channels, one would have $T_{i \to c} = N$, which is an upper limit for the direct transmission probability (32). We expect that an external magnetic field deflects the injected electron beam away from the collector, causing a sudden reduction in $T_{i \to c}$ (10). It would be of interest to study experimentally the effects of a magnetic field on the collector voltage in the geometry of Fig. 3, and in different four- or two-terminal configurations (10). Such a study could provide direct experimental evidence for the horn and barrier related collimation effects originally proposed in Ref. 10. Very recently, it has been argued (33) that horn collimation effects are the primary cause for the quenching of the Hall effect in multi-probe electron wave guides (34), by the enhancement of the transmission probability along the channel, at the expense of that into the side probes (7,25). Barrier collimation effects were not invoked in Refs. (33), but may well be important. The quenching of the Hall effect in the experimental systems thus seems to be in essence a classical and not a quantum mechanical effect (25). The role of collimation in coherent electron focusing is discussed in (35).

4. NON-LINEAR TRANSPORT

The current-voltage characteristic of a semi-classical point contact is expected to be non-linear if the voltage drop across the point contact is no longer small compared to the Fermi energy. This is a consequence of the Fermi-Dirac statistics, which shows up as an interdependence of the Fermi velocity and the electron gas density ($mv_F = \hbar k_F = \hbar(2\pi n)^{1/2}$ in 2D, with k_F the Fermi wave vector). For a finite voltage drop across the point contact, electrons in an energy interval eV contribute to the current, and one has to integrate over velocity magnitude and direction

$$ J = \frac{1}{2}D(E_F)W \int_{E_F-eV}^{E_F} (\frac{2E}{m})^{1/2}dE \, \frac{1}{\pi} \int_{-\pi/2}^{\pi/2} \cos \phi \, d\phi \quad . \tag{13} $$

The resulting current is no longer linear in the applied voltage

$$ I = eD(E_F)\frac{W}{\pi} \frac{2}{3}(\frac{2}{m})^{1/2} [E_F^{3/2} - (E_F - eV)^{3/2}] \quad . \tag{14} $$

This result applies for $eV \leq E_F$. For $eV \ll E_F$ Eq. (5) is recovered. For $eV \geq E_F$ this argument predicts a current limited by a saturation value reached at $E_F = eV$. This saturation is a consequence of our assumption that the entire voltage drop is across the point contact, with neglect of any accelerating fields outside the point contact region. At large voltages this is probably not a realistic assumption, and the actual electric field distribution should be determined self-consistently. However, the fact that non-linear transport can arise, as a consequence of the degeneracy of the electron gas, is true regardless of the sophistication of the model used, and is a distinguishing feature between the physics of Sharvin point contacts and classical Knudsen flow.

Non-linear transport through metallic point contacts has been widely investigated because of the possibility to observe phonon related structure in the second derivative of the I-V characteristics. This is known as point contact spectroscopy (9). Since typical phonon energies are of the order of 30 meV and the Fermi energy in a metal is several eV, the non-linearity of Eq. (14) does not play a role in metals. In a 2DEG, where E_F is typically 10 meV the situation is reversed, and these effects have so far obscured possible structure due to inelastic scattering processes. The arguments of this section (still neglecting self-consistency) are readily generalized to the quantum case. The associated breakdown of the conductance quantization of point contacts has been discussed elsewhere (36,37).

The effect of an abrupt potential barrier of height E_o in the constriction can be taken into account heuristically, by replacing E_F by $E_F - E_o$ in Eq. (14). A realistic description would have to include the detailed shape of the barrier, and its dependence on the applied bias voltage. Also, it is important to note that the constriction is electro-statically defined by means of an external gate voltage, referred to the potential of one of the ohmic contacts. An appreciable voltage drop across the sample will therefore also affect the effective gate potential, and thereby the shape and barrier height of the constriction .

Due to the presence of a barrier in the constriction, the point contacts can be used as hot-electron injectors. The non-equilibrium velocity distribution of the injected electrons has been studied in an electron focusing experiment (35,38). One could exploit this by using the geometry of Fig. 5 as a novel *hot-electron transistor*. Williamson and Molenkamp [39] have proposed, as an example, that current amplification can be realized by arranging that the point contact spacing is of the order of a few interaction lengths. Avalanche electron multiplication can then occur in the region between the point contacts. If the second point contact is just pinched-off, the hot electrons created in that proces are transmitted to region 4..6, causing an excess replenishing current in region 3-7.

5. CONCLUDING REMARKS

As we have tried to show, both the trajectory and the mode picture give valuable insight in the physics of quantum ballistic transport. Specifically, we have proposed (10) that the presence of a reduced carrier density in narrow constrictions, and the flaring of their opening into a horn, has a dramatic effect on ballistic magneto-transport effects, because of the associated collimation of the injected electron beam. An exploitation of these effects, and of hot-electron injection by means of quantum point contacts, is likely to lead to a new class of ballistic electron devices. We conclude by listing a few areas deserving further attention. It would be of interest to model the shape of the potential landscape around the point contact, and at the 2DEG boundary, on the basis of a self-consistent solution of Poisson's equation. Charge-transfer effects between the narrow constriction and the wide 2DEG regions can be expected to play a role in strong magnetic fields. The role of spin has not been discussed in this paper, but there are significant indications that it causes interesting anomalies in high magnetic fields, possibly due to a discrepancy in the value of the g-factor in the point contacts and in the wide 2DEG regions (4). Non-linear transport clearly deserves further experimental and theoretical study. It is obvious that the physics of hot-carrier transport in layered structures ("vertical" transport) and the device concepts developed in that field (such as hot-electron spectroscopy, resonant tunneling hot-electron devices) (40) may be readily translated into similar concepts and devices in the parallel ballistic transport regime discussed here. Finally, we note that quantum ballistic transport is not a branch of "mesoscopic" physics, even though it deals with sub-micron structures (16).

The experimental realization of a device with *predictable* ballistic quantum interference effects therefore remains an important challenge.

REFERENCES

1. *Physics and Technology of Submicron Structures* H. Heinrich, G. Bauer and F. Kuchar, eds. (Springer-Verlag, Berlin, 1988).
2. Yu.V. Sharvin, Zh.Eksp.Teor.Fiz. **48**, 984 (1965) [Sov.Phys.JETP **21**, 655 (1965)]; see also V.S. Tsoi, Pis'ma Zh.Exp.Teor.Fiz. **19**, 114 (1974) [JETP Lett. **19**, 70 (1974)].
3. B.J. van Wees *et al.* Phys. Rev. Lett. **60**, 848 (1988); Phys. Rev. B **38**, 3625 (1988).
4. H. van Houten *et al.* Europhys. Lett. **5**, 721 (1988); C.W.J. Beenakker, H. van Houten and B.J. van Wees, *ibid.* **7**, 359 (1988); H. van Houten *et al.* Phys. Rev. B., to be published.
5. D.A. Wharam *et al.* J.Phys.C **21**, L209 (1988).
6. R. Landauer, IBM J.Res.Dev. **1**, 223 (1957); Z.Phys.B. **68**, 217 (1987).
7. M. Büttiker, Phys. Rev. Lett. **57**, 1761 (1986); IBM J. Res. Dev. **32**, 317 (1988).
8. M. Knudsen, *Kinetic Theory of Gases* (Methuen, London, 1934).
9. A.G.M. Jansen, A.P. van Gelder and P. Wyder, J. Phys. C **13**, 6073 (1980).
10. C.W.J. Beenakker and H. van Houten, Phys. Rev. B., to be published.
11. N.S. Kapany, in J. Strong, *Concepts of Classical optics* (Freeman, San Francisco, 1958).
12. L.D. Landau and E.M. Lifshitz, *Mechanics* (Pergamon, Oxford, 1976).
13. H. van Houten, B.J. van Wees and C.W.J. Beenakker, in Ref. 1.
14. D.S. Fisher and P.A. Lee, Phys. Rev. B. **23**, 6851 (1981); see also A.D. Stone and A. Szafer, IBM J.Res.Dev. **32**, 384 (1988).
15. L.I. Glazman *et al.* Zh.Eksp.Teor.Fiz. **48**, 218 (1988) [JETP Lett. **48**, 238 (1988)].
16. Y. Imry, in *Directions in Condensed Matter Physics,* Vol. 1, edited by G. Grinstein and G. Mazenko (World Scientific, Singapore, 1986) page 102.
17. E.G. Haanappel and D. van der Marel, Phys. Rev. B, to be published.
18. A. Szafer and A.D. Stone, Phys. Rev. Lett. **62**, 300 (1989).
19. G. Kirczenow, Solid State Comm. **68**, 715 (1988).
20. A. Kawabata, unpublished.
21. B.J. van Wees *et al.* unpublished.
22. D.A. Wharam *et al.* J.Phys.C. **21**, L887 (1988).
23. B.J. van Wees *et al.* Phys. Rev. Lett., **62**, 1181 (1989).
24. C.W.J. Beenakker, H. van Houten and B.J. van Wees, Superlattices and Microstructures **5**, 127 (1989).
25. C.W.J. Beenakker and H. van Houten, Phys. Rev. Lett. **60**, 2406 (1988); F.M. Peeters, *ibid.* **61**, 589 (1988).
26. P.H.M. van Loosdrecht *et al.* Phys. Rev. B. **38**, 10162 (1988).
27. H. van Houten *et al.* Phys. Rev. B **37**, 8534 (1988).
28. N.F. Ramsey, *Molecular Beams* (Clarendon, Oxford, 1956).
29. M. Büttiker, Phys.Rev.B. **38**, 9375 (1988); *ibid.* **38**, 12724 (1988).
30. S. Washburn *et al.* Phys. Rev. Lett. **61**, 2801 (1988); R.J. Haug *et al.* Phys. Rev. Lett. **61**, 2797 (1988); S. Komiyama *et al.* unpublished.
31. G. Timp *et al.* in Ref. 1; Y. Takagaki *et al.* Solid State Comm. **68**, 1051 (1988).
32. The series resistance of point contacts has been studied numerically by S. He and S. Das Sarma, unpublished.
33. H.U. Baranger and A.D. Stone; A.M. Chang and T.Y. Chang; Ford *et al.*, unpublished.
34. M.L. Roukes *et al.*, Phys. Rev. Lett. **59**, 3011 (1987); C.J.B. Ford *et al.*, Phys. Rev. B **38**, 8518 (1988); see also G. Timp, in *Mesoscopic Phenomena in Solids,* P.A. Lee, R.A. Webb and B.L. Altshuler, eds. (Elsevier Science Pub., Amsterdam) to be published.
35. C.W.J. Beenakker, H. van Houten and B.J. van Wees, Festkörperprobleme/ Advances in Solid State Physics, **29**, U. Rössler, ed. (Pergamon/Vieweg, Braunschweig, to be published).
36. L.P. Kouwenhoven *et al.* subm. to Phys. Rev. B.
37. L.I. Glazman and A.V. Khaetskii; P.F. Bagwell and T.P. Orlando; unpublished.
38. J.G. Williamson *et al.* unpublished.
39. J.G. Williamson and L.W. Molenkamp, unpublished.
40. A. Palevski *et al.*, unpubl.; see also J.R. Hayes and A.F.J. Levi, IEEE J. Quant. Electr. **QE22**, 1744 (1986); M. Heiblum, *et al.* Phys. Rev. Lett. **56**, 2854 (1986).

QUANTUM BALLISTIC TRANSPORT IN HIGH MAGNETIC FIELDS

B. J. van Wees, L.P. Kouwenhoven, E.M.M. Willems, and C.J.P.M. Harmans

Delft University of Technology, 2600 GA Delft, The Netherlands

J.G. Williamson

Philips Research Laboratories, 5600 JA Eindhoven, The Netherlands

1. INTRODUCTION

Quantum point contacts (QPCs) in a two-dimensional electron gas (2DEG) have proven to be valuable tools for the study of quantum ballistic transport. As their most striking property QPCs show conductance quantization in zero magnetic field[1,2]. The application of a magnetic field results in a gradual transition from this electric quantization (due to the lateral confinement of the carriers) to magnetic quantization[1]. In this paper we focus on the transport through QPCs in quantizing magnetic fields. It will be shown that QPCs can be used to study the electron transport in the wide two-dimensional electron gas (2DEG) to which they are attached. This transport can be studied by using the QPCs as controllable current and voltage probes, which can selectively inject current into specific Landau levels (magnetic edge channels to be more precise) or selectively detect electrons in specific Landau levels.

In the second section we give a basic description of quantum ballistic transport in magnetic edge channels. Section 3 deals with the recently discovered anomalous integer quantum Hall effect[3,4], which can be seen as a consequence of the selective population and detection of edge channels combined with the absence of scattering between these edge channels. Section 4 discusses experiments which show that inter-edge channel scattering can be weak, even on macroscopic length scales[5].

In the last section we show the observation of discrete electronic states in a one-dimensional electron interferometer[6]. We have combined the one-dimensional nature of transport in magnetic edge channels with the controllable transmission of QPCs to fabricate a one-dimensional electron interferometer. In this way we studied the transition from one-dimensional transport to transport through discrete, zero-dimensional states.

2. TRANSPORT IN HIGH MAGNETIC FIELDS

In high magnetic fields the description of electron transport in a 2DEG is relatively simple. The motion of electrons with the Fermi energy E_F, relevant for the transport, is determined by their guiding energy $E_G = E_F - (n-1/2) \hbar\omega_c$ (We ignore spin-splitting for the moment). Electrons belonging to different Landau levels (having different n) flow along different equipotentials $V(x,y)$, which are determined by the condition:

$$-eV(x,y) = E_G \qquad (1)$$

We can model the electrostatic potential of the 2DEG by a flat part in the bulk, and a rising part near the edges. This implies that condition (1) is usually satisfied near the edge of the 2DEG, and one may speak about transport through edge channels. This picture is obviously highly simplified. However, it gives valuable insight into the basic phenomena of transport in high magnetic fields.

Equation (1) also shows that a potential barrier V_B in the 2DEG will reflect those edge channels for which $E_G < V_B$ and will transmit those edge channels with $E_G > V_B$. Such a potential barrier can be formed by a QPC, which is defined by a split-gate on top of the 2DEG. These gates define a saddle-shaped potential, of which the barrier height can be controlled by the applied gate voltage. The two-terminal conductance of a QPC in high magnetic fields can now be written as:

$$G_{2t} = 2e^2/h \ (N + T) \qquad (2)$$

in which N illustrates the number of fully transmitted edge channels, and T the (partial) transmission of the upper edge channel. Quantization of the two-terminal conductance occurs in those magnetic field and gate voltage intervals for which T=0. When used as a current probe, attached to a wide 2DEG, a QPC will inject current into only those edge channels which are transmitted through the QPC. Similarly, when used as a voltage probe, a QPC only measures the occupation of those edge channels which are transmitted through the QPC.

3. ANOMALOUS INTEGER QUANTUM HALL EFFECT

We have investigated the quantum Hall effect in the ballistic regime[3] in the geometry shown in fig. 1. Gates A, B and C define two adjacent point contacts, with separation 1.5 μm. This is within the elastic mean free path ($\approx$ 10 μm in our device). Contacts 2 and 4 are current probes, contacts 1 and 3 measure the Hall voltage. Fig. 1. illustrates the current flow in edge channels for the case of two occupied (spin-degenerate) Landau levels. When the gate voltage is such that both edge channels are transmitted through the QPCs, their presence does not affect the electron transport and the regular quantization, corresponding to two occupied Landau levels, is expected: $G_H = 4e^2/h$.

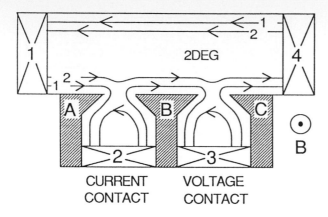

Fig. 1   Current flow in magnetic edge channels, resulting
in an anomalous quantization of the Hall conductance.

Fig. 1 illustrates the case where both QPCs only transmit a single edge channel. In this case the current QPC only injects current into the first edge channel. *In the absence of scattering between the edge channels in the region between the QPCs,* the total current I will enter the voltage QPC. A voltage will build up to compensate this current with an equal out-going current. Since only one channel is available, this voltage is given by $V_H = I\,h/2e^2$, which gives the result $G_H = 2e^2/h$. We therefore have the surprising result that the Hall conductance may show quantized plateaux which are not related to the number of occupied Landau levels in the 2DEG. This can be understood by the fact that edge channels which are neither populated by the current contact, nor detected by the voltage contact, are irrelevant for the electron transport. However, a necessary condition is the absence of scattering between edge channels, to prevent current "leaking away" into initially not populated edge channels. A detailed analysis, given in Ref. 3, shows that when the QPCs do not transmit an equal number of edge channels, the quantization is determined by the QPC which has the largest conductance.

Fig. 2 shows a comparison between the measured two-terminal conductances of the QPCs and the Hall conductance. As can be seen the Hall conductance closely follows the two-terminal conductances, showing a quantized plateau at $2e^2/h$, despite the presence of two occupied Landau levels. The agreement with the theoretical description given above proves that the QPCs do indeed act as selective probes, and that inter-edge channel is extremely weak in this device.

This anomalous QHE will be destroyed when inter-channel scattering is introduced. This is in contrast with the regular QHE, which is not affected by scattering between adjacent edge channels[7] (provided they are located at the same boundary of the 2DEG). Because all edge channels are occupied equally in this case, the scattering rate from edge channel A to edge channel B will be compensated by an equal scattering rate from channel B to A.

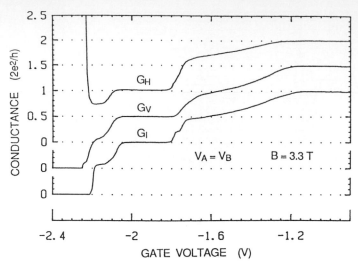

Fig. 2 Comparison between the measured conductances G_I and G_V of the probes with the Hall conductance G_H, showing the anomalous QHE (from Ref.3)

4. ABSENCE OF SCATTERING BETWEEN EDGE CHANNELS ON MACROSCOPIC LENGTH SCALES.

The previous section raises the question on what length scales an unequal occupation of edge channels can persist. We have obtained information about the scattering processes between edge channels by studying the Shubnikov-de Haas (SdH) resistance oscillations, using a QPC as a controllable voltage probe[5].

In the Landauer formulation of electron transport[8](longitudinal) resistance arises from the backscattering of electrons. In the presence of a quantizing magnetic field this backscattering takes place from one edge channel to another edge channel on the opposite edge of the 2DEG, which carries current in opposite direction. Since the SdH oscillations occur each time a Landau level crosses the Fermi energy, we can describe this resistance as scattering between the two opposite edge channels belonging to the upper Landau level.

Fig. 3 shows how this resistance can be suppressed by the selective detection of edge channels by a QPC. This is illustrated for the case of two occupied Landau levels. Bulk contact 2 injects current into the two available channels. This current flows towards contact 4 along the upper edge of the 2DEG. If the magnetic field is such that SdH resistance occurs, part of these electrons will be scattered into the n=2 left-going edge channel. When the QPC transmits both edge channels, a finite current will enter the QPC, and a finite voltage (and resistance) is measured between contacts 4 and 5. In fig. 3 the QPC only transmits the first edge channel, into which initially no electrons have been scattered. In this case only a voltage is measured when scattering is present between the n=2 and n=1 edge channels.

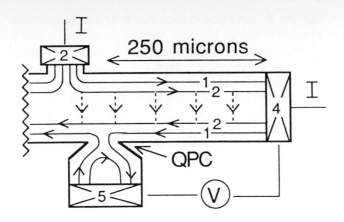

Fig. 3 Suppression of SdH oscillations due to selective detection of edge channels
(from Ref. 5)

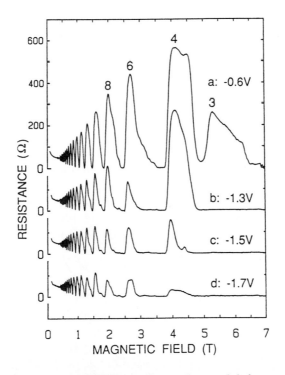

Fig. 4 Measured SdH oscillations. Curves b,c, and d show suppressed SdH
oscillations (from Ref.5)

Experimental results are presented in Fig. 4. It has been indicated which Landau level is responsible for the SdH resistance. At $V_G = -0.6V$ the QPC transmits all edge channels, and a regular SdH trace is observed. When the number of transmitted edge channels is reduced by a decreasing gate voltage, a dramatic suppression of the resistance occurs for fields above 1 Tesla.

These results show that an unequal occupation of edge channels, induced by the SdH oscillations, can persist on length scales larger than 250 μm. Because of this non-locality of the electron transport on macroscopic length scales, we have to draw the important conclusion that the resistance of a macroscopic 2DEG in high magnetic fields can not be defined without specifying the properties of the probes.

In relation to the quantum Hall effect, our results show that contacts do not only play an important role in small samples[7], but also in macroscopic devices. Because of the lack of equilibration between edge channels, a set of non-ideal (not coupling equally to all available edge channels) contacts will produce deviations from exact quantization. The fact that the Hall resistance is usually quantized to an extreme accuracy therefore warrants an investigation into the physics of contacts.

5. OBSERVATION OF DISCRETE ELECTRONIC STATES IN A ONE-DIMENSIONAL ELECTRON INTERFEROMETER.

An elementary way to study discrete, zero-dimensional (0-D) states is to construct a one-dimensional electron interferometer. The insertion of two (impenetrable) barriers into a one-dimensional conductor will lead to the formation of zero-dimensional states. These will be formed at those energies for which constructive interference occurs between the electron waves travelling inside the barriers. By making these barriers partially transparent, these states can then be studied by transport measurements. The conductance will show maxima due to resonant transmission, each time the energy of a 0-D state coincides with the Fermi energy E_F.

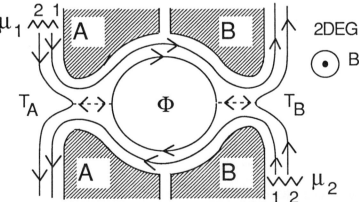

Fig. 5   Current flow in the disk. The second edge channel forms a 1D interferometer (from Ref. 6)

We have constructed such an interferometer by taking advantage of the one-dimensional nature of transport through magnetic edge channels. In Fig. 5 the gate pairs A and B define a disk-shaped 2DEG (diameter 1.5μm), to which QPCs are attached. In this device the one-dimensional conductor is formed by a magnetic edge channel inside the disk, and the barriers are formed by the QPCs. Their transmission can be controlled individually by the applied gate voltage. This device has the additional advantage that the 0-D states can be swept through the Fermi energy by changing the magnetic flux which penetrates the disk.

Fig. 5 shows the current flow in the device for the case of two occupied spin-split Landau levels. In this example the first edge channel is fully transmitted through the device, and gives a fixed contribution e^2/h to the conductance. A one-dimensional interferometer is formed by the second edge channel. In general the conductance of the device can be written as the sum of quantized conductance of the N fully transmitted edge channels and the conductance of a one-dimensional interferometer[9,10]:

$$G_D = \frac{e^2}{h} \left(N + \frac{T_A\, T_B}{1 + (1-T_A)\,(1-T_B) - 2\,\sqrt{(1-T_A)(1-T_B)}\,\cos(\vartheta)} \right) \qquad (3)$$

in this expression T_A, T_B denote the partial transmission of the upper edge channel through the QPCs, and ϑ gives the phase acquired by an electron in one revolution around the disk. Note that the relation between this phase and the flux Φ is given by $\vartheta = 2\pi\, e/h\, \Phi$. When T_A and T_B are zero, discrete 0-D states will be formed at those energies for which ϑ equals a multiple of 2π.

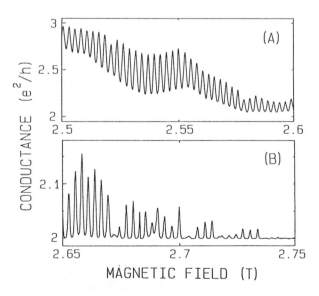

Fig. 6 Resonant transmission through discrete electronic states. (from Ref. 6)

Fig. 6a. shows the oscillations due to the third edge channel, showing the transition from the $3e^2/h$ to the $2e^2/h$ plateau. Extremely regular oscillations are observed, showing a modulation of the channel conductance up to 40%. Fig. 6b. shows this resonant transmission through discete electronic states in the region where the transmission of the third edge channel through both QPCs is low. As a result the transmisssion of the third edge channel through the device is low, except when a 0-D state coincides with the Fermi energy. This shows up in Fig. 6b. as narrow spikes.

6. CONCLUSIONS

In this paper we have shown that quantum ballistic transport in high magnetic fields can be described in terms of edge channels, not only in small devices, but also in a macroscopic 2DEG. The lack of scattering between these edge channels, combined with the edge channel selection of the QPCs, has resulted in the observation of an anomalous quantum Hall effect and the suppression of SdH resistance oscillations. The one-dimensional nature of the transport through edge channels has been used to fabricate a device in which controlled interference occurs between edge channels. This creates possibilities to study elementary electron transport in more complex devices which are made up of one-dimensional conductors.

ACKNOWLEDGEMENTS

We wish to thank C.W.J. Beenakker and H. van Houten for valuable discussions, C.E. Timmering, and M.E.I. Broekaart for making the devices, S. Phelps at the Philips Mask Centre, C.T. Foxon, J.J. Harris at Philips Redhill, and the Delft Submicron Centre for their contribution to the fabrication of the devices. We acknowledge the Stichting F.O.M. for financial support.

REFERENCES

1. B.J. van Wees *et al.* Phys. Rev. Lett. 60, 848 (1988); Phys. Rev. B38, 3625 (1988)
2. D.A. Wharam *et al.* , J. Phys. C21, l209 (1988)
3. B. J. van Wees *et al.* , Phys. Rev. Lett. 62, 1181 (1989)
4. H. van Houten *et al.* , Phys. Rev. B , to be published.
5. B. J. van Wees *et al.* , Phys. Rev. B , to be published ; S.Komiyama *et al.* , preprint
6. B. J. van Wees *et al.* , submitted to Phys. Rev. Lett.
7. M. Büttiker, Phys. Rev. B38, 9375 (1988)
8. R. Landauer, IBM J. Res.Dev. 1, 223 (1957); Z.Phys. B68, 217 (1987)
9. U. Sivan, Y. Imry, and C. Hartzstein, preprint; U. Sivan, and Y. Imry, Phys. Rev. Lett. 61, 1001 (1988), I.K. Jain, Phys. Rev. Lett. 60, 2074 (1988)

NEW THEORETICAL RESULTS ON BALLISTIC QUANTUM TRANSPORT: QUENCHING OF THE HALL RESISTANCE AND QUANTIZED CONTACT RESISTANCE

A. Douglas Stone and Aaron Szafer

Applied Physics, Yale University, New Haven, Connecticut.

Harold U. Baranger

AT&T Bell Laboratories, Holmdel, New Jersey.

1. INTRODUCTION

In this paper we summarize some new theoretical results relating to two novel quantum transport effects recently discovered in ballistic systems(high mobility 2D and quasi-1D GaAs heterostructures): the conductance steps, "quantized" to the value of e^2/h(per spin), observed in 2D systems separated into two regions by a constriction[1,2], and the disappearance or "quenching" of the *low-field* Hall resistance in thin wires[3-6]. In analyzing these phenomena we stress the general principle that in ballistic systems the device geometry can alter and in principle control the electron momentum distribution at the fermi surface to create highly non-equilibrium populations, such as collimated beams, which will then be "detected" by the voltage probes in a manner very different than an equilibrium distribution[7].

2. THE LANDAUER FORMULA AND CONSTRICTION RESISTANCE

We begin by considering the quantized constriction resistance[1,2], an effect that illustrates how the device geometry can act as a filter, but with very different filtering effects, depending on whether the constriction is joined adiabatically or abruptly to the wide regions. The basic experimental facts are the following: 1) A 2D electron gas is divided into two regions by a narrow gate with a lithographically created gap; as the gate voltage is varied, the shape of the constriction is varied (both its width and its potential difference from the 2D region). 2) The low-temperature resistance decreases as the constriction size is increased in a series of steps; after subtracting off a series resistance the height of the steps in conductance is found to be e^2/h (per spin) both in the presence and absence of a magnetic field. 3) Subtraction of a series resistance is sufficient to yield steps of height e^2/h in two, three and four-probe measurements (in zero magnetic field), showing that the basic effect is insensitive to the nature of the measuring probes as long as they are in the wide regions.

The physical origin of this effect can be easily understood by considering the Landauer formula for an ideal two-probe measurement[8,9]. An ideal two-probe measurement is one in which the sample is attached between two perfect reservoirs with electrochemical potentials, μ_1 and $\mu_2 = \mu_1 + eV$ respectively (where V is the applied voltage) and these reservoirs serve both as current source and sink *and* as voltage terminals. In the energy interval eV between μ_2 and μ_1 electrons are injected into right-going states emerging from reservoir 1, but none are injected into left-going states emerging from reservoir 2. Thus there is a net right-going current proportional to the number of states in the interval $\mu_1 - \mu_2$, given by

$$I = e \sum_i^{N_c} v_i \frac{dn_i}{d\varepsilon} eV \sum_j^{N_c} T_{ij} = (e^2/h \sum_{i,j}^{N_c} T_{ij})V \equiv \frac{e^2}{h} g V \tag{1}$$

where N_c is the number of propagating channels *in the sample*, v_i is the the longitudinal velocity for the ith momentum channel at the fermi surface, T_{ij} is the transmission probability from j to i, and we have used the fact that for a quasi-1D density of states, $dn_i/d\varepsilon = 1/hv_i$. Eq. (1) yields an expression for the two-probe conductance, g,(measured in units of e^2/h). It should be noted that exactly the same result is obtained from linear response theory when the two reservoirs are replaced by semi-infinite perfect leads at fixed potentials[10].

In the ballistic limit, in which there is no scattering at all in the sample, $T_{ij} = \delta_{ij}$ and $g = N_c$; if the width of the sample is varied so that N_c changes, then g will change in perfectly sharp (θ-function) steps of unit height (at T=0). As first understood by Imry[8], this effect can be interpreted as an ideal contact resistance between the sample and the reservoirs. However the possibility of experimental systems with components behaving very much like perfect reservoirs was be no means clear, and the observability of this ideal contact resistance was not predicted beforehand. Thus, the observation of the effect demonstrated that under certain circumstances wide regions of the sample behave much like the perfect reservoirs envisioned by this argument, and the "two-probe" formula, Eq. (1) is physically-relevant despite its rather idealized assumptions. The properties of a perfect reservoir appealed to in deriving Eq. (1) were the following: 1) It is initially in equilibrium at electrochemical potential μ and this equilibrium is negligibly disturbed by the current flow. 2) Particles entering the reservoirs never return without loss of phase memory. 3) The connection between the reservoir and the sample generates no additional resistance.

To realize properties 1) is not difficult, all that is needed is to make the contacts at which the voltage is measured sufficiently wide compared to the "sample" which is dominating the resistance. However, properties 2) and 3) raised more subtle questions. Certainly there would be a region of size the inelastic length around the constriction which would be phase-coherent with the "sample". Moreover, in general there would be an impedance mismatch at the orifice, which would generate an additional resistance which is largest at threshhold for propagation of a new mode (for reasons to be discussed below), and this would not simply add a constant resistance independent of the width of the constriction. If the connection between the wide and narrow regions were very gradually tapered(adiabatic), these effects would be very small[9,11]; however the experimental system was not fabricated with any attempt to match smoothly the wide regions to the constriction. This raised an important theoretical question: to what extent is gradual, approximately adiabatic matching between the wide and narrow regions essential for observing the effect. We thus decided to study the opposite limit of an abrupt interface[12].

3. ABRUPT CONSTRICTION: THE MEAN-FIELD APPROXIMATION

Many of the essential physical ideas in this limit are illustrated by the solution of the simpler problem of an "infinite constriction", i.e. transmission through a single sharp interface between a very wide (width W) and a narrow (width W') system with walls defined by hard-wall boundary conditions(see insert to Fig. 1). For this idealized system the conductance is given by $g = \sum_n T_n$ where $T_n \equiv \sum_w T_{nw} = \sum_w T_{wn}$, and T_{nw} is the transmission probability to scatter from a mode w in the wide region to a mode n in the narrow region. It is clear from general arguments that each T_n should vary continuously from zero when the fermi energy ε_F is equal to the threshhold energy ε_n for propagation of mode n, to unity far above threshhold. Since $T_{nw} = T_{wn}$ (from time-reversal symmetry) T_n can equally be thought of as the total transmission probability for a wave *emerging* from the constriction

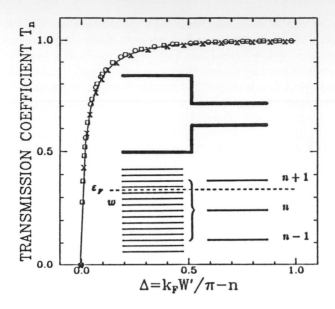

Fig. 1. $T_n(\Delta)$ for WN geometry. Solid line is T_3 in the MFA (Eq. 2), squares are $T(\Delta)$ (Eq. 3). Circles and crosses are numerical results for T_3 and T_4, demonstrating scaling property of exact $T_n(\Delta)$. Inset shows schematically the WN geometry and the threshhold energies for the corresponding regions; brace indicates the channels w contributing to T_n in the MFA.

in mode n, which is clearly going to approach unity for energies far above threshhold. However at, or very near threshhold, such a wave must be totally reflected back into the mode n $(T_n(\varepsilon_n) = 0)$, a result which can be proved from the unitarity relations and the continuity of the wavefunctions with energy[12]. This result is obvious based on the analogy to the transmission coefficient $T_{1D} = 4Kk/(K+k)^2$ above a one-dimensional potential step(where K is the wavevector before the step, k after) since the electron entering the constriction at threshhold will typically need to give up longitudinal energy, and will effectively encounter a barrier at the interface. Thus we expect each T_n to turn on smoothly at ε_n and rise to unity over some interval; in order for $g = \sum_n T_n$ to exhibit sharp steps this turn-on must be rather rapid.

It is possible to calculate T_n in this geometry to arbitrary accuracy by solving the wave-function and derivative matching conditions at the inter-face numerically[12-14]; we were also able to obtain a very good approximate solution analytically[12], which yields further physical insight. The basis of the approximation is the fact that the strongest coupling between modes

w and a given n occurs when the transverse wavevectors q_w, q_n are equal (where $\varepsilon_n \equiv \hbar^2 q_n^2/2m$), and this coupling decays rapidly once $q_w > q_{n+1}$ or $q_w < q_{n-1}$; we thus replace the actual coupling by a uniform coupling for $q_{n+1} > q_w > q_{n-1}$(we refer to this as a *mean-field approximation*(MFA)). In the MFA the initially coupled matching equations decouple and yield the simple result

$$T_n = 4K_n k_n/[(K_n + k_n)^2 + J_n^2].\qquad(2)$$

where $(K_n + iJ_n) = (W'/W)\sum_w^{(n)} k_w$ and the superscript n indicates that only modes with $q_{n-1} < q_w < q_{n+1}$ are summed. Note the exact formal analogy between T_n in the MFA and T_{1D} with the wavevector K replaced by a complex wavevector whose real part K_n is an average longitudinal wavevector for all the propagating modes w with $q_{n-1} < q_w \leq k_F$, and whose imaginary part J_n is an average of the decaying modes w with $k_F < q_w < q_{n+1}$. With some further work it is possible to obtain from the MFA an explicit analytic form for T_n in terms of a dimensionless measure of the deviation from threshhold $\Delta = (k_F W'/\pi) - n$ which has the form

$$T_n(\Delta) \approx T(\Delta) = \frac{12\sqrt{\Delta}(1+\Delta)^{3/2}}{((1+\Delta)^{3/2} + 3\sqrt{\Delta})^2 + (1-\Delta)^3}\qquad(3)$$

i.e. all the steps have the same shape when measured in the natural units as shown in Fig. 1. We have extended the MFA to describe the Wide-Narrow-Wide geometry which is closer to the experimental situation. Here two new effects arise: rounding of the steps due to evanescent modes leaking through the constriction, and resonances due to internal reflection from the interfaces; these effects are discussed in some detail in ref. 12. A significant quantitative result obtained is that the point-contact limit can be approached quite closely before the steps disappear, i.e. a constriction many times as wide as it is long can still generate rather sharp steps in g.

The important points demonstrated by the solution of the abrupt model are: 1) The sharp conductance steps survive completely non-adiabatic matching. 2) Nonetheless the impedance mismatch substantially rounds the rise of the steps even at T=0, and still causes a 0.1% deviation of T_n from unity at threshhold for the rise of the next step. The first point appears quite important for explaining the robustness of the effect, and suggests that this effect is a universal feature of ballistic conduction through a constriction.

The essential physical idea of the MFA used to solve the abrupt limit is that the transverse wavevector is conserved at an abrupt Wide-Narrow interface with an uncertainty given by the spacing (in transverse wavevector) of the modes in the narrow region. If this interface is preceded by a gradual widening which reduces the transverse wavevector and increases the longitudinal momentum, then the interface will collimate the "beam" of electrons injected into the wide region in the forward direction, an effect which will be fundamental to our explanation of the quenching of the Hall resistance.

4. QUENCHING OF THE HALL RESISTANCE: FORMALISM

We now discuss the quenching of the Hall resistance, which is very different from the point contact resistance since the geometry of the sample and Hall probes are crucial to the effect. The observation that the low-field Hall resistance, R_H, in thin GaAs wires was suppressed substantially from its 2D value was first reported by Roukes et al[3], and subsequent experiments by Ford et al.[4] and Chang and Chang[5] on gated samples confirmed that the effect was *generic* in the sense that it occurred in differently-fabricated ballistic structures, and persisted over a wide range of carrier densities, typically spanning an interval over which 3-9 subbands were occupied. The experiment of Chang and Chang and a more recent experiment by Ford et al.[6] convincingly demonstrated that the quenching behavior is sensitive to the geometry of the junction to the Hall probes, and can be eliminated[5,6] or even enhanced (i.e. *negative* values of the Hall resistance can be created)[6] by manipulating this geometry.

The natural formalism for treating such a situation, in which correctly describing the sample geometry is crucial for an adequate theory, is the *multi-probe Landauer formula* due to Büttiker[15]. This formula gives the current response of a phase-coherent system connected to N_L reservoirs where any two can serve as current source and sink and voltage can be applied(or induced) between any two. As shown by Büttiker, an argument exactly analogous to that leading to Eq. (1) yields the result

$$I_m = e^2/h \sum_{n}^{N_L} T_{mn} V_n \qquad (4)$$

where I_m is the total current into lead m, V_n is the voltage applied at lead n, and T_{mn} is the total transmission probability (summed over all channels at the fermi surface) for an electron injected at lead n to be collected at lead m. In the case $N_L = 2$ this formula reduces exactly to Eq. (1), for four or more probes, if the T_{mn} are known, it can be inverted to yield the Hall resistance. Although a natural extension of the reasoning leading to Eq. (1), it represents a conceptual departure from the original Landauer formula $g = T/R$ and its multi-channel generalization[16], in that it treats all the leads equivalently and does not require an ideal non-invasive voltage measurement. It is also very appealing physically because it naturally generates the observed reciprocity symmetries for resistance measurements in a magnetic field[15]. Finally it provides a fermi surface expression for the Hall resistance, R_H which is valid in arbitrary magnetic field. This latter property was quite striking in view of the fact that the Hall resistance is more commonly expressed in terms of all the states below the fermi surface. We have proved that Eq. (4) can be derived rigorously in linear response theory for an arbitrary magnetic field[17]. The proof is rather involved and requires addressing several issues which to our knowledge are not treated in the previous literature. However we will not attempt to describe this derivation here; we simply note that given the results of ref.s 10 and 17 an exact equivalence has been established between the Kubo and Landauer approaches as long as Eq. (4) is used to describe the N_L-probe measurement.

5. QUENCHING OF THE HALL RESISTANCE: PHYSICAL MECHANISM

Using this formalism, given a model for the confining potential, we can calculate the Hall resistance of any ballistic structure numerically by means of a recursive Green function method[18]. Earlier theoretical work[19], loosely based on Eq. (4), had hypothesized that the quenching behavior was related to the transition between transport via edge states in high field, to traversing states in low field. This approach suggested a natural cross-over scale in the behavior of R_H which depended on the width of the wires; however no explicit argument was given to prove that R_H *should be quenched* below this crossover field. Since the geometry of the junction never enters in this approach, it is hard for such an approach to account for the results of ref.s 4-6; and we believe that the microscopic calculations presented below rule out an explanation based simply on length scale arguments.

The first model we studied consisted of two uniform wires with width defined by infinite "walls" joined to form a symmetric "square cross" (the same model was considered recently by Ravenhall et. al. [20]) We found that the Hall resistance for this structure *was not generically reduced from its 2D value*; instead it was an oscillatory function of energy (or carrier density) on energy scales of order the sub-band spacing at T=0, and with a mean value very close to the classical 2D value. Although there were some regions of energy in which R_H was suppressed well below its 2D value, these regions only occurred in the first two sub-bands, i.e. below the experimental carrier densities, and even these "quenched" regions were rapidly eliminated by a reasonable degree of temperature-averaging. A similar result was obtained for a model where the hard-wall confinement is replaced by harmonic con-

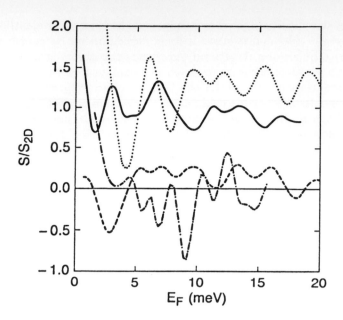

Fig. 2. Plot of average slope of low-field Hall resistance normalized to the 2D value vs. fermi energy ($T \approx 4K$, wire width $\approx 100nm$). Solid line is for hard-wall "square" cross, dotted is harmonic "square" cross; energy of first subband is $\approx 0.5meV$(hard-wall), $\approx 3meV$(harmonic)). Dashed line is hard-wall "graded" cross, dot-dashed is harmonic "graded" cross with realistic grading length.

finement, as first considered by Kirczenow[21]. The results for both models are shown as the top two traces in Fig. 2 ; these results clearly demonstrate that *the quenching of the Hall resistance is not a generic feature of narrow ballistic wires* itself (even with strongly-coupled probes), *but must depend on special properties of the potential at the junction.*

The important physical effect which has been left out of the two models discussed above is the collimation effect due to the (practically unavoidable) widening of the confining potential at the junction. If one plots the equipotential lines for the confining potential, they must run essentially parallel to the x and y axes far away from the junction, and then turn through a 90° angle as they pass through the junction. The hard-wall and harmonic models above have the special and unphysical feature that the equipotential lines run exactly parallel to one axis until they reach a certain distance from the junction, and then abruptly turn 90°. In real systems one expects a gradual rotation over a length comparable to the size of the junction region itself (in the narrowest wires). If this is so, electrons approaching the junction will initially experience an adiabatic widening of the confining potential, which ultimately becomes an abrupt widening as the junction is reached. As discussed above, such an interface will result in collimation of the electrons in the forward direction.

It is easy to see that such a collimation effect is likely to reduce the low-field Hall resistance. The multi-probe Landauer formula for any potential with 90° rotational symmetry gives the expression

$$R_H = (h/e^2)\frac{T_R - T_L}{[2T_F(T_F + T_R + T_L) + T_R^2 + T_L^2]} \tag{5}$$

where T_R, T_L, T_F are the right, left and forward transmission probabilities respectively, and it is assumed that the field is oriented so the Lorentz force is pushing the electrons to the right as they enter the junction. One sees that R_H can be reduced in two ways, either by reducing the asymmetry (caused by the field) between left and right transmission, or by reducing the total sideways transmission, $T_R + T_L$. If the incident electrons are collimated one certainly expects the latter effect, since the forward transmission is enhanced; it is also possible that collimation will reduce the asymmetry, since in the limit where the entire beam is focused forward one expects no asymmetry below a threshhold field. In fact our results indicate that collimation has both effects, and both contribute to the quenching.

In the bottom two traces of Fig. 2 we show the behavior of R_H vs. energy for two structures with this slow grading, one in which the confining potential consists of hard-walls which are slowly widened into the junction, the other in which the confining potential is harmonic, but its steepness is decreased as the junction is approached. One sees that in contrast to the "square" models in which R_H oscillates around its classical value, R_H in the "graded" models oscillates around zero over the entire range of energies studied, corresponding roughly to the occupation of 1-6 sub-bands. For the graded harmonic model, the effective width of the potential is doubled over a distance equal to the initial width, showing that a realistic amount of grading can give rise to generic quenching. We have also studied abruptly widened models (where one does not expect a collimation effect), and found no generic quenching, and have considered other graded models with different types of grading and different degrees of hardness of the confining potential. All of the graded models show generic quenching, whereas none of the square models do. A related result is that analysis of the transmission coefficients of the square cross, which does not show quenching, reveals that the modes with large forward momentum make a very small contribution to R_H; the classical value of R_H is arising almost completely from the high transverse momentum modes that would be unoccupied in the graded models. All of these results combined point conclusively to the collimation effect as the mechanism for the quenching of the Hall resistance in these ballistic systems.

ACKNOWLEDGEMENTS

We acknowledge useful discussions with M. Büttiker, A. Chang, H. van Houten, Y. Imry, M. Roukes, and G. Timp. Work at Yale was supported in part by NSF Grant No. DMR-8658135 and the AT&T Foundation.

REFERENCES

1. B.J.van Wees et al. (1988). Phys. Rev. Lett. 60,848.
2. D.A.Wharam et al. (1988). J. Phys. C21,L209.
3. M.L. Roukes et al. (1987). Phys. Rev. Lett. 59,3011.
4. C.J.B. Ford et al. (1988). Phys. Rev. B38, 8518.
5. A.M. Chang and T.Y. Chang, submitted to Phys. Rev. Lett.
6. C.J.B. Ford et al., submitted to Phys. Rev. Lett.
7. The same point has been stressed by C. Beenakker and H. van Houten, (Phys. Rev. B, to be published); and Y. Imry, (private communication).
8. Y.Imry, (1986). In "Directions in Condensed Matter Physics" (G. Grinstein and G. Mazenko, eds.), p. 101. World Scientific Press, Singapore.
9. R. Landauer (1987). Z. Phys. B68, 217.
10. D.S. Fisher and P.A. Lee (1981). Phys. Rev. B23, 6851.
11. L.I. Glazman, G.B. Lesovick, D.E. Kmelnitskii, R.I. Shekhter (1988). Pys'ma v ZhETF,v.48, No.4.
12. A. Szafer and A.D. Stone (1989). Phys. Rev. Lett. 62, 300.
13. G. Kirczenow (1988). Sol. St. Commun. 68, 715.
14. E.G. Haanappel and D. van der Marel, Phys. Rev. B, to be published.
15. M. Büttiker (1986). Phys. Rev. Lett. 57, 1761.
16. M. Büttiker et al. (1985). Phys. Rev. B31, 6207.
17. H.U. Baranger and A.D. Stone, submitted to Phys. Rev. B; A.D. Stone and A. Szafer (1988). I. B. M. Journal of Research and Development 32,384.
18. H.U. Baranger and A.D. Stone, submitted to Phys. Rev. Lett.
19. C.W.J. Beenakker and H. van Houten (1988). Phys. Rev. Lett. 60, 2406.
20. D.G. Ravenhall, H.W. Wyld and R.L. Schult (1989). Phys. Rev. Lett. 62, 1780.
21. G. Kirczenow, submitted to Phys. Rev. B.

THEORETICAL CONSIDERATIONS FOR SOME NEW EFFECTS IN NARROW WIRES

Yoseph Imry

Department of Nuclear Physics
Weizmann Institute of Science
Rehovot, 76100, Israel

Abstract

The conditions to observe quantized conductance of a ballistic point contact are discussed. These conditions are satisfied in the adiabatic model of Glazman et al.. The corrections to the adiabatic picture for a graded constriction are exponentially small in the smoothness parameter. The collimation effect for an adiabatic horn radiating into space is explained and used to further substantiate the argument of Baranger and Stone for the quenching of the Hall resistance in a graded cross geometry. The intrinsic Hall effect in a wire, which is a well-defined and measurable quantity, should not quench.

1. INTRODUCTION AND SUMMARY

In this paper, a theoretical background will be provided for some interesting effects that have recently been discovered in very narrow ("quasi 1D") conductors. We shall mainly discuss the ballistic regime, where the effective length of the system is shorter than the elastic scattering length. Some remarks will be made concerning the effect of elastic scattering. Inelastic, phase breaking [1], scattering will be assumed to be weak enough so as to be irrelevant. Likewise, the temperature will be taken to be low enough so that the relevant conduction electrons are effectively monoenergetic.

The quantization of the conductance of a ballistic point contact[2,3] will be discussed first. In section 2 we shall demonstrate how this quantization follows from the two-terminal version[4] of the Landauer formula, and emphasize the crucial assumptions necessary for this treatment. This will be discussed in the framework of a model where the channel is weakly coupled to the particle reservoirs. We believe that this model helps clarify the physics of the assumptions involved. To get accurate quantization, however, the above coupling has to be ideally strong, in the sense that, e.g., an electron wave moving along the channel must have probability unity to get out into the reservoir. Glazman et al[5] have recently shown how such smooth transport along and out of the channel is obtained in the "adiabatic" case where the constriction is formed by a slowly changing width of the channel. Exact quantization of the constriction conductance is thus obtained in the adiabatic limit. This limit and the first corrections to it[6] are discussed in section 2. The corrections turn out to be exponentially small in the inverse of the small adiabaticity parameter. We believe that this may well be the main physical reason for the approximate quantization of the ballistic point contact conductance observed in practice. The adiabaticity is due[5] to the electrostatic confinement being smeared in space compared with the lithographic dimensions. Limitations of the two-terminal conductance picture are discussed by Landauer[7]. Elastic scattering in the channel is detrimental to the ideal smooth transport along it. Thus, increasing elastic scattering will provide increasing corrections to accurate quantization. Numerical results[8]

379

show that once the channel length is larger that the elastic mean free path, the universal conductance fluctuations become fully developed. Their size is comparable to the channel opening effect which is still discernible (and, in fact, dominant after ensemble or energy-averaging[8]). In section 3 we also briefly discuss the electron focussing[10], provided by the combined effect of an adiabatic constriction suddenly[11] opening into a wide space. This collimation may be relevant to the Hall effect quenching discussed next.

Since the Hall effect is understood classically as due to the sideways Lorentz force on the moving electrons, one may expect changes in the Hall effect when the lateral confinement by the "walls" of the wire dominates over the Lorentz force. Indeed, Roukes et al[12] and Ford et al[13] found the Hall effect to be "quenched" over significant ranges of the parameters in very narrow high mobility wires. A simple perturbation theory treatment[14] of the "intrinsic" Hall voltage in a narrow ideal wire (discussed in section 4) shows, however, that the intrinsic Hall effect is not quenched and that it is given essentially by the classical result, for a wire which is still wider than the appropriate screening length. By the "intrinsic" Hall voltage we mean in this case the electrostatic potential or the electrochemical potential developed across the wire, both measured by noninvasive, weakly coupled, probes[4,7,14,15]. Spatial averaging over scales larger than the electron wavelength and the screening length is assumed. For a wire narrower than the screening length (but supporting at least one conduction channel) one may talk only about the electrostatic potential difference, which should show quenching. However, we do not believe this limit to be appropriate for the present experiments. Similar results of no quenching with weakly coupled probes had been obtained by Peeters[15]. It has also been found by numerical calculations in Ref.14 that adding disorder to the system will not bring about quenching, so it must be due to something else.

The Hall-effect measurements under discussion[12,13] were not done with weakly-coupled voltage probes, but with the conventional cross geometry in which current and voltage terminals are approximately equivalent. The natural way to treat these experiments is, thus, using Büttiker's[16] multiprobe Landauer-type formula, which is a generalization of the two-probe formula mentioned above. This formula contains the Onsager symmetries appropriate to the multiterminal case[17]. In this formulation the Hall resistance is proportional to the difference between the probabilities T_R and T_L (see Eq.16) of the current-carrying electrons to go into the right- or left- voltage probes, respectively. Beenakker and van Houten[18] were the first to attempt an explanation of the quenching of the Hall resistance using this formulation. They noted that for wire widths much smaller than the size of the edge states[19], T_R and T_L can be physically expected to become equal, and thus the Hall effect will quench. However, the conditions for quenching according to this argument don't agree in detail with experiment. Also, computer simulations[20,21] have not confirmed the quasiclassical prediction of $T_R = T_L$ in wide ranges of the parameters, for sharp boundaries of the cross region. Recently, Baranger and Stone[21] noted that an alternative way to obtain Hall effect quenching is via $T_R = T_L = 0$. This should happen due to the boundaries of the cross region being graded, as in the adiabatic picture discussed above. According to Ref.21, the physical reason for the vanishing of T_R and T_L is that the gradual broadening of the wire selects the low transverse quantum numbers and high longitudinal momenta, which are less likely to turn around the corner into the voltage probe. We believe that this argument has to be strengthened by the focussing considerations[10] discussed in section 3. Those issues are summarized in section 5. Recent experiments[22,23] highlight the relevance of details of the confinement potential near the cross region for the quenching.

2. GENERAL CONSIDERATIONS FOR BALLISTIC POINT CONTACT CONDUCTANCE QUANTIZATION AND A RESONANT-TUNNELING MODEL

To understand this quantization we follow the discussion of Ref.4. The two-terminal Landauer[24-27] type conductance between two particle reservoirs having zero temperatures and chemical potentials μ_1 and $\mu_2 (\mu_1 > \mu_2)$ and connected by a narrow wire, is defined as the ratio of the current between the reservoirs to their voltage $(\mu_1 - \mu_2)/e$. G_2 is given by

$$G_2 = \frac{Ie}{(\mu_1 - \mu_2)} = \frac{e^2}{\pi\hbar}\sum_{ij} T_{ij} \tag{1}$$

The narrow wire is assumed to have ideal sections (see Fig.1a) connected to the reservoirs and straddling an elastic scatterer. T_{ij} is the transmission probability from channel j on the LHS of the scatterer to channel i on its RHS. The physical assumptions involved in obtaining (1) are that a) each reservoir keeps all conduction channels coming out of it full (i.e. in equilibrium with it); b) an electron wave moving along the wire into a reservoir gets absorbed into it and thermalizes there. Note that for an electron with energy below the chemical potential of the absorbing reservoir, a better way of describing what happens is that because of the Pauli principle this electron can not be absorbed into the reservoir, but is rather reflected back. However, the net current from the reservoir is the same as if the electron got absorbed and another one was emitted. Since electrons are indistinguishable, the two descriptions are equivalent. The third important physical assumption, explained and emphasized in the original paper by Landauer[25] is that electrons are emitted (or reflected) from the reservoirs incoherently (i.e. with no phase relationship to one another). It can be demonstrated[28] that the irreversible ohmic dissipation associated with (11) occurs vie the thermalization of carriers in the reservoirs.

The above picture can be regarded as a reasonable description of the narrow contact situation of Refs.2 and 3 (except that the contact is not long enough to treat its longitudinal states as a continuum, but this can be approximately handled by using a voltage which is not too small). The real, nontrivial, assumptions are those of the ideal coupling of the reservoirs to the narrow wire. Non-ideal impedance mismatch might cause coherent reflections in the wire-reservoir interface, that are neglected in the above picture. Below, we shall consider a model which highlights the importance of those assumptions. However, whenever (1) is valid, and if the wire is "ballistic" (i.e. $T_{ij} = \delta_{ij}$), one finds[4] (with spin degeneracy)

$$G_2 = \frac{e^2}{\pi\hbar}N_c \tag{2}$$

Where N_c is the number of conduction channels supported by the ideal portion of the narrow wire. The finite resistance (2) should not[4] be regarded as due to the wire itself (which is "ideal") but to the whole arrangement. It may be interpreted[4] as due to the (idealized, by assumption!) contacts of the wire with the reservoirs. Non-ideal contacts (due to impedance mismatch) will yield values for G_2 smaller than (3), even without impurity scattering.

To expand on the last point we consider the resonant tunneling-type model, depicted in Fig. 1b. On the contacts between the narrow wires (taken to be ideal) and reservoirs, barriers with transmission probabilities $T_1, T_2 \ll 1$ are placed. Since all transmission probabilities between reservoirs 1 and 2 are symmetric, we take, without loss of generality $T_1 \gtrsim T_2$. The quantized states of the closed wire (when $T_1 = T_2 = 0$) become resonant states whose widths are given by $\hbar/\tau_i$, where

$$\tau_i \sim \frac{L}{v_i \max(T_1, T_2)} \tag{3}$$

where L is the length of the wire and v_i the longitudinal velocity of the the channel. Since L is finite the longitudinal states are discrete, to make connection with the Landauer picture we take (still, of course, with $\mu_1 - \mu_2 \ll \mu_1$)

$$\mu_1 - \mu_2 \gg \frac{\hbar v_i}{L} \left(\gg \frac{\hbar}{\tau_i} \right) \tag{4}$$

In this case, all the resonances between μ_2 and μ_1 are fed by 1. Each transverse channel i leads to a current given by $(e/\tau_i)(\mu_i - \mu_2)\frac{L}{\pi \hbar v_i}$ where in τ_i we have to take in this case T_2 in Eq. (3) (denoted by $T_{2,i}$) for channel i since decay into reservoir 1 is blocked. Thus, we find for the total current between 1 and 2

$$I = \frac{e}{\pi \hbar} \sum_i T_{2,i}(\mu_1 - \mu_2) \ll \frac{e}{\pi \hbar} N_c(\mu_1 - \mu_2) . \tag{5}$$

So, the conductance of a ballistic point contact which is poorly matched to the reservoirs is much less than the ideal quantized value. We believe that this example shows very vividly that the assumptions of ideal matching of the wire to the reservoirs in indeed crucial. The question of matching of the "wire" to the reservoirs was first discussed by Landauer[29], who also suggested improving this matching by "flaring" the interface and making it smooth. This forms the subject of the next section. Before discussing that we remark that the linear transport regime where $(\mu_1 - \mu_2)$ satisfies inequalities opposite to those of (4) is also of interest. It is in this latter limit where the conductance can be used for a spectroscopy of the resonances.

3. THE ADIABATIC CONSTRICTION AND THE SMALL CORRECTIONS TO IT

Glazman et al[5] were the first to construct an adiabatic[29] model for a constriction, which clearly demonstrates the quantization of the constriction conductance. Imagine a ballistic (disorder-free) 2D wire, or electronic waveguide having a width $d(x)$ yielding a smooth constriction (see Fig.2). This is obtained by taking $d(x)$ to be a symmetric smooth function with $lim_{x \to \pm \infty} d(x) = d$, $d(0) = d_0 \ll d$ and changing slowly from d to d_0 over a scale $L \gg d, d_0, \lambda$. One now makes a Born-Oppenheimer type separation of the "slow" longitudinal variable x and the "fast" one y. The y-problem is just a square well having energies

$$E_n(x) = \frac{\pi^2 \hbar^2}{2m[d(x)]^2} n^2 \tag{6}$$

As is familiar from the usual separation, $E_n(x)$ plays the role of an additional potential for the mode n along the x direction. This potential depends on n and has the shape of a barrier with a maximum proportional to $(n/d_0)^2$. For a given E_F, a finite number of modes will have their energy above the barrier. This number, n_0 is the integral part of $k_F d/\pi$. For large L (adiabatic limit) tunneling below the barrier and reflections above it are negligible and the x-problem is well approximated by the WKB method. It follows that n_o modes have $T_{ij} = \delta_{ij}$ and all the others have zero transmission, hence

$$G_2 = \frac{e^2}{\pi \hbar} n_0 , \tag{7}$$

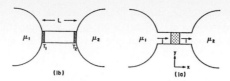

Fig.1:   a) The general Landauer geometry, as explained in the text.

b) The Landauer geometry with an ideal wire but with weakly transmitting barriers to the reservoirs - a "resonance tunneling" model.

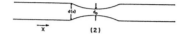

Fig.2:   An adiabatic constriction with slowly varying $d(x)$

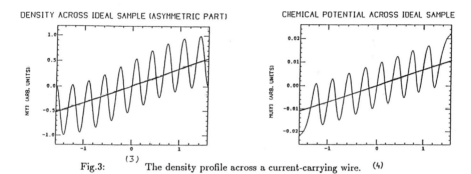

Fig.3:   The density profile across a current-carrying wire.

Fig.4:   The chemical potential profile across a current-carrying wire, according to Eq.(13).

(the straight line is after averaging over the oscillations on the scale of λ)

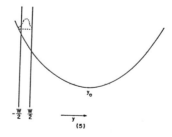

Fig.5:   The Landau parabola and the wire for a skipping edge state. The dashed lines schematize the ground state and its wavefunction for $l_s \sim w$.

which establishes quantization of G_2 in this limit. Glazman et al[5] also evaluated the small corrections due to reflection or transmission in each mode and found them to be exponentially small (except very near the points were modes are switched on and off):

$$\delta G = \frac{e^2}{\pi \hbar} \left[1 + exp(-z\pi^2 \sqrt{2R/d}) \right]^{-1}, \tag{8}$$

where R is the radius of curvature of the constriction and $z = k_F d/\pi - n$. It is of interest now to check how do corrections to the adiabatic approximation affect the T_{ij} for $i \neq j$, in order to establish the adequacy of this approximation for the conductance. This calculation was done by Yacoby and the author[6]. One writes the wavefunctions as sums of products of transverse and longitudinal components and obtains a differential equation, which is best analyzed in terms of the variable $u = x/L$ and whose terms depends on different powers of $1/L$. By making an asymptotic analysis of this equation around the Airy equation in the large L limit, one finds that the corrections to the adiabatic approximation involve integrals of products of adiabatic WKB-type functions. These integrands have quickly varying phases, and analyzing them by steepest descent and stationary phase methods, one arrives at the surprising result that all the corrections to T_{ij} and R_{ij} are exponentially small, as in Eq. 8. It also follows[6] that one does not need the whole constriction to be adiabatic, it is sufficient that adiabaticity holds only until the channel opens up to be far enough from threshold. We believe that these are the reasons why the adiabatic picture may be valid in practice and the conductance quantization holds approximately.

Another interesting aspect of the adiabatic picture is that for a wave moving adiabatically along a guide, the mode number, n, is conserved. For a guide whose width is slowly increasing, this implies that the transverse energy gets smaller and smaller, while the longitudinal one increases continuously. Thus, the motion is more and more in the forward direction. If such an adiabatic horn suddenly opens up into space, it turns out (see eg. Ref.11 in the present context) that the radiation emitted conserves approximately the transverse momentum. Thus, a wave packet is radiated out with transverse wave numbers in the interval $\pm \pi/d_{max}$(for $n = 1$, where d_{max} is the final opening of the horn. It turns out[10] that this wave packet describes a wave which is collimated in space and at large distances reaches an angular width of about $\pi k d_{max}$. Among the applications of this collimating effect there may be[21] the quenching of the Hall effect as discussed in section 5.

4. INTRINSIC HALL EFFECT OF A NARROW WIRE

This section is based on Ref.14. Let us consider again a narrow[20] wire through which a current is driven as in Fig. 1. A magnetic field B is applied perpendicularly to the place of the wire. The electron confinement in the wire is taken to be due to an infinitely deep potential well for $\frac{w}{2} \leq y \leq \frac{w}{2}$. In the Landau gauge, the vector potential is given by $\vec{A} = (-yB, 0, 0)$. The effect of B is through the "Landau parabola"

$$V_L(y) = \frac{1}{2} m\omega_c^2 (y - y_0)^2 \tag{9}$$

where $\omega_c = eB/mc$, $y_0 = l_H^2 k_x$, $Bl_H^2 - \hbar c/e$. k_x is the wavenumber of the electron's running wave in the x-direction. The combined effect of the confinement potential and V_L leads to discrete transverse states $|n>$ with energies E_n. Both $|n>$ and E_n vary with

384

B. Ref.14 used elementary perturbation theory for small B to evaluate the low order effects of B on $|n>$ and E_n. We note also that at the Fermi level, $E_n + \hbar^2 k_x^2/2m = E_F$. From this one may evaluate the change in the electron density across the wire due to the magnetic field (see Fig.2). This change is clearly proportional to the variations of the electrostatic potential and the electrochemical potential (the latter is the Hall voltage) across the wire. By the intrinsic Hall voltage we mean the latter difference, as measured by noninvasive voltage probes in the manner of Ref.30 (see also Refs. 4,27,28 and 31). This measurement should include proper averaging over length scales larger than the electron wavelength, λ, and the screening length λ_s. Self-consistent screening will be discussed at the end of this section.

It should be emphasized that this noninvasive measurement is not like the conventional Hall measurement with the cross geometry, where the probes are strongly coupled to the system (although recent measurements[22] are approaching the weak coupling in some sense). The cross geometry measurements have to be handled by using the full formulation[16], which will be briefly discussed in section 5. That does not mean, however, that the weak coupling measurement is impossible. Criticisms of the latter [32] are based on the neglect of important correlations among transmission probabilities from the voltage probe to the current probes. These issues will be fully discussed in Ref.31.

A generalization of the single-channel expression of Ref.30 to many channels yields for the chemical potential at a "site" (whose dimensions must be larger than $\lambda, \lambda s), \mu(x)$:

$$\mu(x) = \frac{\mu_1 \sum_i |\Psi_i^l(x)|^2/v_i + \mu_2 \sum_i |\Psi_i^r(x)|^2/v_i}{\sum_i (|\Psi_i^l(x)|^2 + |\Psi_i^r(x)|^2)/v_i} \tag{10}$$

v_i being the longitudinal velocity of the ith channel at the Fermi-level, $\Psi_i^l(x)$ and $\Psi_i^r(x)$ being the scattering states emanating from the LHS and RHS reservoirs, respectively at x. The $|\Psi|^2$ values are averages over spatial ranges larger than λ and λ_s. This weighted average of μ_1 and μ_2 is of the same nature as the voltages defined in Ref.26. It can be easily shown to agree with the general formulation of Refs.16 and 32 in the weak coupling limit, at least in the single channel case.

Using the perturbation theory wavefunctions and Eq.(10) one finds the chemical potential variation across the wire, as shown in Fig.3. By taking the difference in the limits of Eq.13 on the two edges of the wire, and using a L'Hospital limiting procedure, we find that the difference in μ across the system is equal to the classical Hall voltage times a coefficient of order unity. Thus, the intrinsic Hall effect is not quenched. This agrees with the result of Peeters[15].

We have calculated the Hall voltage perturbatively for $B \rightarrow 0$. It is of interest to estimate the range of B for which this is valid. Let us concentrate on "skipping" edge state, i.e. those for which the center of the Landau parabola, y_0, is for outside the wire $y_0 \gg w/2$ (see Fig.5). Approximating the section of the parabola inside the wire by a straight line one finds for the size of the nth edge state [19] bound near one (and neglecting the other) wall:

$$l_s^{(n)} \sim \left(\frac{l_H^2}{k_n}n^2\right)^{1/3} \tag{11}$$

For a field so strong that $l_s^{(n)} \ll w$, this state is bound on one wall and does not feel the other one. One would expect that the condition for the validity of perturbation theory is the opposite inequality:

$$l_s^{(n)} \gg w \tag{12}$$

It is interesting that this condition does not follow from the naive one that the perturbation matrix elements be much smaller than the energy level differences. It does follow, however, from a closer examination which notes that the perturbation contributions to a given state from e.g. those immediately above and below it tend to almost cancel. Taking this into account, we do find (12) as the condition for the validity of perturbation theory.

The above calculations were done for an ideal system with no impurity scattering. Will some disorder scattering bring about the quenching? There seems no particular reason for this to be the case. Inserting a scatterer in the sample and evaluating its effect numerically using the Green's function method was attempted in Ref.14. The scattering influences quantitatively the Hall effect but does not lead to a generic quenching.

Until now we discussed how the Lorentz force changes the density across the wire and hence the chemical potential for noninteracting electrons. With Coulomb interaction, the accumulated positive and negative charges on the two sides of the wire will create a Coulomb field which will tend to cancel the effect of the Lorentz force. This is in fact the classical explanation of the Hall effect. Such self consistent screening was generally treated by Landauer[24,25]. The final result is that on scales larger than λ and λs, such screening does occur and restores charge neutrality. So that in the real situation the electrostatic potential is changed, the local Fermi level (proportional to the density in 2D) stays constant and the electrochemical potential (sum of the two) follows the space variation of the chemical potential calculated naively in the noninteracting system. A short calculation demonstrates that this indeed happens also in our case. For a wire narrower than λ_s, the electrostatic potential difference does not fully develop and one may get some "quenching" of the Hall effect, which is not the one discovered in Ref.12,13 (for example the thickness below which this happens is independent of B). The initial quenching experiments[12,13] did not address the intrinsic Hall effect but employed the conventional cross, four-terminal geometry for the Hall effect. This is discussed in the next section.

5. QUENCHING OF THE HALL EFFECT IN THE SYMMETRIC CROSS GEOMETRY

For a symmetric-cross geometry the four-terminal formulation[16] yields the following expression for the Hall resistance:

$$R_H \equiv \frac{V_H}{I} = \left(\frac{\pi\hbar}{e^2}\right) \frac{T_R - T_L}{2T_F(T_F + T_R + T_L) + T_R^2 + T_L^2} \tag{13}$$

Where T_L, T_R and T_F are respectively the probabilities for a current-carrying electron to turn to the left branch of the cross, turn to its right branch, or to continue forward. Beenakker and van Houten[18] proposed as a mechanism for quenching that T_R may become equal to T_L if inequality (12) holds. This is reasonable since in the quasiclassical picture the electron is equally likely then to bounce off the two walls of the wire. Thus, if the openings leading to the L and R branches are equivalent, it will follow classically that $T_L = T_R$. This is to be contrasted, however, to the case where the opposite inequality holds and the electron skips, for example, on the right edge, so that T_R="1" (N for N conduction channels) and $T_L = 0$ (the latter situation leads to a quantized Hall effect). It is not clear, however, that this quasiclassical argument will really hold in the fully quantum mechanical situation, due to diffraction effects at the entrances to the voltage probes. In fact, $|\Psi|^2$ larger near one edge of the wire might lead to a larger transmission to the lead on the same side. Nevertheless, one may feel that the Beenakker-van Houten argument should play some role. It was noted in Ref.14 that the condition (12) depended on n, so that according to it different channels should quench, for example, at different values of B. This does not seem to happen generically. Also, the width dependence of the quenching field does not appear to agree with (12). Computer calculations[20,21] show a strongly fluctuating R_H for a straight wire cross, with an average closer to the classical value than to generic quenching.

Quite generally, one expect interference effects to decrease in a gradual "adiabatic" geometry, so that quenching might occur in a graded junction. Baranger and Stone[21] took up this point and noted that the vanishing of <u>both</u> T_L and T_R can also lead to quenching of R_H. They suggested that this could happen since, as discussed in section 3, the adiabatic broadening of the wire before getting into the cross region selects a larger ratio of longitudinal to transverse momentum and energy. This should lead to a smaller probability of going sideways.. We point out that the more complete treatment of collimation[11] given in section 3, where the adiabatic "horn" radiates into a sudden opening, may strengthen the argument of Ref.21. It is also important to point out[33] that the vanishing of the T_R and T_L at $B = 0$ is not sufficient for quenching. It is the asymmetric part $(T_R = \alpha B \quad ; \quad T_L = -\alpha B)$ which is relevant. It will take further work to fully establish quenching of the Hall effect.

6. CONCLUSIONS

The adiabatic picture[29,5] appears to be the crucial ingredient to justify the assumptions needed to establish[4] the quantization of the ballistic point contact conductance. The smallness[5,6] of the corrections to this approximation explains why this is observed[2,3] in real devices. The collimation effect[11] obtained in the adiabatic picture may well be relevant to explain the quenching of the Hall effect in the cross geometry. Experiments[22,23] controlling what happens in the cross region will be crucial for a final understanding of quenching. At the same time, we emphasize that the intrinsic Hall effect is not quenched. This is consistent with the results of Ref.21 where only the voltage probes coupling is weakened. However, experiments with controlled weak-coupling which is uniform among the channels are needed to establish this point. This will also test the general idea of the possibility of noninvasive unbiased voltage measurement[27,30,31].

ACKNOWLEDGEMENTS

Much of this work is based on joint work with A. Devenyi, I. Kander, U. Sivan and A. Yacoby. Illuminating discussions with M. Büttiker, O. Entin-Wohlman, R. Landauer, U. Sivan and A. Yacoby are gratefully acknowledged. This research was supported by grants from the Fund for Basic Research administered by the Israeli Academy of Sciences and by the Minerva Foundation, Munich, The Federal Republic of Germany.

REFERENCES

1. A. Stern, Y. Aharonov and Y. Imry, General conditions for the loss of quantum interference, preprint, 1989.
2. B.J. van Wees, H. van Houten, C.W.J. Beenakker, J.G. Williamson, L.P. Kouwendhoven, D. van der Marel and C.T. Foxon, Phys. Rev. Lett. **60**,848 (1988).
3. D.A. Wharam, M. Pepper, H. Ahmed, J.E.F. Frost, D.G. Hasko, D.C. Peacock, D.A. Ritchie and G.A.C. Jones, J. Phys. **C21**, L209 (1988).

4. Y. Imry, Physics of mesoscopic systems, in : Directions in Condensed Matter Physics, G. Grinstein and G. Mazenko, eds, Memorial Volume in honour of S.-k Ma, World Scientific, Singapore (1986), p.101-163.

5. L.I. Glazman, G.B. Lesovik, D.E. Khmel'nitskii, and R.I. Shekhter, Pis'ma Zh. Exsp. Teor. Fiz. **48**, 218 (1988) [Sov. Phys. JETP Lett. **48**, 238 (1988)]; A. Kawabata, preprint, O. Entin-Wohlman and M. Ya Azbel, preprint.

6. A. Yacoby and Y. Imry, preprint (1989).

7. R. Landauer, those proceedings and to be published.

8. I. Kander and Y. Imry, in preparation.

9. H. van Houten, these proceedings.

10. N. Lang, A. Yacoby and Y. Imry, in preparation.

11. G. Kirczenow, Sol. St. Comm. **68**, 715 (1988); A. Szafer and A.D. Stone, Phys. Rev. Lett. **62**, 300 (1989); E.G. Haanappel and D. Van der Marel, preprint (1988) Y. Avishai and Y.B. Band, preprint (1988); N. Garcia, J.J. Saenz and R. Casero, to be published.

12. M.L. Roukes, A. Scherer, S.J. Allen, Jr., H.G. Craighead, R.M. Ruthen, E.D. Beebe and J.P. Harbison, Phys. Rev. Lett. **59**, 304 (1988).

13. C.J.B. Ford, T.J. Thornton,R. Newbury, M. Pepper, H. Ahmed, D.C. Peacock, D.A. Rithcie, J.E.F. Frost and G.A.C. Jones, Phys. Rev. **B38**, 8518 (1988).

14. A. Devenyi, M.Sc. thesis, Weizmann Inst., unpublished (1988); A. Devenyi and Y. Imry, in preparation.

15. F.M. Peeters, Phys. Rev. Lett. **61**, 580 (1988).

16. M. Büttiker, Phys. Rev. Lett. **57**, 1761 (1986).

17. H.B.G. Casimir, Revs. Mod. Phys. **17**, 343 (1945).

18. C.W.J. Beenakker and H. van Houten, Phys. Rev. Lett. **60**, 2406 (1988).

19. R.E. Prange and T.W. Nee, Phys. Rev. **168**, 779 (1968).

20. D.G.Ravenhall, H.W. Wyld and L.R. Schult, preprint (1989); G. Kirczenow, preprint (1989).

21. H.V. Baranger and A.D. Stone, these proceedings, and preprint (1989).

22. A.M. Chang and T.Y. Chang, these proceedings, and preprint (1989).

23. C.J.B. Ford, S. Washburn, M. Büttiker, C.M. Knoedler and J.M. Hong, these proceedings, and preprint (1989).

24. R. Landauer, IBM J. Res. Dev. **1** (1957).

25. R. Landauer, Phil Mag. **21**, 863 (1970).

26. M. Büttiker, Y. Imry, R. Landauer and S. Pinhas, Phys. Rev. **B31**, 6207 (1985).

27. O. Entin-Wohlman, C. Hartzstein and Y. Imry Phys. Rev. **B34**, 921 (1986).

28. U. Sivan and Y. Imry, Phys.Rev. **B33**, 551 (1986).

29. R. Landauer, Z. Phys. **B68**, 217 (1987).

30. H.L. Engquist and P.W. Anderson Phys. Rev. **B24**, 1151 (1981).

31. U. Sivan and Y. Imry, to be published in: Mesoscopic Phenomena in Solids, P.A. Lee. R. Webb and B.L. Altshuler, eds, North Holland (1989)

32. M. Büttiker, Phys.Rev. **B35**, 4123 (1987).

33. A. Yacoby, private communication.

THE LOW-FIELD HALL EFFECT IN QUASI-BALLISTIC WIRES

C.J.B. Ford

IBM Thomas J. Watson Research Center
Yorktown Heights, New York 10598, USA

Results on the Hall effect in narrow quasi-ballistic channels are presented: for particular cross geometries, the Hall voltage can be enhanced, quenched or actually negative. These results can be explained qualitatively in terms of a simple trajectory picture. The "last plateau" is seen for all geometries and appears to arise when transmission into a voltage probe dominates over the straight-through transmission.

I. INTRODUCTION

Within the past few years, several anomalous features have been observed in the Hall resistance R_H in narrow, high-mobility devices in which electrons can propagate distances many times the channel width without undergoing collisions, and in which the Fermi wavelength is not much less than the channel width. R_H is approximately zero ("quenched") over a range of magnetic field B around $B = 0$, rather than rising linearly with B as expected classically (1,2,3). As B is further increased, R_H rises steeply above the classical line, and then levels off (forming a "last plateau"), before rejoining the classical line and forming quantum Hall plateaux. The origin of these two features was unclear, and only recently have theoretical models begun to match the experimental results. A first attempt at explaining the quenching (4) suggested that a Hall voltage would not build up until the electron wavefunction was confined to one side of the wire by the Lorentz force, forming an edge state. However, poor quantitative agreement was found with experiment (2).

More recent work has focused on the role of the Hall probes used to make the measurements; these are generally channels similar to the current-carrying channel and meeting it at right angles. A calculation for the case of probes which did not disturb the main channel found that R_H would not be quenched (5), so other groups have calculated the scattering coefficients in the cross region quantum-mechanically by direct numerical solution of the Schrödinger equation for model potentials with strongly-coupled probes. When substituted into the multi-probe resistance formula (6) they give R_H. A perfect square cross with a square-well or parabolic potential produced quenches only for specific values (7), or ranges (8), of the Fermi energy E_F, contrary to experimental results (2), which showed a quench over a wide range of E_F. On filling more than two sub-bands (formed by the transverse confinement), the quench rapidly disappeared, in contrast to early devices (1,2) which probably had up to 9 occupied sub-bands when quenching was observed.

Another group (9) found the same behaviour, so they introduced a more realistic geometry, allowing for the broadening of the channel in the region around the cross. Quenching was found for E_F in a wide range, due to the collimating effect of the widening. As the wire widens the sub-band energy decreases, so that for an electron at the Fermi energy, the transverse momentum k_{tr} decreases and the longitudinal momentum k_L increases to keep the electron's energy constant, (assuming adiabatic transport in which electrons stay in the same sub-band without a change in energy) (10). Thus when the electron enters the region open to the probes, its wavefunction propagates in a limited range of angles around the straight-through direction. Without this collimation effect, the wavefunction would spread out rapidly if k_{tr} were large relative to k_L, which is the case when E_F lies just above the bottom of a sub-band. That seems to be why the quench comes and goes as E_F is varied in the simulations of a perfect cross. When the electrons are collimated, a finite magnetic field is required to get enough transmission into the voltage probes to produce a significant Hall voltage.

II. RESULTS AND DISCUSSION

In this paper, measurements of devices made using an etching technique (12) will be discussed. Reactive ion etching is used to remove the 300Å thick GaAs cap layer of a GaAs-$Al_xGa_{1-x}As$ heterostructure except over the channel. A gate is then deposited over the whole device. The etched regions do not conduct, confining the two-dimensional electron gas (2DEG) formed at the heterojunction to a narrow channel. The channel width and the carrier concentration n_s decrease as the gate voltage V_g is reduced towards threshold. Scattering off the walls is expected to be specular in all the devices, since the depletion regions smooth out lithographic irregularities. These particular devices were too narrow to conduct at zero V_g due to edge depletion. The threshold voltage was around 0.35V, varying between devices and each time a particular device was cooled down to 4K. All the samples were made simultaneously on the same wafer to minimise the possibility of variation between different devices, except due to the intentional changes in the geometry. The lithographic width was $0.29\mu m$, and the electrical width (estimated from magnetic depopulation of the 1D sub-bands (13)) was in the range 0.20-$0.09\mu m$ depending on V_g. Measurements were made at 4.2K, where the elastic and inelastic lengths were expected to be considerably greater than the size of the cross region.

Earlier cross regions fabricated with split gates (2) had corners rounded both by the deficiencies in the lithography and by the smoothing effect of the electrostatic confinement, so the centre of the cross was considerably wider than the leads, and probably somewhat irregular. To investigate in a controlled way the effect of having a cross region without perfectly square corners, this set of devices (17) included a widened cross, as shown inset in Figure 1a. The result was surprisingly clear: instead of quenching, the R_H actually became *negative* (Figure 1a, solid curve). As on each of the samples measured, there was a "normal" (nominally-perfect) cross (radius of curvature of the lithographic corners $\sim400\text{Å}$) in series with the widened cross, connected to it by a $6\mu m$ long channel of the same width. The corresponding Hall resistance is shown dashed in the same figure. R_H was quenched as expected, in contrast to the behaviour of the widened cross. A second nominally-identical device showed very similar behaviour, as shown offset vertically in Figure 1a, indicating that the effect is not due to random impurity scattering or lithographic imperfections, but rather to the shape of the cross.

This can be explained intuitively in terms of semi-classical trajectories, which give an indication of the behaviour in a magnetic field of the transmission coefficients in a quantum-mechanical treatment to be described later. The incoming electron trajectory (moving at a fairly small angle to the forward direction due to the collimation described earlier), is bent towards the "correct" voltage probe by the Lorentz force, but in the widened cross the electron is reflected off the diagonal wall, as shown inset in Figure 1a, and is reflected into the "wrong" probe, giving rise to a negative R_H. Eventually B is large enough to bend the electron into the right probe. In the case of the normal cross, at low fields the curvature is not sufficient for the electron to enter the voltage probe (15,16), so it just bounces off the wall and continues

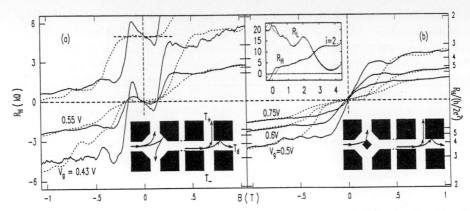

Fig. 1. (a) R_H vs B for several values of V_g. The solid and dotted lines are for the widened and normal cross, respectively (on the same sample). The trace offset vertically by 5kΩ shows corresponding results for a different, nominally identical, sample. (b) R_H vs B for various values of V_g. The solid and dotted lines are for the widened cross with a central dot and the normal cross on the same sample. Inset: high field R_L and R_H for $V_g = 0.5$V. Inset to each figure: device geometries, and the electron paths as discussed in the text.

along the channel (see inset to Figure 1a). Hence the Hall voltage is quenched at low B.

Another geometry investigated was similar to the widened cross, but with a dot etched out of the centre (see inset to Figure 1b). Here the Hall resistance became *enhanced* over the classical Hall value, rather than quenched (see Figure 1b). The dot appears to scatter electrons into the probes, and any field is enough to cause increased scattering into the "correct" probe, leading to an enhanced R_H. In the normal crosses, R_H was always quenched for V_g below a certain value. Thus the quenching was independent of the position of E_F relative to the bottom of a sub-band, in contrast to the results of calculations for perfect crosses described earlier.

Quantum-mechanically, for a four-fold symmetric cross, $R_H = (h/2e^2)(T_+ - T_-)/[2T_d(T_d + T_+ + T_-) + T_+^2 + T_-^2]$, where T_d is the transmission coefficient directly through the cross, and T_+ and T_- are those into the probes (T_+ for the probe favoured by the Lorentz force) (6,7,8,9). In the normal cross, T_+ and T_- are small compared with T_d at low B, hence the quench. In the widened cross, T_- becomes greater than T_+ due to the reflection into the "wrong" probe, giving a negative R_H, and in the widened cross with the dot, T_+ is greater than it would be without the obstacle, enhancing R_H.

The "last plateau" was apparent in all the devices (Figure 1b). In earlier devices defined with split gates, the last plateau was clear even with many (up to 15) occupied sub-bands (which are equivalent at high B to Landau levels) (2,18). The usual quantum Hall plateaux were seen at high fields, then as B decreased the plateaux became blurred out, with R_H fluctuating about the

392

classical line. The last plateau seems to occur when the magnetic field is strong enough to bend most electrons into the usual voltage probe. Then $T_+ \gg T_d \gg T_-$, and the above formula yields $R_H \simeq (h/2e^2)/(N + T_d)$, where N is the number of spin-degenerate sub-bands occupied. If T_d were zero, there would be a staircase of plateaux as in the quantum Hall effect. However, when T_d is non-zero, (due mainly at these fields to scattering between opposite edges of each probe), the height of the plateau is reduced. Experimentally, the value of T_d found by comparing this formula with the plateau height fluctuates between 0 and $\lesssim 2$ along the plateau for each N (even for $N \simeq 15$). (It is assumed that since the plateau is fairly flat while B increases by a factor of two, T_d goes nearly to zero sometimes, rather than fluctuating between some value $t > 0$ and $t + 2$. Thus the maximum value on the plateau is close to $h/2Ne^2$.) As B increases so that magnetic confinement dominates, N decreases, but $T_d \sim 1$ still, so that there are fluctuations on the quantum Hall plateaux. Since the plateaux are about the same length as the field scale of the fluctuations, no distinct plateaux are seen until higher fields, where T_d is smaller and the plateaux are long enough for the average value to be apparent. Reducing channel width or temperature makes interference effects due to multiple reflections and bound states in the cross region more likely as the phase coherence length becomes much larger than the size of the cross, and these give rise to larger fluctuations in T_d and hence in R_H. The last plateau will be discussed further elsewhere (17).

III. SUMMARY AND CONCLUSIONS

In conclusion, it is found that the low field Hall effect in narrow high-mobility devices depends on the precise geometry of the cross region and can be understood simply by considering electron trajectories: when the cross approximates to a perfect cross, R_H becomes quenched, whereas if the wires widen considerably around the cross, the Hall resistance actually becomes negative, due to electrons being reflected into the "wrong" probe. If an obstacle is added in the centre of the widened cross, reflections off it assist the magnetic field in bending the electron trajectories into the "correct" voltage probe, causing an enhanced Hall resistance. These effects occur over a wide range of carrier concentrations (and hence Fermi energy), and are therefore insensitive to the position of E_F relative to the bottom of a sub-band. The last plateau has approximately the appropriate quantum Hall value, but it is reduced since there is still some direct transmission from one current lead to the other.

ACKNOWLEDGMENTS

I would like to thank my collaborators S. Washburn and M. Büttiker (in particular for suggesting including a central gate and for pointing to the interpretation of the results in terms of the simple trajectory picture). J.M. Hong grew the material and C.M. Knoedler carried out the RIE. I am also grateful to A.B. Fowler, D.P. Kern, K.Y. Lee and S. Rishton for helpful advice and discussions.

REFERENCES

1. M.L. Roukes, A. Scherer, S.J. Allen, Jr., H.G. Craighead, R.M. Ruthen, E.D. Beebe and J.P. Harbison, Phys. Rev. Lett. **59**, 3011 (1987).
2. C.J.B. Ford, T.J. Thornton, R. Newbury, M. Pepper, H. Ahmed, D.C. Peacock, D.A. Ritchie, J.E.F. Frost and G.A.C. Jones, Phys. Rev. B, **38**, 8518 (1988).
3. A.M. Chang and T.Y. Chang, unpublished.
4. C.W.J. Beenakker and H. van Houten, Phys. Rev. Lett. **60**, 2406 (1988).
5. F.M. Peeters, Phys. Rev. Lett., **61**, 589 (1988).
6. M. Büttiker, Phys. Rev. Lett., **57**, 1761 (1986) and IBM J. Res. Dev. **32**, 317 (1988).
7. D.G. Ravenhall, H.W. Wyld and R.L Schult, Phys. Rev. Lett. **62**, 1780 (1989).
8. G. Kirczenow, Phys. Rev. Lett., **62**,1920 (1989) and unpublished.
9. H.U. Baranger and A.D. Stone, unpublished.
10. C.W.J. Beenakker and H. van Houten, unpublished.
11. C.J.B. Ford, T.J.Thornton, R.Newbury, M.Pepper, H.Ahmed, D.C. Peacock, D.A.Ritchie, J.E.F. Frost, and G.A.C. Jones, Appl. Phys. Lett. **54**, 21 (1989)
12. T.P. Smith, III, H. Arnot, J.M. Hong, C.M. Knoedler, S.E. Laux, and H. Schmid, Phys. Rev. Lett. **59**, 2802 (1987).
13. K.-F. Berggren, G. Roos and H. van Houten, Phys. Rev. B **37**, 10118 (1988).
14. M.L. Roukes, T.J. Thornton et al, unpublished.
15. Y. Takagaki, K. Gamo, S. Namba, S. Ishida, S. Takaoka, K. Murase, K. Ishibashi and Y. Aoyagi, Sol. St. Comm. **68**, 1051 (1988).
16. Y. Avishai and Y.B. Band, unpublished.
17. C.J.B. Ford, S. Washburn, M. Büttiker, C.M. Knoedler and J.M. Hong, Phys. Rev. Lett. (to be published).
18. C.J.B. Ford, Ph.D. thesis, Univ. of Cambridge, England.

EVIDENCE FOR AN INHOMOGENEITY SIZE EFFECT IN MICRON SIZE GaAs/AlGaAs CONSTRICTIONS

A.Sachrajda, D.Landheer, R.Boulet, J.Stalica
and T.Moore[1]

Physics Division
National Research Council
Ottawa,Canada

1. INTRODUCTION

The breakdown of the Quantum Hall Effect (QHE) has been studied in wide samples by several groups[1]. Recently structures have been observed[2-4] in the breakdown curves of the QHE in narrow structures ($\approx$ micron) during both magnetic field sweeps at constant high current density and current sweeps at constant field. Several groups[3-5], ourselves included, originally interpreted the results in terms of an inter-Landau level Zener mechanism. In this paper we provide evidence that the actual origin of the structures is the spatially dependent breakdown of the Q.H.E.. A simple model shows that the structures become detectable when the size of an inhomogeneity (a region which has separate breakdown characteristics) is comparable to the width of the sample. Results are presented which demonstrate the dramatic effect of illumination on the breakdown characteristics. Measurements were performed at 1.2K.

2. EXPERIMENTAL RESULTS

The inset in figure (3) illustrates our sample geometry. It is based upon that used by Bliek et al.[2] and makes use of the fact that breakdown is related to the current density so that contacts need not be mounted on the constriction itself. Contacts mounted on the same side of the constriction were used to confirm that the breakdown was only occuring in the constriction at the currents used in the measurements. In this paper the results presented were made on a sample with a constriction with an electrical width of 1.5 microns and a length of 10 microns. The

[1]Present address: BNR, Ottawa, Canada

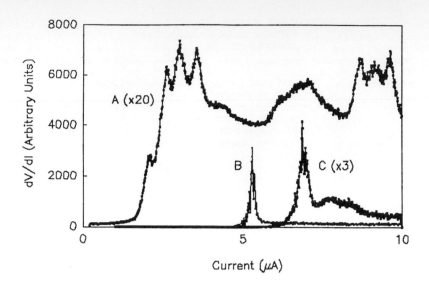

Fig.1   Three different current sweeps,A at 7.5T B at 8.9T and C at 8.4T (see text).

wafer had a mobility of 97,000 cm^2/Vs in the dark and up to 200,000 cm^2/Vs after illumination and a carrier concentration of 3.7.10^{11}cm^{-2} before illumination. In the Zener model current sweeps and field sweeps are different in that the ratio eV$_H$/$\hbar\omega_c$ (the maximum energy the electron can gain from the Hall electric field to the cyclotron energy) changes only by a few per cent in a field sweep but can change easily by up to twenty times in a current sweep. All of our samples, however, showed as many structures in field sweeps as in current sweeps. It was therefore decided to measure the field and current dependencies of individual structures using the numerical derivative of the current and field sweeps to track the structures. Curve A in figure (1) is a typical derivative of a current sweep. Figure (2) illustrates the position of the structures obtained from current sweeps. Where a line connects the points is where it is clear that the points refer to the same structure only shifted in field and current. Field sweeps gave a similar result. Also shown are the actual breakdown currents. Based upon data such as figure (2), in which it can be seen that the individual structures follow a similar curve to the actual breakdown curve only shifted in field and current, it is suggested that the structures are a manifestation of spatially dependent breakdown. A simple model was developed, Sachrajda et al.[6], in which spatially dependent breakdown was treated by dividing the sample into field and current dependent resistors,each resistor having a field and current dependence consistent with the behaviour of individual structures as shown above. The model showed that to reproduce the structures the inhomogeneities had to be placed in series i.e. the inhomogeneities had to span the width of the constriction. The same model also showed that even if only a small spread in carrier density exists between the voltage probes the measured critical current is dramatically reduced. Curve B in figure (1) shows a typical sweep after the sample has been illuminated for a short time with a red LED. The multiple structures disappeared. This suggests that a non-uniform occupation of deep traps (those responsible for persistent photoconductivity) causes the inhomogeneities. On illumination the traps are emptied and if the traps are uniformily distributed the above model trivially predicts that only a single breakdown structure should occur and that the critical current should increase (as it does). The properties (including the probability of occupation of the traps at low temperatures) will be influenced by various crystal non-uniformities, including alloy composition fluctuations, dopant atom or defect clustering, statistical fluctuations of the dopant atoms, layer thickness fluctuations etc... Curve C in figure (1) shows a typical current sweep after a slight relaxation of the persistant photoconductivity which occured at least once after each illumination. A second structure can be seen suggesting that the sample has become more inhomogeneous. Figure (3) shows the increase in the critical current as a function of magnetic field after successive bursts of illumination. It can be seen that the increase is sharper than that predicted by the Zener breakdown model. Points A and B are at approximately the same field.

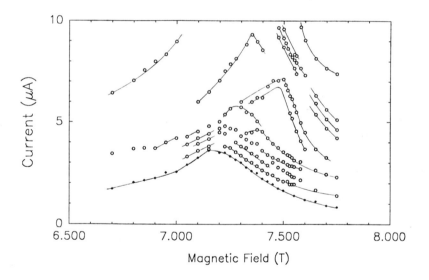

Fig. 2. The current and magnetic field dependences of individual structures. "♦" are the actual critical currents.

Point A was taken immediately after illumination while point B was taken after illumination to a higher field and a relaxation of the carrier density to this field value. The difference in critical current confirms the importance of inhomogeneities in critical current measurements. Several models[7] exist for a spatially inhomogenous 2DEG in a magnetic field.

After several illuminations the breakdown curves became fairly noisy due to a switching of the longitudinal voltage between discrete voltage levels. At some fields the breakdown was dominated by only two voltage levels. The transition region between the two was typically very narrow (approx. 0.1 microamps at 5 microamps or 20 Gauss at 11 Tesla). Figure 4 shows an oscilloscope trace at such a transition region. The voltage switches between the two levels every few milliseconds, but while it is in a particular level, spikes occur towards the second level. Switching between the levels only occurs if the spike makes it to the second level. This is similar to an effect seen by Cage et al.[8] on a wider sample which was interpreted by them as a switching between different current flow states. We have also observed this kind of effect in wider samples and are currently investigating it.

3. CONCLUSIONS

Breakdown measurements provide a useful characterization technique for the uniformity of samples on the length scale of

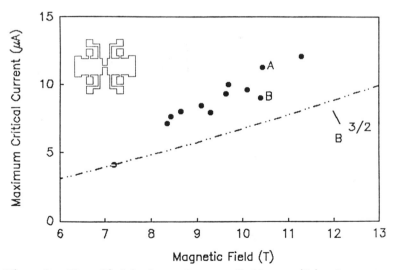

Fig. 3. The field dependence of the critical current (see text). The inset illustrates the geometry used (not drawn to scale). The line illustrates the field dependence expected from the Zener Breakdown model.

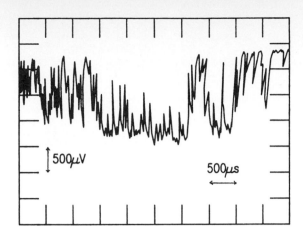

Fig. 4. Voltage level switching at 5 microamps and 10.96T.

the width of the sample. Structures in the breakdown observed by us and others in micron sized constrictions are evidence that inhomogeneities on this length scale exist in these samples. Illumination experiments confirmed the important role that deep traps play in creating these non-uniformities in the 2DEG.
After repeated illumination the breakdown became a switching between discrete voltage levels together with voltage spikes from the levels.

REFERENCES

1. G.Ebert, K.Von Klitzing, K.Ploog and G.Weimann, J. Phys. C 16 5441 (1983), S.Komiyama, T.Takamasu, S.Hiyamazu and S.Sasa Sol. St. Comm. 54 479 (1985)
2. L.Bliek,G.Hein,V.Rose,J.Niemeyer,G.Wiemann and W.Schlapp in High Magnetic Fields in Semiconductor Physics ed. G.Landwehr, Springer series in solid state sciences p.113
3. J.R.Kirtley, Z.Schlesinger, T.N.Theis, F.P.Milliken, S.Wright and L.F.Palmeteer, Phys. Rev. B.34 ,1384
4. A.Sachrajda, M.D'Iorio, D.Landheer, P.Coleridge, T.Moore, C.Miner, A.SpringThorpe to be published in the proceedings of the conference on The Application of High Magnetic Fields in Semiconductor Physics, Wurzburg 1988.
5. L.Eaves and F.Sheard, Semicond. Sci. Technol. 1, 346
6. A.Sachrajda, D.Landheer, R.Boulet and T.Moore submitted for publication.
7. V.Gudmundsson and R.R.Gerhardts p.67, Johnston and L.Schweitzer p.71 in "High Magnetic Fields in Semiconductor Physics" Springer Series in Solid-State Sciences 71.
8. M.Cage,R.Dziuba,B.Field,E.Williams,S.Girvin,A.Gossard,D.Tsui and R.Wagner Phys. Rev. Let. 51, 1374

A NEW TECHNIQUE TO PRODUCE SINGLE CRYSTAL EPITAXIAL NANO-STRUCTURES.

S.D. Berger, H.A. Huggins, Alice E. White, K.T. Short and D. Loretto
AT&T Bell Laboratories, Murray Hill, NJ 07974

I. INTRODUCTION

High dose ion implantation into silicon followed by annealing can lead to the formation of buried stoichiometric silicide layers for a variety of implanted species. Of particular interest here is the implantation of cobalt ions. It has been shown that with an implant dose of $\sim 2 \times 10^{17}$ cm^{-2} and an anneal temperature of the order 1000^0 C a single-crystal oriented layer of metallic cobalt disilicide is formed within the bulk silicon. This technique is known as mesotaxy [1]. The process is illustrated in figure 1 by reference to Rutherford Backscattering Spectroscopy (RBS) of the material both post-implant (fig. 1a) and post-anneal (fig. 1b). The as-implanted cobalt distribution as a function of depth is the usual skewed Gaussian and shows some degree of crystalline order. However, after annealing the profile shows a dramatic sharpening and increase in crystallinity. Furthermore, transmission electron microscopy (TEM) reveals that the silicide layer and the silicon are uni-axial and the interfaces are abrupt and relatively smooth.

Electrical measurements have shown that the resistivity of the layers produced by mesotaxy are superior to those grown by vapor deposition techniques and within a factor of 2 of the bulk material [1].

Using a combination of electron beam lithography and reactive ion etching we have fabricated masks which confine the implant dose laterally. In this way we are able to produce discrete structures of CoSi$_2$ with nanometer dimensions.

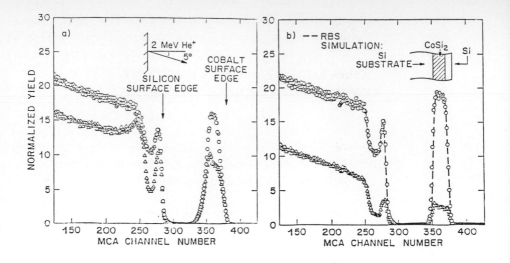

Figure 1. RBS spectra illustrating the mesotaxy process for a planar geometry: a) post-implant and b) post-anneal. Notice how the cobalt peak contracts during the anneal.

Preliminary experiments have shown that using the above technique single-crystal oriented wires can be successfully formed. The wires, which are buried ~60 nm below the (001) silicon surface, show electrical integrity over many microns of length with a cross section of ~100 nm.

II. FABRICATION

Basically the fabrication process involves using a mask to confine the lateral extent of the cobalt implant while subsequent annealing leads to the formation of the oriented silicide structure.

To produce the implant mask 500 nm of SiO_2 was grown onto a (100) Si wafer. The thickness of mask required is determined by the distribution of the implanted ions around their mean penetration depth. In our case we used a value of range + 5σ, where σ is the standard deviation of the distribution. PMMA is spun onto the wafer and patterned using a JEOL JBX-5D electron-beam lithography machine. After development a thin layer of Cr (~20 nm) is shadow evaporated onto the surface. The Cr layer acts as an etch mask for reactive ion etching (RIE) of the oxide layer using CHF₃. Following the RIE the Cr and PMMA are lifted off in acetone. In this way the e-beam pattern is transferred into the SiO_2.

Cobalt ions are then implanted into the wafer at an accelerating voltage of 170 kV and a dose of 2×10^{17} cm^{-2} with the substrate held at 350° C. The implant mask is removed and the wafer annealed at 1000° C for half an hour.

The process is illustrated schematically in figure 2 below.

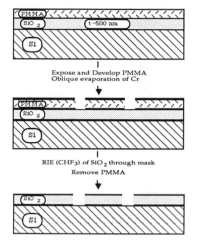

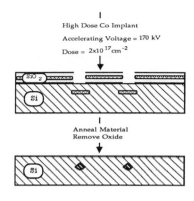

III. STRUCTURAL STUDIES

Arrays of wires lying along the 110 direction in the Si wafer were produced and studied in an electron microscope. In figure 3 we show an electron micrograph of part of an array. Diffraction patterns were taken from individual wires as shown in figure 4. The patterns all showed 002 type diffraction spots which is a clear indication of the presence of $CoSi_2$ since this type of spot is forbidden in the Si structure. Furthermore, from the sharpness and position of the diffraction spot we can infer that the silicide is both single-crystal and uni-axial with the Si. Figure 5 shows a detail of a wire imaged using a 002 spot. The wire can be seen to have abrupt edges which demonstrates that the mesotaxy process does indeed operate in-plane in addition to the contraction already observed along the implant direction. Also visible in the image are two dense dislocation bands running parallel with the wire. A cross-section of a wire is shown in figure 6, again imaged with a 002 type diffraction spot. The cross-section is a truncated rhomboid facetted on the 111 planes. The dislocation bands can be seen to run from the wire to the silicon surface. We believe that these are formed during the anneal and serve to relieve the stress of the system.

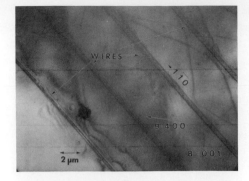

Figure 3. Bright field electron micrograph of part of an array of wires.

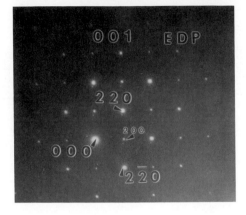

Figure 4. Electron diffraction pattern from a single wire.

IV. ELECTRICAL STUDIES

Thus far we have only checked for electrical continuity of the wires. Individual wires were fabricated as interconnects between two contact pads separated by 10 mm. Room temperature resistance measurements between the pads showed that of 23 wires measured so far, all were continuous.

V. CONCLUSION

Clearly, the ability to grow these buried epitaxial metallic structures with abrupt interfaces opens the way for many exciting experiments. Since the implant masks are produced by e-beam lithography we have considerable flexibility concerning the shape, size and crystallographic orientation of the final structures. In this way we hope to study both the growth mechanisms and transport properties of the structures as a function of these variables.

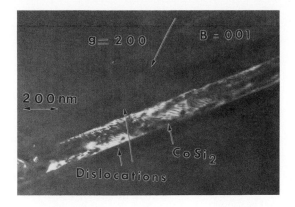

Figure 5. Plan view dark field image of a CoSi$_2$ wire.

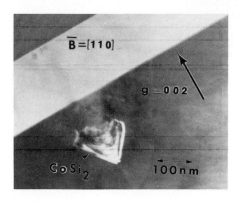

Figure 6. Dark field image of a cross-section of a CoSi$_2$ wire.

REFERENCES

[1] Alice E. White, K.T. Short, R.C. Dynes, J.P. Garno and J.M. Gibson, Appl. Phys. Lett., 50, 95, 1987.

DOUBLED FREQUENCY OF THE CONDUCTANCE MINIMA IN ELECTROSTATIC AHARONOV-BOHM OSCILLATIONS IN ONE-DIMENSIONAL RINGS[1]

M. Cahay[a], **S. Bandyopadhyay**[b] **and H. L. Grubin**[a]

[a]**Scientific Research Associates, Inc.**
Glastonbury, Connecticut 06033

[b]**Department of Electrical and Computer Engineering**
University of Notre Dame
Notre Dame, Indiana 46556

We predict the existence of *two* different sets of conductance minima in the conductance oscillation of a one-dimensional ring due to the electrostatic Aharonov-Bohm effect. The two sets of minima arise from two different conditions and effectively *double* the frequency of the conductance troughs in the oscillations. This makes the frequency of the troughs *twice* that predicted by the Aharonov-Bohm effect. We discuss the origin of this feature along with the effects of temperature and elastic scattering. We also compare it with the magnetostatic Aronov-Al'tshuler-Spivak effect and point out the similarities and differences.

I. INTRODUCTION

Oscillatory conductance due to the electrostatic Aharonov-Bohm effect has been predicted for a variety of ring structures along with potential device applications of that effect. In this paper, we point out an intriguing feature in the conductance oscillation of a one-dimensional ring due to the electrostatic Aharonov-Bohm effect. Unlike in the magnetostatic effect, the conductance in the electrostatic effect reaches its minimum under two *different* conditions which gives rise to *two*

[1]The work at SRA was supported by the Air Force Office of Scientific Research under contract no. F49620-87-C-0055. The work at Notre Dame was supported by the same agency under grant no. AFOSR-88-0096 and by an IBM Faculty Development Award.

distinct and independent sets of conductance minima in the oscillations. One set of minima arises from the usual destructive interference of transmitted electrons and the other arises from constructive interference of reflected electrons. The minima in each individual set recur in the oscillations with the periodicity predicted by the Aharonov-Bohm effect, but the separation between two adjacent minima (belonging to the two different sets) is smaller than and unrelated to the Aharonov-Bohm periodicity. In the following Sections, we establish this feature and discuss various issues related to it.

II. THEORY

The conductance G of a one-dimensional structure in the linear response regime is given by the two-probe Landauer or Tsu-Esaki formula [1]

$$ G = \frac{e^2}{2hkT} \int dE \ |T_{total}(E)|^2 \ sech^2(\frac{E - E_F}{2kT}) \ , \qquad (1) $$

where $T_{total}(E)$ is the transmission coefficient of an electron with incident energy E through the entire structure (i.e. from one contact to the other), T is the temperature and E_F is the Fermi level.

The problem of calculating the conductance G is essentially the problem of calculating T_{total}. The quantity T_{total} can be found from the overall scattering matrix for the structure. For a ring structure, the overall scattering matrix is determined by cascading three scattering matrices [2] representing propagation from the left lead of the ring to the two interfering paths, propagation along the paths, and propagation from the paths to the right lead. For simplicity, we will represent the first and the last of these scattering matrices by the so-called Shapiro matrix which is defined in Ref. 3.

A. Ballistic Transport

In the case of ballistic transport, cascading the aforementioned three scattering matrices (according to the prescription of Ref. 2) yields the overall scattering matrix and the transmission T_{total} [1,4] as

$$ T_{total} = \frac{\epsilon[(t_1 + t_2) - (b - a)^2 t_1 t_2 (t_1' + t_2')]}{[1 - t_1(a^2 t_1' + b^2 t_2')][1 - t_2(a^2 t_2' + b^2 t_1'] - a^2 b^2 t_1 t_2 (t_1' + t_2')^2} $$
$$ (2) $$

where ϵ, a and b are the elements of the Shapiro matrix[2], and t and r stand for transmission and reflection amplitudes within the two interfering paths. The subscripts '1' and '2' identify the corresponding path and the unprimed and primed quantities are associated with forward and reverse propagation of the electron.

---

[2] For a definition of these elements, see Ref. 1, 3 or 4.

In the presence of an external potential V inducing the electrostatic Aharonov-Bohm effect, t_1, t_2, t_1' and t_2' transform according to the following rule [4]:

$$\begin{pmatrix} t_1 \rightarrow \hat{t}_1 & t_1' \rightarrow \hat{t}_1' \\ t_2 \rightarrow \hat{t}_1 e^{i\phi} & t_2' \rightarrow \hat{t}_1' e^{i\phi} \end{pmatrix} , \tag{3}$$

where the quantities with the "hats" represent the transmission amplitudes in the absence of the external potential V, and ϕ is the electrostatic Aharonov-Bohm phase-shift between the two paths induced by V and given by

$$\phi = \frac{e}{\hbar} V < \tau_t > = \frac{\sqrt{2m^*E}}{\hbar} [\sqrt{1 + \frac{eV}{E}} - 1]L \tag{4}$$

Here $< \tau_t >$ is the harmonic mean of the transit times through the two paths which depends on V and the kinetic energy E of the electrons, m^* is the electron's effective mass and L is the length of each path.

Using the transformations given by Equation (3) in Equation (2) and assuming that in the absence of the external potential V the two paths are identical in all respects (i.e. $\hat{t}_1 = \hat{t}_2$ and $\hat{t}_1' = \hat{t}_2'$), we obtain

$$T_{total}(\phi) = \frac{e\hat{t}_1(1 + e^{i\phi})(1 - (b - a)^2 \hat{t}_1 \hat{t}_1' e^{i\phi})}{D(\hat{t}_1, a, b, \phi)} , \tag{5}$$

where the denominator D is a function of $\hat{t}_1$, a, b and ϕ.

We find from the above equation that $T_{total}(\phi)$ vanishes and hence the conductance (see Equation (1)) reaches a minimum whenever

$$\phi = (2n+1)\pi , \quad i.e. \ when \quad \frac{\sqrt{2m^*E}}{\hbar} [\sqrt{1 + \frac{eV}{E}} - 1]L = (2n+1)\pi \tag{6}$$

This gives the usual conductance minima (which we call the *primary* minima) associated with destructive interference of transmitted electrons.

However, we find from Equation (5) that $T_{total}(\phi)$ also vanishes whenever

$$(b - a)^2 \hat{t}_1 \hat{t}_1' e^{i\phi} = 1 \tag{7}$$

From the unitarity of the Shapiro matrix (see Ref. 4) it can be shown that $b - a$ differs from unity only by a constant phase factor, i.e

$$b - a = e^{i\nu} \tag{8}$$

Now since in ballistic transport $\hat{t}_1 = \hat{t}_1' = e^{ikL}$ (where k is the electron's wavevector in either path in the absence of the external potential V), Equation (7) really corresponds to the condition

$$2k_1 L + \phi + 2\nu = \frac{\sqrt{2m^* E}}{\hbar}[\sqrt{1 + \frac{eV}{E}} + 1]L + 2\nu = 2m\pi \quad (9)$$

Whenever Equation (9) is satisfied, another set of conductance minima should appear in the oscillations since the numerator of $T_{total}(\phi)$ goes to zero and the conductance should fall to a minimum unless the denominator of $T_{total}(\phi)$ also happens to go to zero at the same time. It is easy to see that the denominator of $T_{total}(\phi)$ vanishes whenever ϕ is an even multiple of π. Hence, unless Equation (9) is satisfied only by those values of ϕ that are even multiples of π (which requires $2(k_1 L + \nu)$ to be also an even multiple of π), the conductance of the structure should reach a minimum whenever ϕ satisfies Equation (9). This gives rise to an additional set of minima which we call the *secondary minima*. Actually, the secondary minima always occur unless $2(k_1 L + \nu)$ is an even or an odd multiple of π. The latter case is not proved here for the sake of brevity, but is proved in Ref. 4.

B. Diffusive Transport

In the case of diffusive transport, $T_{total}(\phi)$ can again be found from the prescription of Ref. 2, except that now we have to evaluate it numerically. We have calculated the conductance G vs. the electrostatic potential V for both ballistic and diffusive transport. The results are displayed in Fig. 1. The secondary minima are not washed out by elastic scattering in the weak localization regime. However, they begin to wash out with the onset of strong localization and with increasing temperature. The effect of temperature has been discussed in Ref. 4. Note also the interesting feature exhibited by the secondary minima; they become more and more pronounced in the higher cycles of oscillations (increasing V) unlike the primary minima. This implies that in an experimental situation, even if the secondary minima cannot be observed in the first few cycles, they could show up in the later cycles.

III. DISCUSSION

Before concluding this paper, we briefly discuss the origin of the secondary minima. Equation (9), which predicts the existence of the secondary minima in the ballistic case, physically represents the condition that an electron reflected *around* the ring interferes constructively with itself at its point of entry into the ring. This minimizes the conductance by maximizing the reflection. Such a phenomenon can be viewed as some kind of "coherent backscattering", but it is not exactly similar to the magnetostatic Aronov-Al'tshuler-Spivak (AAS) effect which also involves backscattering, but specifically involves interference of two backscattered *time-reversed paths*. Conductance modulation due to the interference of time-reversed

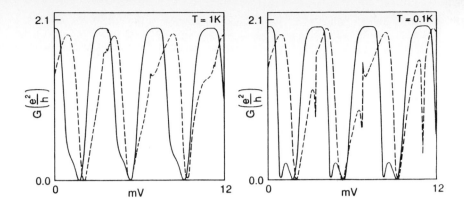

Fig. 1. Electrostatic A-B oscillations in a 1-d ring. The length of each path is 5000 Å. The carrier concentration is 1.55 x 10^6 cm^{-1} and the parameter ϵ = 0.5. The solid curve is for ballistic transport and the broken curve is for diffusive transport. In the latter case, there are 10 elastic scatterers in each path arbitrarily located. Strong localization would have set in if there were 33 scatterers in either path. In both ballistic and diffusive transport, the secondary minima are bleached out much more rapidly than the primary minima as the temperature is increased.

paths cannot occur in the electrostatic case since the time reversed paths always interfere constructively and an external electrostatic potential cannot change that[3]. However, in spite of this basic difference, there is undeniably the superficial similarity between the two effects in that they both double the frequency of the conductance troughs in the oscillations.

REFERENCES

1. Bandyopadhyay, S., and Porod, W. (1988a). Appl. Phys. Lett. _53_, 2323. ; Bandyopadhyay, S., and Porod, W. (1988b). Superlattices and Microstructures, to appear.
2. Cahay, M., McLennan, M., and Datta, S. (1988). Phys. Rev. B15., _37_, 10125.
3. Shapiro, B. (1983). Phys. Rev. Lett. _50_, 747.
4. Cahay, M., Bandyopadhyay, S., and Grubin, H. L. (1989). to appear Phys. Rev. B15 (Rapid Communication).

---

[3]This happens because the phase-shifts suffered by an electron in traveling along opposite directions have the *same* sign in the electrostatic case, but *opposite* signs in the magnetostatic case.

MESOSCOPIC CONDUCTANCE FLUCTUATIONS IN DISORDERED METALS

Adrianus Pruisken and Ziqiang Wang

Department of Physics, Columbia University
New York, New York 10027

I. Introduction

Mesoscopic conductance fluctuations has become one of the most important aspects in nanostructure condensed matter physics [1]. The experimentally observed sample specific but reproducible variations in the conductance as a function of chemical potential and magnetic field are known to be a manifestation of statistical fluctuations about the impurity averaged transport parameter. The problem of statistics in conductance distributions has received considerable attention lately [1,2,3]. In particular, the concept of Universal Conductance Fluctuation (UCF) [1,2] seems to be a commonly accepted explanation of the experimental and numerical results. The question still remains as how to reconcile the anomalous mesoscopic conductance fluctuations in quantum transport with the renormalization group (RG) theory of localization phenomena [4].

Recently, the precise meaning of the statistics in conductance distributions has been studied in the framework of the nonlinear σ model approach to localization. It turns out that the fluctuations about the ensemble averaged transport parameters are described by higher dimensional (irrelevant) operators in the theory which become important when the system *approaches* scaling [5]. The results, however, fundamentally disagree with the existence of UCF and indicate that the universal feature of the fluctuations arises solely due to a mishandling of infrared divergences in perturbation theory.

II. Extended Nonlinear σ Model Representation

Quite clearly, any theory on electronic disorder can not possibly be taken to be complete unless there is a way of analyzing the fluctuations about the ensemble averaged quantities. To this end, as well as for the general interest in understanding the effect of irrelevant operator insertions, the following extended theory has been studied

$$\mathcal{L} = -\frac{1}{8t_o} \int d^D r \mathrm{Tr} \partial_\mu Q \partial_\mu Q + \frac{H_o^2}{4t_o} \int d^D r \mathrm{Tr} Q g_o + \int d^D r \sum_{i=1}^{10} \lambda_i{}^o O_i[Q]. \quad (1)$$

Here, the first two terms represent the ordinary Grassmannian nonlinear σ model with unitary symmetry

$$Q(r) = T^{-1}(r) g_o T(r); \qquad T(r) \in U(2m); \qquad g_o = \begin{pmatrix} \mathbf{1}_m & 0 \\ 0 & -\mathbf{1}_m \end{pmatrix}. \quad (2)$$

The additional dimension four operators O_i are given by

$$
\begin{aligned}
O_1 &= \mathrm{Tr} \partial_\mu Q \partial_\mu Q \mathrm{Tr} \partial_\nu Q \partial_\nu Q; & O_2 &= \mathrm{Tr} \partial_\mu Q \partial_\nu Q \mathrm{Tr} \partial_\mu Q \partial_\nu Q; \\
O_3 &= \mathrm{Tr} \partial_\mu Q \partial_\mu Q \partial_\nu Q \partial_\nu Q; & O_4 &= \mathrm{Tr} \partial_\mu Q \partial_\nu Q \partial_\mu Q \partial_\nu Q; \\
O_5 &= \tfrac{1}{2} (\epsilon_{\mu\nu} \mathrm{Tr} Q \partial_\mu Q \partial_\nu Q)^2; & O_6 &= \mathrm{Tr} \partial_\mu^2 Q \partial_\nu^2 Q - \mathrm{Tr}(\partial_\mu Q \partial_\mu Q)^2; \\
O_7 &= H_o^2 \mathrm{Tr}\{Q, g_o\} \partial_\mu Q \partial_\mu Q; & O_8 &= H_o^2 \mathrm{Tr}\{Q, g_o\} \mathrm{Tr} \partial_\mu Q \partial_\mu Q; \\
O_9 &= H_o^4 \mathrm{Tr}\{Q, g_o\} \mathrm{Tr}\{Q, g_o\}; & O_{10} &= H_o^4 \mathrm{Tr}\{Q, g_o\}\{Q, g_o\}.
\end{aligned}
$$

$$(3)$$

The totality of Eqs. (1)-(3) constitutes the lowest order, non-trivial extension of the conventionally studied theory of localization with unitary symmetries. The derivation closely follows the methodology of Refs.[7,8] and the application to localization theory follows the replica limit ($m = 0$ number of field components).

The renormalization of the extended nonlinear σ model has been studied by employing the background field method in $2 + \epsilon$ dimensions [5,6]. The renormalization of the λ fields involves a 9×9 RG matrix Z_{ij}

$$\lambda_i^o = \mu^{D-4} Z_{ij} \lambda_j \quad (4)$$

where μ is an arbitrary momentum scale introduced in dimensional regularization. The Z_{ij} has been computed in the minimal subtraction scheme and the result to one loop order is

$$Z_{ij} = \delta_{ij} + \frac{t}{D-2} \Lambda_{ij} \quad (5)$$

where

$$
\Lambda = \begin{pmatrix}
0 & 2m & \frac{3}{2} & 1 & 0 & 0 & 0 & 0 & 0 \\
0 & -4m & -1 & -2 & 0 & 0 & 0 & 0 & 0 \\
8 & 4 & 0 & 0 & -4 & 0 & 0 & 0 & 0 \\
-8 & -4 & -m & -2m & 0 & 0 & 0 & 0 & 0 \\
0 & 0 & -2 & 0 & 0 & 0 & 0 & 0 & 0 \\
0 & 0 & 0 & 0 & 0 & -4m & 4m & 0 & 16 \\
-4 & -6 & -3m & -2m & 2 & 0 & -2m & 0 & 0 \\
-4m & -2m & -\frac{3}{2} & -1 & 0 & 0 & 0 & -2m & 0 \\
0 & 0 & 0 & 0 & 0 & \frac{1}{4} & -\frac{1}{2} & -2m & -4m
\end{pmatrix}. \quad (6)
$$

The complete list of RG functions for the t, H and λ fields becomes [5,6]

$$\beta(t) = \epsilon t - 2mt^2 - 2(m^2 + 1)t^3$$

$$\gamma(t) = -2mt + \mathcal{O}(t^3) \tag{7}$$

$$\beta_i(t, \lambda_i) = \frac{\partial \lambda_i}{\partial \ln \mu}|_{t_o, \lambda_i{}^o} = (4 - D)\lambda_i - t\Lambda_{ij}\lambda_j .$$

The critical exponents y_α of the λ fields are expressed in terms of the eigenvalues ω_α of the RG matrix (7),

$$y_\alpha = 4 - D - \omega_\alpha t_c \tag{8}$$

with

$$\omega_1 = 4; \qquad \omega_2 = 0; \qquad \omega_{3,4,5} = -2m;$$

$$\omega_6 = -4; \qquad \omega_7 = -4m; \qquad \omega_{8,9} = -2(2m \pm 1).$$

III. AC Conductance and RMS Conductance Fluctuations

The results of the RG functions (7) do not provide a complete knowledge of the singularities in the transport parameters in the application to localization problems. The remaining question is how the exact expression for the conductivity (Kubo formula) of disordered metal translates into the language of nonlinear σ model. For instance, it has been shown in Ref.[6,9] that the impurity averaged conductivity $\langle \sigma_{xx} \rangle$ can be expressed as an invariant correlation of the σ-model Noether current

$$\sigma_{xx} = \sigma_{xx}^o + \frac{(\sigma_{xx}^o)^2}{8mnDL^D} \int d^D r \int d^D r' \mathrm{tr} \langle Q(r)\partial_\mu Q(r)Q(r')\partial_\mu Q(r') \rangle, \tag{9}$$

where the expectation on the *rhs* is with respect to the bare theory of Eq.(1). The result (9) has been exploited to derive a universal scaling law from the nonlinear σ model [6]

$$\sigma_{xx} = \mu^\epsilon t^{-1} R(t, H^2) \tag{10}$$

where R is given through the closed form expression

$$\frac{H^2}{R^\kappa} = \left(\frac{t_c}{t} \right)^{2/\epsilon} \left(1 - \frac{1 - t/t_c}{R^{1/\epsilon\nu}} \right)^{2/\epsilon} . \tag{11}$$

In this relation, the renormalized parameters t^{-1} and H^2 play the role of CPA conductivity and frequency respectively and it is helpful to think of the renormalization point μ^{-1} as the mean free path for elastic scattering. The critical point t_c ($\beta(t_c) = 0$) in $2 + \epsilon$ dimensions corresponds to the "mobility edge" which separates the metallic phase ($t < t_c$) from the insulating ($t > t_c$) phase. The exponents in (11) are $\kappa = (D - \gamma(t_c))/\epsilon$ and $1/\nu = -\beta'(t_c)$. Equation (11) is the generalization of the DC ($H = 0$) result first obtained by Wegner [10] and is completely analogous to the Widom scaling form for the equation of states in critical spin systems [11].

As far as the conductivity fluctuations are concerned, one can follow a procedure which is similar to the one outlined above for the ensemble averaged conductivity. However, the role of the λ fields is quite different and the matter can only be adequately formulated in a background field formalism applied to the extended theory of (1)-(3). Crudely speaking, the fluctuations about the ensemble averaged quantities (as well as other nonlinear transport parameters) are determined by the behavior of the λ fields at large distances, i.e. they are "irrelevant" in the context of the renormalization group. It has been shown [5] that the conductance fluctuations can be written as a linear combination of 9 independent quantities λ'_α,

$$\lambda'_\alpha = \mu^{D-4}\lambda_\alpha \mathcal{N}_\alpha(t, H^2, \lambda_\alpha) \tag{12}$$

where the form of $\mathcal{N}_\alpha$ is determined by the RG and is given by

$$\mathcal{N}_\alpha(t, H^2, \lambda_\alpha) = \left(1 - \frac{t}{t_c}\right)^{-\omega_\alpha t_c \nu} \mathcal{F}\left(H^2\zeta(t), \lambda_\alpha \eta(t)\right), \tag{13}$$

with

$$\zeta(t) = \left(\frac{t}{t_c}\right)^{2/\epsilon}\left(1 - \frac{t}{t_c}\right)^{-(D+\gamma(t_c))\nu}; \quad \eta(t) = \left(\frac{t}{t_c}\right)^{1-2/\epsilon}\left(1 - \frac{t}{t_c}\right)^{y_\alpha \nu}. \tag{14}$$

In these equations, the λ fields actually stand for the diagonalized version of the original variables in (7). Although Eqs.(12)-(14) clearly demonstrate the importance of the λ fields, one encounters serious difficulties in the attempt to extract explicit scaling forms of $\mathcal{N}_\alpha$ from perturbation theory. For instance, a detailed computation shows that at low temperatures (t)

$$\mathcal{F}(x, y) = 1 + \frac{C_\alpha}{x^{1-\epsilon/2}y} - \frac{\omega_\alpha}{\epsilon}x^{\epsilon/2}. \tag{15}$$

This result indicates that perturbation theory in the presence of the λ fields is actually carried out about the point $\mathcal{N}_\alpha(0, \infty)$ or, more precisely, about the theory with $\lambda_\alpha = \infty$. This drawback should not be a complete surprise since in perturbation theory one deals with ultraviolet singularities of the model. Hence, the result (15), as such, does not provide any information about the behavior of the system at large distances (i.e. $\lambda \sim 0$), a statement which can be carried over directly to the naive formulation of the problem based on impurity diagrams [1-3] (which, by the way, ignores the presence of the λ terms altogether).

The possibility of explicit computations in the regime of physical interest ($\lambda \sim 0$) has been discussed in Ref.[5]. In this reference, a closed form expression was found for the $\mathcal{N}_\alpha$ which is analogous to the result (11) and valid for large λ. This expression, together with the fact that the theory does not describe a phase transition as a function of finite λ, allows one to draw conclusions about the regime of $H, \lambda \sim 0$. The implications for the localization problem can be formulated as follows. By translating the infrared regulator H in terms of the scale factor $b = \mu L$, where L corresponds to the system size, the final result for the conductance ($\Sigma = L^\epsilon \sigma$) fluctuation becomes

$$\langle (\Sigma_{\mu\nu} - \langle \Sigma_{\mu\nu} \rangle)^2 \rangle \propto L^{D-4} \sum_\alpha \theta_\alpha \lambda'_\alpha = \sum_\alpha \theta_\alpha b^{2z_\alpha} \lambda_\alpha^2 \left(b^{z_\alpha} \lambda_\alpha + C_\alpha \right)^{-1} \quad (16)$$

where θ_α and C_α are of order unity. The exponent z_α equals $D-4$ in the metallic regime ($t = 0$). At the mobility edge, $z_\alpha = -y_\alpha$ which is dominated by the smallest y_α. Just as is the case for the parameters t, H and the renormalization point μ, the values for the λ fields are determined by the short distance properties of the disordered electronic system and are non-universal.

Acknowledgment This work has been supported in part by a grant from NSF, grant No. DMR-8600009. One of the authors (AMMP) acknowledges receipt of an Alfred P. Sloan Fellowship.

References

1. P.A. Lee, A.D. Stone and H. Fukuyama, Phys. Rev. **B35** 1039 (1987). and references therein;
2. P.A. Lee and A.D. Stone Phys. Rev. Lett. **55** 1622 (1985).
3. B.L. Altshuler, V.E. Kravtsov and I.V. Lerner, Sov. Phys. JETP **64** 1352 (1986); JETP Lett. **43** 441 (1986).
4. E. Abrahams, P.W. Anderson, D.C. Licciardelo, and T.V. Ramakrishnan, Phys. Rev. Lett. **42** 673 (1979)
5. A. Pruisken and Z. Wang, CU-TP-146 (1989), submitted to Phys. Rev. Lett.
6. A. Pruisken and Z. Wang, Nucl. Phys. **B[FS]**, to be published.
7. A. Pruisken and L. Schäfer, Nucl. Phys. **B200** 20 (1982);
8. A. Pruisken, Nucl. Phys. **B235** 277 (1984).
9. A. Pruisken, Nucl. Phys. **B290** [FS20] 61 (1987).
10. F. Wegner, Z. Phys. **B38** 207 (1979).
11. B. Widom, J. Chem. Phys. **43** 3892; 3898 (1965).

Chapter 7

Related Fabrication and Phenomena

RESOLUTION LIMITS OF ELECTRON BEAM LITHOGRAPHY
and
METHODS FOR AVOIDING THESE LIMITS

A.N.Broers

Cambridge University Engineering Department,
Trumpington Street, Cambridge, CB2 1PZ, England

I. INTRODUCTION

The development of electronic devices that rely on electron tunneling, and the requirement for extremely precise structures to verify theoretical predictions for electron behaviour in solids, have introduced the need for devices with dimensions approaching 1 nm. The resolution of present electron beam lithography methods is inadequate for this task. Structures smaller than 10 nm have been produced by a variety of methods, particularly by the direct sublimation of various compounds, but none of these methods has been usable with a practicable etching or deposition process (1,2,3,4).

This paper describes experiments in which the resolution of the electron beam lithography techniques that have proved useful for the fabrication of electrically testable devices, was measured at higher accelerating voltage than before. In particular, the resolution of PMMA and hydrocarbon vapour resist, known as contamination resist, were measured at 350 kV with a JEOL 4000EX transmission electron microscope (TEM) that was modified to produce a minimum beam diameter of 0.4 nm. The beam was controlled with a computer pattern generator. The resist results are similar to those obtained at lower voltages (50 kV) except that at 350 kV thicker resists can be used and the edge definition in the resist patterns appears to be slightly better. Some of the results have been reported previously(5).

Resist layers exposed at 350 kV were used to mask metal films from ion milling and nitride films from reactive ion etching. With vapour resist, continuous metal lines as small as 5nm wide have been made. This is smaller than has been produced before at lower voltage with similar processes(6). The better resolution is thought to be due to the higher aspect ratio resist profiles produced at 350 kV. We have also selectively closed gaps in silicon nitride films to a minimum dimension of 3 nm by writing vapour resist into the gap.

The use of a TEM rather than a scanning transmission microscope (STEM) for lithography is shown to have considerable advantages.

NANOSTRUCTURE PHYSICS
AND FABRICATION

421

These benefits derive from the projection system which can be used to form a greatly magnified image of the beam thereby making correction of astigmatism and measurement of beam diameter easy even for beam diameters below 1 nm. The fidelity of the computer driven pattern can also be assessed by viewing the pattern directly at high magnification (see for example figure 1). In addition, the immediate availability of high quality TEM observation is of great value when examining samples that are readily distorted by electron irradiation.

II. ELECTRON PROBE

The probe system consists of an LaB$_6$ cathode electron gun, the three TEM condenser lenses and the upper half of the objective lens. For beam writing the four lenses are adjusted so that the Gaussian image of the source at the sample has a diameter of about 0.2 nm. The condenser lenses are also adjusted so that the entrance pupil of the objective lens is filled in an optimum manner. When this is done, the diffraction, aberration and image disks are matched and the minimum total beam diameter is produced.

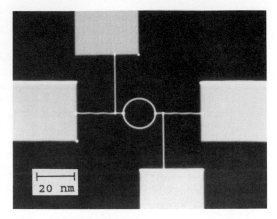

Figure. Highly magnified image obtained with the TEM projection system as electron beam 'writes' pattern for 16 nm diameter ring.

The upper half of the objective lens acts as the final probe-forming lens. The present pole-piece dimensions are not ideal for this purpose, because the lens is designed for transmission electron microscopy with the lower half of the lens field acting as the objective lens, but the aberration coefficients (C$_s$ = 2.6 mm, C$_c$ = 2.8 mm) are still low enough to produce a beam diameter of 0.5 nm for a current of 10^{-12} Amp. For beam writing, the objective lens and the

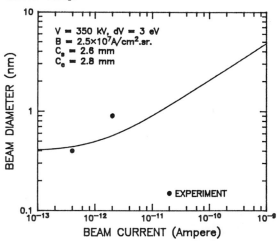

Figure 2. Beam diameter versus current for electron probe used in resist measurements.

projection system are adjusted so that the sample and the beam focus coincide and their images are in focus on the phosphor screen. Accurate focusing of the beam on the sample is achieved by eliminating the halo in the beam image that occurs due to electron scattering and diffraction when the beam focus does not coincide with the sample.

To calculate the beam current versus beam diameter (defined here as the diameter containing 80 % of the beam current) it is necessary to know the brightness of the beam and the energy spread. The brightness was measured at the sample to be about 2.10^7 Amp/cm^2.ster. and the energy spread was assumed to be less than 3 eV based on previous data for LaB$_6$ cathode electron guns operating at similar total currents(7). Beam diameter versus beam current is shown in figure 2 together with experimental data.

Lithography patterns were generated with an IBM PC pattern generator which produces patterns containing rectangles, circles and triangles. The beam is deflected with double deflection coils located between the third condenser lens and the objective lens.

III. PMMA RESIST EXPOSURE DISTRIBUTION

Figure 3 shows part of the test pattern used to measure the resolution of PMMA. The pattern was written in a 50 nm thick layer of PMMA supported on a 100 nm thick Si$_3$N$_4$ membrane. The resist has been shadowed at 45° with 3 nm of PtPd to enhance contrast and to indicate the resist thickness. The sizes of the pattern shapes were chosen so that the minimum 'written' linewidth (3.2 nm) was about four times smaller than the half-width of the fundamental exposure distribution, and the largest

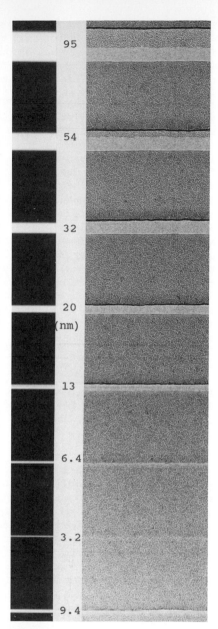

Figure 3. Portion of the test pattern used to measure the resolution of PMMA. On the left is the directly recorded image (obtained without a sample); on the right is the developed and shadowed (at 45°) PMMA resist pattern.

linewidth was about 10 times larger than the half-width (118 nm). For each experiment, the pattern was repeated at seven different exposure doses. The lightest dose was below that needed to open up the largest shapes. The heaviest dose above that needed for the resist to develop through to the substrate in the site of the narrowest line.

Accurate measurement of the width of resist lines below about 50 nm is very difficult because of electron irradiation effects. The method used here to analyze the results and derive the effective exposure distribution, avoids the need to measure the actual width of the resist lines. Only the written linewidth and the exposure dose required for complete development of the line, are required. The method has been described in detail elsewhere(9). The fractional exposure

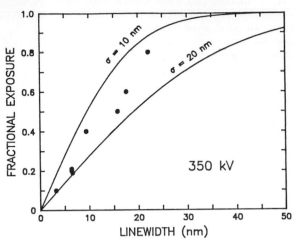

Figure 4. Fraction of large area dose received at the center of lines of different width.

(fraction of the large area exposure dose received at the center of each line) data for 350 kV exposure are plotted in figure 4. They suggest an exposure distribution with a width of about 15 nm and therefore a minimum useful linewidth for closely spaced patterns of about 20 nm (9).

As has been discussed previously, the distribution arises not because of any limitations of the electron beam system, the resolution of which far exceeds the distribution width, but because of the interaction range of the electrons with the resist molecules. The molecular size of the resist does not appear to play a role either, although resolution measurements made to date have been for molecular sizes (molecular weight for PMMA of 20,000 to 500,000) that very much exceed the resolution of the electron beam system and it has been suggested that the resolution might improve for much lower molecular weights.

IV. LIFT-OFF SHADOWING OF PMMA PATTERNS

Figure 5 shows metal patterns produced by evaporating about 2 nm of PtPd onto developed PMMA patterns and then lifting off the metal on top of the resist by dissolving the resist with acetone. The metal produces a 'shadow' image which provides higher contrast in the TEM than the original resist pattern and allows smaller features to be observed.

The metal patterns obtained when the resist was exposed at a dose just below the correct dose, contain hair-line breaks. This is shown in the ring pattern shown in figure 5(A). The ring in figure 5(B) was exposed at a slightly heavier dose and is continuous (to the extent that the thin metal layer is continuous). It is possible that the hair-line breaks are due to single molecules of PMMA bridging the lines. They appear in all areas of the pattern and are predominantly orientated normal to the line. This would be expected because it is the shortest path across the line and therefore molecules oriented in this way would have the least probability of being exposed. The breaks have a width of 2 - 3 nm. If this explanation is correct, it suggests that at least some of the molecules in the resist are stretched out. After exposure, the smaller fragments are dissolved away in the developer and broken molecules left protruding into the exposed area must 'curl up' against the edge of the lines to produce the clean edges observed in the developed patterns.

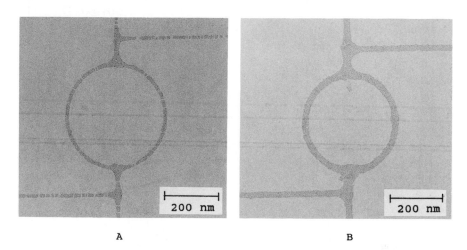

A   B

Figure 5. (A) Lift-off pattern at exposure dose slightly below optimum showing fine breaks, (B) Lift-off pattern at optimum dose.

These explanations are obviously highly speculative and needs further verification but the resolution of the lift-off shadowing method is high enough to make it possible to observe molecular-scale phenomena.

V. VAPOUR OR CONTAMINATION RESIST

The hydrocarbons that condense onto surfaces in vacuum systems form a negative electron resist which acts as a satisfactory mask for ion etching. This 'contamination resist' process was developed in the 1960's and has been used to make a variety of devices(8,9). 350 kV exposure yields very high aspect ratio (up to 50:1) resist lines because of reduced lateral scattering.

The exposure distribution for contamination resist cannot be measured in the same way as that for conventional resists such as PMMA. This is because the rate of build up of resist depends on the hydrocarbon supply in the immediate vicinity of the beam and this varies with the exposure rate and with the pattern density. In our instrument most of the contamination was carried in on the sample and was not deposited in situ. When this is the case there is a steady decrease in the rate of writing in any given region of the sample making it impossible to use the resist pattern used for PMMA. Instead, the width of the exposure distribution was estimated from the linewidths obtained with ion milled metal patterns. For 20 nm thick AuPd layers the minimum linewidths are between 5 nm and 15 nm. This suggest that the resist has a width of about 15 nm. This is similar to the width of the distribution in PMMA and it seems likely that the mechanisms that delocalize the exposure are the same.

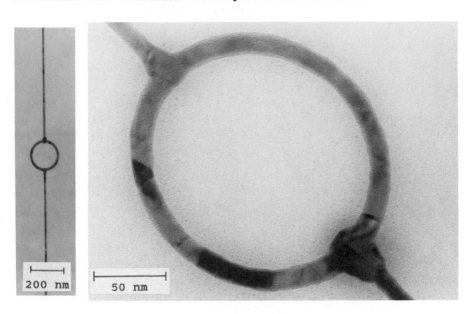

Figure 6. AuPd ring fabricated on Si_3N_4 membrane using contamination resist and ion milling.

Figure 7. 5 nm wide AuPd wire on Si_3N_4 membrane fabricated with contamination resist and ion milling. Wire is about 20 nm thick.

Figure 6 shows an approximately 20 nm thick AuPd ring produced by ion milling through a contamination resist mask. The diameter of the ring is 160 nm and the linewidth is about 15 nm. Linewidth variations are less than ± 1 nm. Figure 7 shows a AuPd wire which has been reduced in width by 'over-etching' to about 5nm.

VI. GAP-CLOSING

The 30 times degradation between the written linewidth and the developed resist linewidth need not provide a fundamental limit to the size of the structures that can be produced, provided the edge definition of the patterns is adequate. If this is the case, it should be possible to narrow the spacing in the resist pattern by uniform deposition on either side of the written or etched lines. The minimum dimension then depends on the edge sharpness and on the uniformity of the etching and/or deposition techniques and not on the initial linewidth. In particular it should be possible to produce gaps in wires and other structures by controllably closing wider openings by depositing or growing material on either side of the gap. It may also be possible to form narrower lines by depositing or etching through the narrowed gaps. Controlled over-etching can also be used to produce narrower lines, as was the case for the wire shown in figure 10.

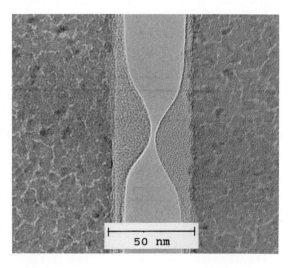

Figure 8 shows an example of gap in a silicon nitride membrane that has been controllably closed to 3 nm. This dimension could not have been obtained by resist exposure. The protrusions on either side of the initially 50 nm wide gap were produced by scanning an electron beam across

Figure 8. Gap in Si_3N_4 membrane closed with contamination writing to a width of 3 nm.

the gap and allowing contamination to build up on either side. The process can be monitored with a resolution of better than 0.5 nm by observing the gap and the scanning electron beam in the TEM mode. When the desired width is achieved, the beam diameter at the sample is greatly increased and the deposition ceases. The sample can subsequently be coated with metal or semiconductor and the coating layer patterned to produce a wire that intercepts the gap at the narrow constriction.

VII. DISCUSSION

The measurements reported here confirm that there is a fundamental limit to the width of an electron beam exposed resist line of about 15 nm. This width is approximately the same at 350 kV as it is at lower voltages and is also approximately the same in vapour resist as it is in liquid spin-coated resist although in practice narrower isolated lines can be produced with vapour resist. The limit is evidently set by the range of the interaction between the electrons and the resist molecules and is not a function of the resolution of the electron optical system, which can be 20 times higher. There is evidence that, with PMMA, the resist molecules are stretched out in the resist rather than coiled up and for the molecular weight range of 20,000 to 500,000 considered here, this indicates that the narrowest lines are smaller than the length of the molecules. The lines must therefore develop out by the developer removing partial molecules.

All of the experiments reported here were made on thin membranes where the influence of backscattered electrons is negligible. For many practical devices, thin membranes cannot be used and backscattered electrons must be considered. As the accelerating voltage increases the area from which the backscattered electrons emerge increases and by 350 kV it is more than 100 μm in diameter. The actual width of the distribution at 350 kV has not been measured. For voltages above about 50 kV, where the width is about 8 μm, the area is so large compared with the dimensions of the structures to be made that the backscattered electrons can generally be ignored. They merely create a faint background fog to the exposure. This would not be the case for a device requiring a dense array of features over an area larger than the backscatter range but at present most devices are small compared to the backscatter range particularly at 350 kV. Large area devices may be needed in the future when many individual devices must be integrated together.

A more significant problem with bulk substrates is that it is more difficult to examine the structures and to position the electron beam with respect to previously fabricated structures in order to make multi-layer devices. Backscattered or secondary electrons must be collected and a scanning image formed. Scanning images do not have the resolution of transmitted images (their resolution can be limited by the same interaction phenomena that limit the resolution of resist exposure) and are less efficient in their formation resulting in larger electron dose to the sample. They can nevertheless be used to achieve resolution down to about 2 nm and to position the beam with the same accuracy. This allows dimensions of 15-30 nm to be usefully employed. It is possible to place backscatter electron detectors inside the pole-piece gap of the immersion probe lenses of the type used in the measurements reported here and obtain a satisfactory surface image.

The limits identified in this paper prevent dimensions below about 15 nm being achieved in useful devices, but to date this limit has not really been a constraint. The smallest dimensions in electrically tested devices is generally about 30 nm, and most device features are 50 nm or greater, so there remains a factor of at least 2 improvement available to workers prepared to spend enough time to reach the fundamental limits of conventional electron beam resist processing. The limits can be reached at any accelerating potential above about 50 kV with higher voltages being better. The best overall compromise, considering cost, appears to be 100 kV.

To reach dimensions below 15 nm in useful devices, it will be necessary, either to perfect complementary processing methods for the direct sublimation processes mentioned in the introduction, or to rely on the uniform deposition of material on either side of written or etched lines in order to define narrower features. The controlled closing of a gap to 3 nm is presented as an illustration of the potential for this approach.

ACKNOWLEDGEMENTS

The author would like to recognize and thank Roger Koch who designed and installed the pattern generator on the TEM fabricator, Arthur Timbs who designed the vibration isolation system on which the instrument is mounted and Mark Lutwyche who prepared many of the samples.

REFERENCES

1. A.N.Broers, J.J.Cuomo, J.Harper, W.Molzen, R.B.Laibowitz, M.Pomerantz, Electron Microscopy 1978, Vol. III, J.M.Sturgess, Ed., Micros. Soc. Canada, Toronto, 1978, p 343.

2. M.Isaacson and A.Muray, J.Vac. Sci. Technol. 19 p 1117-1120, 1981

3. E.Kratschmer and M.Isaacson, J.Vac.Sci.Technol. B4, p 361-364, 1986

4. M.E.Mochel et al Proc. 41st Annual Meeting Electron Microscopy. Soc. of America, San Francisco Press, San Francisco, p 100-101, 1983

5. A.N.Broers, A.E.Timbs and R. Koch, Proc. Microcircuit Engineering 88, to be published 1989.

6. A.N.Broers, W.W.Molzen, J.J.Cuomo, and N.D.Wittels, Appl. Phys. Lett., 29, p 596, 1976

7.   A.N.Broers and H.C.Pfeiffer, Record of 11th Symp. on Electron
      Ion and Laser Beam Technology, R.F.M.Thornley Ed., IEEE N.Y.,
      p 205-207, 1971

8.   A.N.Broers, 1st Int. Conf. on Electron & Ion Beam Sci. &
      Technol., R.Bakish Ed., John Wiley, N.Y., p 191-204, 1964.

9.   A.N.Broers, IBM J.Res. Develop., 32, p 502-513, 1988

QUANTUM DEVICE MICROFABRICATION AT THE
RESOLUTION LIMIT OF ION BEAM PROCESSING

A. Scherer, M.L. Roukes and B.P Van der Gaag

Bellcore
Red Bank, New Jersey 07701

ABSTRACT

We have developed a technique which allows us to substantially decrease the size of electrical devices patterned from high mobility two-dimensional electron gases (2DEG). Our fabrication technique relies on the precise introduction of electrical damage by exposing these shallow GaAs/AlGaAs heterostructures to low energy ion beams. By controlling the ion energy and dose, the lateral spread of the ion-induced electrical damage can be limited to below 20 nm. The optimal ion dose is reproducibly obtained by monitoring the resistance of our devices during the damage process. We have applied this technique to the fabrication of complex self-aligned gated devices with dimensions below 100 nm.

I. INTRODUCTION

If the electrical widths of conducting structures are reduced to approach the Fermi wavelength, quantum confinement effects can be observed in electrical transport. This dimension, λ_F, is ~100 nm for $GaAs/AlGaAs$ two-dimensional electron gases (2DEG) with carrier densities $n_s \sim 10^{11}$ cm^{-2}. Conventional microfabrication methods used to create such narrow structures in 2DEG rely on channel definition by removal of the donor layer [1] or by electrostatic gating [2]. These techniques result in electrical channel widths which are much smaller than the structural (lithographic) dimensions. The large difference between these dimensions severely limits both the complexity of geometries which can be transferred into high-mobility 2DEG and how close adjacent small features may be placed. "Pinched" gate structures, such as point contacts, suffer from the limitation that the electrostatic fields created by the gates act to define the conducting geometry. It is therefore impossible to alter the carrier density in the resulting channels without altering their geometry as well.

To overcome these problems and more fully take advantage of the very high resolution available from electron beam lithography, we have developed a new microfabrication technique. It relies on the controlled spatial modulation of the electrical properties of the 2DEG by selective low energy ion bombardment. Masks are defined to provide selective protection during this process. High-resolution pattern transfer of the mask geometry into the 2DEG is obtained by precise control of the ion beam exposure. This technique produces narrow conducting channels without exposing any free surfaces and thereby avoids serious

431

depletion effects. Device definition has been accomplished by using low-voltage $Ar^+{:}Cl_2$ [3], low-energy He^+ [4], and Ne^+ ion beams [5]. This ion damage patterning technique has permitted us to produce and explore the transport properties of "quantum wires" with *structural* widths well below 100 nm.[6]

In this paper, we describe this microfabrication technique and its application to patterning ultranarrow conducting channels. We present the relationship between the ion beam parameters and the resolution of pattern transfer obtained. Mechanisms responsible for the pattern transfer and their implications on the ultimate resolution obtainable are discussed. Finally, we present recent applications of this technology in which ion damage patterning is used together with electrostatic gating to produce complex quantum devices.

II. PROCEDURE

We use $GaAs/AlGaAs$ 2DEG heterojunction material with an initial mobility of 1×10^6 $cm^2/V{\cdot}sec$ and a carrier density of 3×10^{11} cm^{-2}. Optical lithography is performed to define 100 μm square $Ni/AuGe/Au$ ohmic contacts. These contact pads converge to 20 x 40 μm rectangular fields of view by a Au/Cr lead-frame. SrF_2 masks are then patterned by electron beam lithography using a liftoff procedure. For this study, these masks form a Hall geometry with 100 to 800 nm wide links 6 μm long. After these steps, the samples, now complete except for the final ion beam exposure, are mounted and wirebonded to 28-pin chip carriers. We insert these chip-carriers into a mating socket mounted on a sample introduction rod and wired to permit the samples to be electrically connected to the measurement electronics.

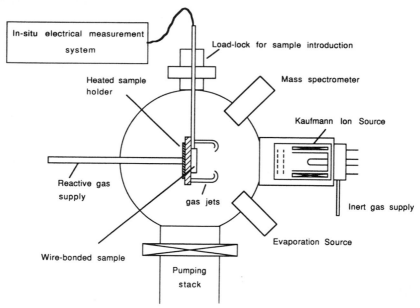

Fig. 1. *Schematic of the ion beam etching system.*

Electrical resistances of the samples in the vacuum system are measured during ion beam exposure using 4-terminal ac measurements. The etching chamber was designed with a load-lock sample introduction system, which allows us to exchange wire-bonded specimens within only a few minutes. A Kaufman ion source is used to generate ion beams of Ne^+, He^+ and Ar^+ with voltages ranging from 100 eV to 1500 eV. The ion flux at the specimen is monitored with a shutter, and varied from 5 to 50 $\mu A/cm^2$. The inert gas flow can be regulated with a mass flow controller to 10 sccm for Ne, 45 sccm for He and 2 sccm for Ar to yield a chamber pressure of $1x10^{-4}$ Torr. Chlorine, used during Ar^+ ion exposure, is introduced close to the sample surface using gas jets (Figure 1).

III. RESULTS AND DISCUSSION

A. Structural vs. Electrical Width

In Figure 2, we present a schematic cross section of a Ne^+ ion-defined 2DEG wire. From this schematic, we can see that the electrical channel width (w_{el}) is smaller than the structural mask width (w_{str}). This difference results from the depletion of electrons beneath each edge of the mask. Figure 2a shows the evolution of the resistance of three narrow wires, $w_{str} \sim$ 100 nm, 300 nm and 500 nm, with increasing exposure to a 200 eV Ne^+ beam. Three regions are observed: below 3×10^{16} $ions/cm^2$, the measured resistance corresponds to the low resistance characteristic of the undamaged 2DEG. At this definition dose (η_{def}), the unmasked regions of the 2DEG are rendered non-conducting and the mask pattern is transferred into the 2DEG. The resistance rises to an initial resistance (R_i), determined by the mask geometry and the the 2DEG sheet resistance. The resistance gradually rises with increasing η as lateral depletion reduces the width of the remaining conducting 2DEG beneath the ion mask. Finally, at the cutoff dose, the channel beneath the mask pinches off completely because of lateral depletion. During this entire ion damage process with Ne^+ at 200 eV, only 10 nm of the cap layer is removed by ion sputtering.

The electrical width of the patterned conductors can be determined approximately using the measured lengths of our devices and the 2DEG sheet resistance. Figure 2b displays w_{el} versus η for the three channels. The electrical width drops rapidly to approach the mask width near η_{def}, and then slowly decreases once a channel is defined. The dose range above η_{def} and below the cutoff dose determines the range in which a conducting structure exists underneath the mask. Clearly, as the size of the conducting structure is decreased, lateral electrical depletion increases in significance. Sub-100 nm structures are extremely sensitive to processing conditions compared to larger structures, and therefore require very precise control over the ion dose delivered. This control can be obtained by directly monitoring their electrical characteristics during ion beam definition.

B. Influence of Ion Beam Parameters

The resolution of the ion damage patterning process depends on the ion energy, mass and dose. Figure 3a shows in-situ resistance measurements obtained when 400 nm wide masks are transferred into the 2DEG by damage at various Ne^+ ion energies. Changes in η_{def} with ion energy demonstrate that fewer ions are required to shut off conduction at 500 eV than at 130 eV. Since the sputtering yield does not change significantly within the energy range, the electrical damage evidently propagates deeper into the heterostructure at higher ion voltages. We also note that the pattern transfer is less directional at higher Ne^+ energies, as the resistance measured immediately after definition increases. This means that at the point of definition, w_{el} is smaller, i.e. lateral depletion is more serious for higher ion energies. If 200 nm breaks are deliberately patterned into the 400 nm masks, the resistance of the

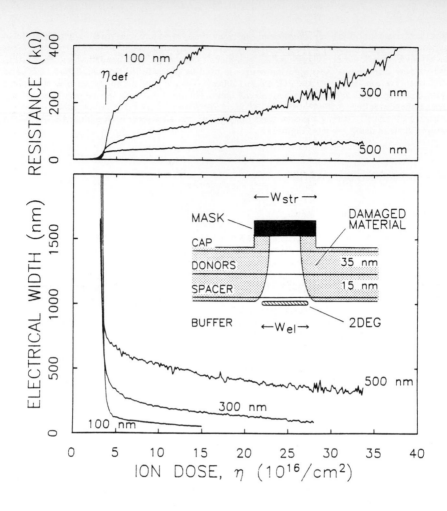

Fig. 2a. *Evolution of the resistance with ion dose, measured during bombardment, for 12 µm long wires with structural (mask) widths, w_{str} as shown.*
Fig. 2b. *Approximate electrical widths deduced from curves of Fig. 1a. The inset shows a schematic cross section of a GaAs/AlGaAs 2DEG wire patterned by selective damage.*

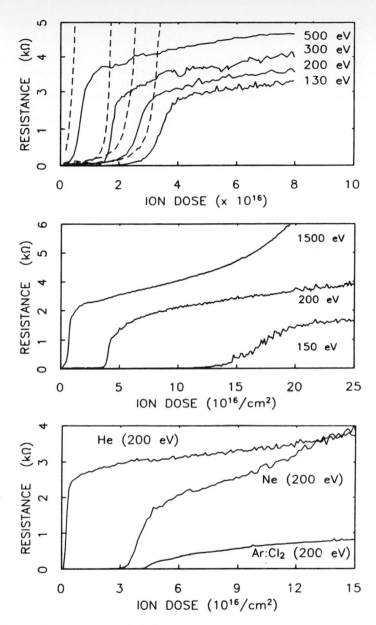

Fig. 3a. *Resistance vs. ion dose for a series of w_{str} = 400 nm wires exposed to different Ne^+ ion energies. The dashed curves show the behavior of wires patterned to include a small break in the mask.*

Fig. 3b. *Resistance vs. dose curves of 400 nm wires exposed to $Ar^+:Cl_2$ at various ion energies. These structures were etched with 2 sccm Ar to form the ion beam and 3 sccm Cl_2 injected close to the sample surface.*

Fig. 3c. *Comparison between the resistance vs. dose behavior of 400 nm wires when exposed to He^+, Ne^+ and $Ar^+:Cl_2$.*

corresponding links is observed to abruptly increase at η_{def}, as is evident from the dashed lines.

A similar set of curves is obtained during $Ar^+:Cl_2$ ion etching at different ion energies, and is shown in Figure 3b. In this case, since the Ar^+ mass is larger, the corresponding electrical damage penetration GaAs/AlGaAs is reduced [7]. In addition, the GaAs etch rate during $Ar^+:Cl_2$ chemically assisted ion etching is considerably higher [8]. Thus, at low Ar^+ ion energies, physical removal of cap and donor layer material plays an important role in determining η_{def}. The definition behavior of 2DEG channels with both Ne^+ and $Ar^+:Cl_2$ differs markedly from that of very light ion species, such as He^+. Even at very low energies where the sputter yield is essentially zero, He^+ generates damage throughout the 2DEG heterostructure, and influences the resistance of the unmasked material at extremely low doses. Figure 3c shows the behavior of 400 nm wide masked 2DEG channels exposed to $Ar^+:Cl_2$, Ne^+ and He^+ at an ion energy of 200 eV. The initial resistance (at η_{def}) of the conducting channels defined is seen to depend very significantly on the ion species, and is highest for the He^+-damaged sample. This results from the larger electrical depletion widths obtained with this gas— He^+-patterned structures with $w_{str} < 300$ nm failed to conduct at low temperatures.

C. Electrical Properties at Low Temperatures

Investigation of quantum transport are carried out at cryogenic temperatures to take advantage of high electron mobilities and to observe confinement effects without thermal smearing. We must therefore adapt our ion damage process to generate sub-100nm structures at low temperatures. We observe experimentally, however, that η_{def} depends on the temperature above roughly 200 K. This can be explained by the existence of low energy, thermally activated, electron traps formed during ion bombardment. In Figure 4, we show the 78 K and 300 K resistance of 400 nm wide 2DEG channels after exposure to 200 eV Ne^+ for several ion doses. For both temperatures, the measurements can be fit assuming that the resistance is generated by the parallel sum of two exponentials. At low doses, electrical conduction through a broad conducting channel dominates, which increases in resistance exponentially due to direct ion bombardment. Above η_{def}, conduction through the masked link determines the resistance. This resistance increases at a slower exponential rate as a result of lateral depletion. From the latter part of this fit, we observe that 400 nm wide structures are even more sensitive to ion bombardment above η_{def} at low temperatures. Sub-100 nm links, which are much more sensitive to lateral depletion, no longer conduct below 200K after exposure to the room temperature definition dose. Therefore, the narrowest structures are successfully fabricated by using lower ion doses at which a parallel conduction path still remains at room temeperature. This parallel conduction through the ion-bombarded 2DEG region "freezes out" below 200K. Low temperature magnetotransport measurements confirm that the resulting devices form well defined high-mobility channels with very narrow electrical widths at 4K.

D. Electrical Damage Mechanisms

The conductivity of a 2DEG, $\sigma = n_s e \mu$, can be reduced to zero by decreasing either the carrier density (n_s) or the mobility (μ). Elsewhere we show that ion beam parameters can be chosen which reduce σ either by predominantly increasing scattering (reducing μ) or by predominantly depleting the 2DEG [9]. In Figure 5 we display further evidence that these two competing mechanisms are operative. In this experiment, unmasked 2DEG with different spacer layer thicknesses are monitored in-situ during overall Ne^+ ion bombardment. The ion dose required to shut off conduction in the 2DEG (η_{def}) is seen to depend on spacer layer dimensions below 20 nm. Above that thickness, η_{def} remains constant. If electrical damage is formed deep enough in the material to locally influence the 2DEG, increased scattering can

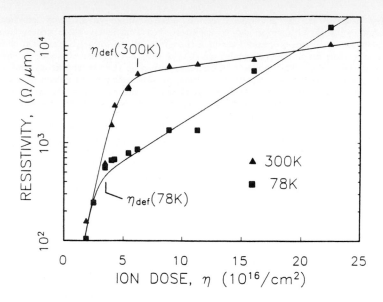

Fig. 4. *Resistivity (Resistance/length) vs. ion dose for a series of $w_{str} = 400$ nm wires, exposed to specific endpoints and measured at 300K and 78K. Each point on the full curve is the average of eight measurements on different length samples. The 300K and 78K definition doses are quite different. A simple expression describing the parallel sum of exponentially increasing resisitivities fits the data over several decades of resistance.*

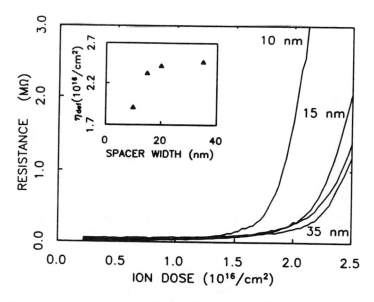

Fig. 5. *Resistance vs. ion dose for broad unpatterned 2DEGs with different spacer layer thicknesses. The inset shows the dependence of the definition dose (η_{def}) on the spacer dimension.*

dramatically decrease σ. In this case, the ion dose required to shut off conduction monotonically increases with spacer layer thickness. For sufficiently thick spacer layers, however, donor electrons can be trapped through the effects of ion-induced damage at or near the surface. This can result in a depopulation of the 2DEG before much local damage is created at heterointerface. This indirect mechanism occuring at or near the surface then dominates η_{def} and reduces σ by depletion, resulting in the observed saturation of η_{def} for large spacer thicknesses. Our studies indicate that this latter mechanism results in extremely high resolution pattern transfer— conducting channels as narrow as 50 nm have been obtained.

E. Determining the Resolution Limit

The resolution obtainable from the ion damage pattern transfer process is limited by lateral depletion. The extent of lateral electrical depletion beneath each edge of the ion etch mask, the "depletion length", can be determined for 200 eV Ne^+ ions at 78 K as a function of ion dose [5]. At each chosen dose, we simultaneously expose Hall bar masks having widths $w_{str} =$ 100, 200, 400 and 800 nm. We characterize each Hall geometry sample by magnetotransport measurements at 4K to ascertain that well-formed narrow channels are defined. By extracting the electrical width from low temperature measurements, and making the simple assumption that a constant dose-dependent depletion length causes a reduction of w_{str} to w_{el} for each of the conducting channels by the same amount, we can deduce an average depletion width for each dose. This depletion width is strongly dependent on the ion dose, and can vary from 15 nm for $\eta = 2.6 \times 10^{16} cm^{-2}$ to 75 nm for $\eta = 5.6 \times 10^{16} cm^{-2}$. Therefore, for 200 eV Ne^+ exposure at the optimum dose, the minimum structural size of a conducting channel at 4 K is 30 nm for the heterojunction we have employed. This minimum width clearly depends not only on the ion dose, but also on the ion mass and energy. Therefore, to improve the resolution below 30 nm, optimization must involve all process parameters, and may require innovations in heterojunction geometry as well.

F. Applications of Ion Damage Patterning

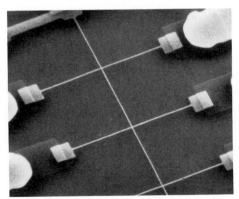

Fig. 6a. *Scanning electron micrographs of a gated 2DEG channel with* $w_{str} = 100$ *nm. The bright lines show the Au gate deposited as a self-aligned ion damage mask.*
Fig. 6b. *SEM micrograph of a series of gated point contacts with* $w_{str} = 200$ *nm and lengths of 200, 400 and 600 nm. The brigtht line shows the Au gate, and the dark areas show the* SrF_2 *mask used to define the undisturbed 2DEG areas.*

Patterning of 2DEG heterostructures by selective ion damage permits the fabrication of high-mobility channels in which the electrical path geometry is controlled by the mask, and the carrier density is not degraded by the fabrication process. If a metallic mask is used as a self-aligned gate and ion damage mask, the carrier density in a narrow conducting channel can be changed with no corresponding change in the conductor geometry. Magnetic depopulation studies [10] in such wires verify that the electrical width remains constant even if the carrier density is modulated significantly. This behavior contrasts with that found with split gates, where both geometry and carrier density in the channel changed with gate bias [11]. Figure 6a shows a gated 100 nm wire in which the geometry is determined by the self-aligned gold wire used as a mask during ion damage patterning. This same self-aligned technique — in the limit of very short length channels — can be used to produce a novel type of point contact. In these new structures, the Fermi energy is swept through subbands occuring at eneries which are fixed by the channel geometry. Figure 6b shows a SEM micrograph of a series of 200 nm wide point contact with 200, 400 and 600 nm lengths, which connect to large undisturbed 2DEG areas. These point contacts show clean quantization even at temperatures as high as 4K, demonstrating the well-defined geometries attainable by ion damage patterning. [10] This technique can easily be extended to create parallel arrays.

IV. CONCLUSIONS

We have developed techniques to routinely and reproducibly fabricate conducting structures in high-μ 2DEGs with dimensions at or below 100 nm. In the course of optimizing this process, we have investigated the limits to the resolution of this pattern transfer process for a range of ion beam parameters. When used together with electrostatic gating, this patterning method allows us to microfabricate ultrasmall devices with new transport properties.

REFERENCES

1. H. van Houten, B.J. van Wees, M.G.J. Heijman and J.P Andre, Appl. Phys. Lett., 49, 1781 (1986) / G. Timp, A.M. Chang, J.E. Cunningham, T.Y. Chang, P. Mankewich, R. Behringer, and R.E. Howard, Phys. Rev. Lett., 58, 2814 (1987).
2. T.J. Thornton, M. Pepper, H. Ahmed, D. Andrews, G.J. Davies,.br Phys. Rev. Lett., 56, p1198 (1986).
3. A. Scherer, M.L. Roukes, H.G. Craighead, R.M. Ruthen, E.D. Beebe J.P. Harbison, Appl. Phys. Lett., 51, 2133 (1987).
4. T.L. Cheeks, M.L. Roukes, A. Scherer, and H.G. Craighead, Appl. Phys. Lett., 53, 1964, (1988).
5. A. Scherer, M.L. Roukes, to be published in Appl. Phys. Lett. (1989).
6. M.L. Roukes, A. Scherer, S.J. Allen, Jr., H.G. Craighead, R.M. Ruthen, E.D. Beebe, and J.P. Harbison, Phys. Rev. Lett., 59, 3011 (1987).
7. S.W. Pang, M.W. Geis, N.N. Efremow, and G.A. Lincoln, J. Vac. Sci. Technol., B3, 389, (1985).
8. G.A. Lincoln, M.W. Geis, S.W. Pang and N.N. Efremow, J. Vac. Sci. Technol., B1, 1043, (1983).
9. M.L. Roukes and A. Scherer, and T.L. Cheeks, unpublished.
10. T. Thornton, M.L. Roukes, A. Scherer, B.P. Van der Gaag, Prodeedings of the NATO Advanced Research Workshop "Science and Engineering of 1- and 0-Dimensional Semiconductor Systems, S.P. Beaumont, ed., Plenum, 1989 (to appear).
11. C.J.B. Ford, T.J. Thornton, R. Newbury, M. Pepper, H. Ahmed, D.C. Peacock, D.A. Ritchie, J.E.F. Frost, and G.A.C. Jones, Phys. Rev. B38, 8518 (1988)

FABRICATION OF NANOMETER FEATURES WITH A SCANNING TUNNELING MICROSCOPE

Edward E. Ehrichs
Alex L. de Lozanne

Department of Physics
University of Texas
Austin, Texas   78712

I.   INTRODUCTION

There is increasing interest in producing structures of nanometer size for use in device fabrication and the probing of physical laws in nanometer sized structures.   As this field is growing, it is important to investigate new techniques for nanometer device fabrication.   Several groups have shown that the STM can be used to modify surfaces on nanometer scales (1-12).   We have developed a technique which allows us to fabricate metal and contamination resist lines with linewidths as small as 10 nm with a Scanning Tunneling Microscope (STM).

II.   WRITING TECHNIQUE

In its ordinary mode of operation, the STM can make spectroscopic and topographic maps of surfaces with angstrom resolution (13).   Typically, the voltage between the tip and sample is maintained between $\pm$ 10 mV and $\pm$ 2 volts.   In this regime, the electrons do not have enough energy to overcome the work function barrier of the tip or sample and exist as free electrons between them.   They can however tunnel between the tip and sample if the tip is sufficiently close to the sample.   For a large class of materials, scanning with the STM does not alter the surface if used properly.

We can use this same STM to write very small features on surfaces.   We evacuate the microscope chamber and then introduce an organometallic or other metal-containing gas which is easily decomposed.   The voltage between tip and sample is increased in a controlled way until the microscope is in the field emission mode.   At this point, the electrons can acquire enough energy to decompose molecules of the gas in an adsorption layer under the tip.   The metal from the decomposed gas is deposited under the tip onto the substrate. The pattern that is formed can be controlled by scanning the

NANOSTRUCTURE PHYSICS
AND FABRICATION

tip over the surface in the desired pattern. The deposition process ends when the voltage between the tip and sample is reduced back to the tunneling regime. At this point, any surface modifications can be observed by taking a picture of the surface in the normal operating mode with the STM.

The STM that we use has a standard tripod piezo configuration which has been described elsewhere (4). The microscope is operated in a vacuum chamber with a base pressure of 10^{-8} torr. Most of our work has been carried out on n-type silicon (111) surfaces with resistivities from 0.5 to 13 Ω cm. We have found that the process works well on metal surfaces also (5). Silicon has many advantages. It is hard and has a high melting temperature, so it resists surface modification by tip damage or excessive current between the tip and sample. Lightly doped silicon has the advantage that it is conductive enough at room temperature to drain the tunneling and writing currents of the STM and yet it becomes insulating at low temperatures allowing us to measure the conductivity of features written on the surface. We prepare the silicon samples as follows. They are degreased by rinsing in de-ionized water, ultrasonically cleaning in ethanol for 5 min, boiling in trichloroethylene for 10 min, ultrasonically cleaning in ethanol for 5 min, and rinsing in de-ionized water. The oxide layer is then removed by soaking in 49% HF solution for 15 s and rinsing in de-ionized water. The samples are then blown dry with helium, loaded onto the microscope, and the chamber is immediately pumped down to vacuum.

III. FEATURE FABRICATION

The first organometallic gas that we used was dimethyl cadmium. For these experiments, in imaging mode, we biased the sample to -1.7 volts and grounded the tip. Initially, to switch to writing mode, we simply increased the tip to sample voltage to 5 or more volts. With this configuration, we were able to write 25 nm features (4). As the microscope is switched from tunneling to field emission mode, the current increases and the feedback compensates by retracting the tip. It was our feeling that this might limit the resolution of the process, so we decided to use voltage pulses for writing. If the pulses have a low duty cycle and the feedback time constant is much longer than the pulse duration, the tip should be maintained quite close to the sample. Using this process, we wrote two large lines (1μm width) on silicon. The pulse voltage was 8 volts, pulse duration 10 ns, and pulse frequency was 130 Hz. We intentionally made them large so they could be found and analyzed with Auger Electron Spectroscopy (AES). The results of this analysis indicate that there is cadmium in the line area and there is none in an area away from the line. The analysis Also indicates that there is a large amount of carbon in the deposit (1).

At this time, we found that we were able to make deposits without introducing any organometallic gas into the chamber. We made a large line of this material and analyzed it with AES. The results indicated that the deposits were primarily carbon. We concluded that we were observing a process described by Laibowitz and Umbach (14). Surface organic contamination is decomposed leaving a carbon deposit. The deposits can drain the tunneling current during imaging, so they are not insulating.

In order to reduce the amount of surface carbon contamination, we began using argon plasma cleaning on our samples prior to writing (1). This has reduced the surface contamination and allowed us to make our smallest features with the microscope to date. Figure 1 shows a contamination resist grid pattern with a 10 nm line width . These features were made with 4.5 volt 400 ns pulses with a frequency of 13 kHz. The pattern generation was controlled by the computer.

IV. TRANSPORT MEASUREMENTS

Our next goal was to measure some of the transport properties of our features. Most of the features for these experiments were made with the tip negative with respect to the sample. McCord and coworkers have determined that this is important for increasing the metal content of the deposits (2). The obvious approach was to fabricate contact pads on silicon using conventional lithography and connect the pads with STM lithography. The problem is that our microscope has a maximum scan distance of about 15 μm. It would be virtually impossible to load a sample onto our microscope positioned to within a few microns. The solution was to fabricate an interdigitated finger pattern. Each of two contact pads are connected to 25 fingers, 1 mm long and 5.5 μm wide. The entire structure is 2 mm long and 0.5 mm wide. It is easy to get the tip of the microscope somewhere over this pattern. Then, we connect two of the fingers with the STM writing across a 5.5 μm gap in order to connect the two contact pads. We fabricate two identical such patterns on each chip so that one can act as a control.

The first feature whose resistivity we tried to measure was a carbon contamination resist line. The line was about 5.5 μm long by 0.3 μm by 0.3 μm. It was found experimentally

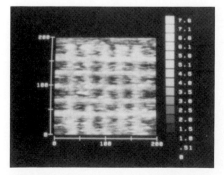

FIG. 1. STM 200 x 200 nm image of contamination resist pattern.

FIG. 2. Optical micrograph of aluminum deposit.

443

that if the turbo pump is accidentally shut off overnight, back streaming oil from the mechanical pump provides abundant material for writing contamination resist lines. As we cooled the chip, the resistance of the pattern with the line never deviated significantly from the control pattern. The resistance was greater than 40 MΩ which gives a resistivity of greater than 65 Ω cm.

We have also made aluminum features. The gas precursor was trimethyl aluminum. So far, the only aluminum feature which has conducted current was a large aluminum blob. Our smaller lines had breaks due to roughness of the substrate. We wrote with 10^{-3} torr trimethyl aluminum and continuous (not pulsed) tip voltage of -30 volts. We did not use voltage pulses so the feed back system retracted the tip and the deposit spread out to be a circle 7 μm in diameter (figure 2). AES analysis indicates that the surface of the deposit contains 10% aluminum, 56% carbon, and 16% oxygen. Figure 3 shows a graph of the log of the resistance vs. inverse temperature for the pattern with the aluminum and the control pattern. The resistance of the control pattern continues to increase as the temperature drops and carriers freeze out of the silicon substrate. The resistance of the pattern with the aluminum assumes a constant value of 7100 Ω as current is conducted through the deposit. Unfortunately, we cannot assign a resistance to the aluminum deposit itself because the resistance of the chrome fingers is on the order of 7 kΩ. We can however give an upper cutoff to the aluminum deposit resistance of 7 kΩ. Assuming a height of about 0.3 μm for the deposit (the maximum distance the feedback can retract the tip), this gives an upper cutoff of the resistivity of less than 0.3 Ω cm.

V. ETCHING

In addition to depositing materials on surfaces, the STM can be used to etch surfaces. In this experiment, we introduced 30 mTorr of tungsten hexafluoride into the chamber. After imaging the silicon surface, we attempted to

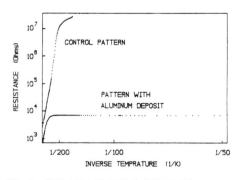

FIG. 3. Resistance of control pattern and aluminum deposit.

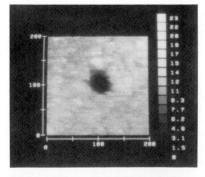

FIG. 4. STM 200 x 200 nm image of etched pit on silicon surface.

write at the center. We used -20 volt pulses with a pulse width of 100 ns, and a frequency of 200 Hz. After attempting to write, we imaged the surface again and found that instead of writing a feature we had etched a pit (figure 4). It is unlikely that the tip itself made the hole because the silicon is very hard and contact between the two always bends and wrecks the tip. We were able to obtain good pictures after etching which indicates that the tip was still in good condition. E-beam etching with tungsten hexafluoride has been observed by Matsui and coworkers (15). They observed that tungsten was deposited on silicon under electron bombardment, but that etching took place if the substrate temperature was increased beyond 50 °C. During the writing pulse, the current density (which is governed by field emission instead of tunneling) may be over 1000 times greater than during tunneling (16). This could cause local heating of the silicon substrate inducing etching.

VI. CONCLUSION

The STM shows great promise for being a useful tool in the fabrication of nanometer structures. It has the ability to deposit metals, contamination resist, and to etch with 10 nm resolution. The process does not rely on any complicated mask scheme, so it is trivial to generate new and different patterns every time. Much further work is required to characterize the process.

REFERENCES

1. E. E. Ehrichs, S. Yoon, and A. L. de Lozanne, Appl. Phys. Lett. 53, 2287 (1988).
2. M. A. McCord, D. P. Kern, and T. H. P. Chang, J. Vac. Sci. Technol. B 6, 1877 (1988).
3. O. E. Hüsser, D. H. Craston, and A. J. Bard, J. Vac. Sci. Technol. B 6, 1873 (1988).
4. E. E. Ehrichs, R. M. Silver, and A. L. de Lozanne, J. Vac. Sci. Technol. A 6, 540 (1988).
5. R. M. Silver, E. E. Ehrichs, and A. L. de Lozanne, Appl. Phys. Lett. 51, 247 (1987).
6. J. S. Foster, J. E. Frommer, and P. C. Arnett, Nature 331, 324 (1988).
7. J. Schneir, R. Sonnenfeld, O. Marti, P. K. Hansma, J. E. Demuth, and R. J. Hammers, J. Appl. Phys. 63, 717 (1988).
8. U. Staufer, R. Wiesendanger, L. Eng, L. Rosenthaler, H. R. Hidber, and H. J. Güntherodt, Appl. Phys. Lett. 51, 244 (1987).
9. D. W. Abraham, H. J. Mamin, E. Ganz, and J. Clarke, IBM J. Res. Dev. 30, 492 (1986).
10. M. A. McCord and R. F. W. Pease, J. Vac. Sci. Technol. B 4, 86 (1986).
11. C. W. Lin, F. R. Fan, and A. J. Bard, J. Electrochem. Soc. 134, 1038 (1987).
12. R. S. Becker, J. A. Golovchenko, and B. S. Swartzentruber, Nature 325, 419 (1987).
13. P. K. Hansma and J. Tersoff, J. Appl. Phys. 61, R1 (1987).
14. R. B. Laibowitz and C. P. Umbach, in Percolation, Localization, and Superconductivity, edited by M. A. Goldman and S. A. Wolf (Plenum, New York, 1984), pp. 267-286.
15. S. Matsui and K. Mori, J. Vac. Sci. Technol. B 4, 299 (1986).
16. Y.Z. Li, L. Vazques R. Piner, R. P. Andres, and R. Reifenberger, J. Appl. Phys. (in the press).

METAL-CHALCOGENIDE PHOTORESISTS FOR HIGH RESOLUTION LITHOGRAPHY AND SUB-MICRON STRUCTURES

A.E. Owen, P.J.S. Ewen, A. Zakery

Department of Electrical Engineering
University of Edinburgh
Edinburgh EH9 3JL
United Kingdom

M.N. Kozicki, Y. Khawaja

Center for Solid State Electronics Research
College of Engineering and Applied Sciences
Arizona State University
Tempe, AZ 85287-6206

I. INTRODUCTION: METAL PHOTO-DISSOLUTION IN CHALCOGENIDES

At least seven distinct photo-induced structural or physico-chemical changes have been observed in amorphous (glassy) chalcogenides when samples in a suitable form are exposed to light or other irradiation viz: photo-crystallization, photo-polymerization, photo-decomposition (in compounds), photo-induced morphological changes, photo-vaporization, photo-dissolution (of certain metals), and light induced changes in local atomic configuration (Owen, Firth and Ewen, 1985). All of the photo-induced phenomena affect chemical solubility to some extent and are, therefore, suitable for photoresist applications. Particularly marked changes in solubility occur, however, on the photo-dissolution of metals and it is with the application of this phenomena that the present paper is concerned.

The normal configuration used to observe the effect is a 0.1 - 1 micron thick amorphous chalcogenide film which has been deposited on (or beneath) a thin metallic (or metal compound) layer, typically 0.01 - 0.05 micron in thickness. Illumination from either side with light (or other sources of irradiation with energy above or slightly below the band gap for the chalcogenide) causes the metal to dissolve rapidly into the amorphous film and migrate through it. The effect has been observed in many chalcogenide systems e.g. As-S, As-Se and Ge-Se, and a variety of metals, metal compounds and alloys have been used as the metal source e.g. Ag, Cu, Au, Zn, Ag_2S, $AgNO_3$ and alloys in the Ag-Cu system. The phenomenon does, however, seem to be unique to the chalcogenide compounds. High concentrations of metal may be driven considerable

447

depths by the photo-dissolution process. In As_2S_3, for example, Ag concentrations of 10 - 40 at. % are reached at depths of 20 micron or more on exposure for about one minute to UV light of intensity 10 - 100 mW/cm² (Kokado, Shimizu and Inoue, 1976). A very important characteristic of photo-dissolution is that for vertical illumination the metal migrates vertically with negligible lateral movement in most cases, notably in As-S compounds. Also, in the case of Ag dissolution, at any rate, it is generally observed that the concentration profile is step-like, implying that the phenomenon is not a straightforward diffusion process (Goldschmidt, Bernstein and Rudman, 1977).

II. METAL-CHALCOGENIDE PHOTORESISTS

The metal photo-dissolution process in chalcogenides possesses the two basic properties required of a high-resolution photoresist technology for applications in VLSI fabrication, nanometer electronics and nanometer scale opto-electronics. As mentioned above, in most metal-chalcogenide combinations there is negligible lateral movement of the metal. It has been demonstrated that the ultimate resolution could be <10 nm (Mizushima and Yoshikawa, 1981). In addition, metal-free ("undoped") chalcogenide compounds are readily soluble in dilute alkali solutions and are rapidly dry etched in plasmas of CF_4 and SF_6. By contrast, the metal-doped chalcogenide is virtually insoluble in any solutions except some very strong acids (e.g. hot concentrated H_2SO_4); it is also etched only very slowly by CF_4 (and SF_6) plasmas. It has recently been shown, however, that Ag-doped As-S compounds can be selectively dry-etched in a plasma of sulfur gas (Belford, Hajto and Owen, 1989). On development, therefore, the contrast is high (>10).

Because of their high contrast and high absorption coefficient in the optical and UV parts of the spectrum, chalcogenide photoresists, unlike organic resists, do not suffer problems with standing wave formation or with matching the modulation transfer function of the exposure system, and so can be used to achieve sub-micron resolution using conventional optical sources. Indeed, as a "bonus" of the mechanics of metal-photodissolution, there is an edge sharpening effect which enhances the resolution beyond what is predicted by diffraction-limited optics (Tai *et al*, 1980). With excimer laser sources metal-chalcogenide photoresists have been shown to have a resolution capability of 0.2 micron and a contrast >10 (Polasko *et al*, 1985). Chalcogenide resists are also sensitive to X-rays, ion beams and electron beams. Their high contrast makes them attractive for use with electron beam sources and some experimental results are given in section III(i).

Chalcogenide resists can be deposited by a variety of techniques such as spin coating from a solution, thermal vacuum evaporation and sputtering. Spin coating is a standard technique in microelectronics and in the commercial fabrication of films for holographic recording; it can be used to produce films up to 100 micron or more in thickness, which is important in some optical applications. Vacuum evaporation and sputtering yield uniform films in the range 0.01- 10 micron. Very thin films (<100 nm), for multi-layer lithography, can also be prepared either by spin coating or vapor deposition methods.

III. EXPERIMENTAL

(i) High Resolution Lithography

Experimental work has been concentrated mainly on base glasses in the As-S system and, for lithographic studies, especially the combination Ag-As$_{33}$S$_{67}$. The reason for the choice of that particular composition is that, with metallic Ag as the photo-dissolution source, the final composition of the fully photo-doped material is a homogeneous amorphous phase roughly in the center of the 3-component Ag-As-S phase diagram and close to the composition AgAsS$_2$ as shown in Figure 1 (Owen, Firth and Ewen, 1985). For ultimate lithographic resolution it is obviously desirable to ensure that the photoresist is a uniform single-phase material; "grainy" materials tend to be limited to dimensions in the order of the grain size. We have already shown that the Ag-As$_{33}$S$_{67}$ photoresist is capable of resolving lines and spaces of 0.4 micron or better, using relatively crude wet etching (developing) methods, in a conventional optical wafer stepper (Firth *et al*, 1985).

Recent experiments have evaluated the application of the same Ag-As$_{33}$S$_{67}$ photoresist in ultra-high resolution electron beam lithography using a converted ISI 100B SEM controlled by a computer based scanning system (Bernstein *et al*, 1988). A grating structure has been successfully fabricated, with lines and spaces of 35 nm, again using wet etching methods. The grating was fabricated on a silicon substrate; a Ag layer approximately 30 nm thick was first deposited and that was overlayed with a vacuum evaporated film of approximate composition As$_{33}$S$_{67}$, 150 nm thick. The films were then exposed at 40 keV with a 5.5 nm diameter beam. The pixel spacing was 22 nm with a current per pixel of 2.7 pA and a time per pixel of 20 ms. This resulted in a line dose of 2.5 x 10^{-8} C/cm, which is approximately 100 times higher than that required for the most sensitive commercially available electron resists. The exposure pattern contained lines separated by progressively smaller spaces. A 20 s develop in a mixture of sodium hydroxide developer (KTI 351), isopropyl alcohol (IPA) and deionized water (1:25:50), was used to produce the minimum dimensions described above.

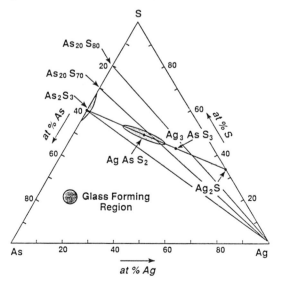

Fig. 1. The three component system Ag-As-S.

(ii) Electronic and Optical Effects

Photodoping with a metal affects all the physical and chemical properties of the base chalcogenide, including electronic and optical properties. Moreover, it is possible to exercise control of the properties of the resultant material simply by varying the amount of metal available for dissolution. Typical results for the optical absorption edge and refractive index are illustrated in Figures 2 and 3 respectively. The base glass in both cases was $As_{30}S_{70}$ and layers of Ag varying in thickness from 50 to 150 nm provided the source. Note that the optical energy gap varies in a smooth fashion from 2.5 eV in the undoped $As_{30}S_{70}$ to approximately 1.9 eV. Similarly, the refractive index varies smoothly; for example, at a wavelength of 632 nm, from approximately 2.35 in the undoped material to 3.10. At a wavelength of 10.6 micron the corresponding change is 2.21 to 2.73. Moreover, the reflectivity of $As_{30}S_{70}$ is extremely high for infra-red wavelengths, with a sharp transition between the optical and infra-red at around 1.1 micron.

IV. CONCLUSIONS

The metal photodissolution effect seems to be unique to amorphous chalcogenide compounds, as are the other photo-induced effects mentioned in the introduction. Certainly the authors are not aware of any other phenomenon or system in which such dramatic <u>but controllable</u> changes can be photo-induced in the chemical and physical properties of a class of materials so readily adapted to a variety of thin- and thick-film fabrication methods. A substantial degree of control can be exercised over the "magnitude" of properties such as refractive index and optical energy gap and, even more significantly in the present context, over the size or "scale" of a given property across the area of a chalcogenide film.

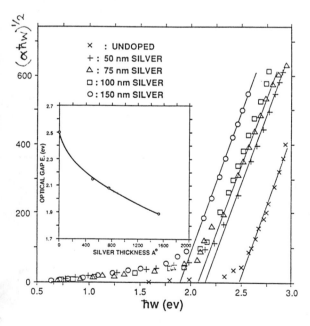

Fig. 2. The optical absorption edge and optical energy gap of 0.5 micron $As_{30}S_{70}$ films photodoped from Ag layers of varying thickness.

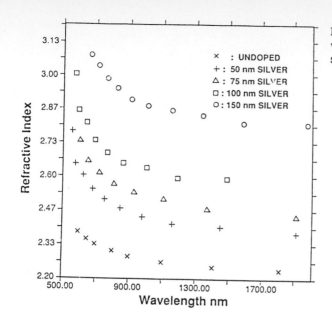

Fig. 3. Refractive index vs. wavelength for the same series of samples shown in Fig. 2.

As demonstrated, the metal photo-dissolution effect has an inherently high spatial resolution, well into the submicron range using optical lithography and on the scale of a few tens of nanometers or less with electron beam exposure. This unusual combination of properties, their easy fabrication in thin- and thick-film form and the ability to selectively etch either the undoped or doped material makes the chalcogenide compounds uniquely appropriate to a variety of microelectronic and nanometer scale applications e.g. high resolution lithography, 3-dimensional and planar waveguide structures in integrated opto-electronics and small geometry "diffraction grating" type structures.

REFERENCES

Belford, R.E., Hajto, E., and Owen, A.E. (1989). To be published in Thin Solid Films.
Bernstein, G.H., Liu, W.P., Khawaja, Y., Kozicki, M.N., Ferry, D.K., and Blum, L. (1988). J. Vac. Sci. Technol., B6, 2298.
Firth, A.P., Ewen, P.J.S., Owen, A.E., and Huntley, C.M. (1985). Proc. SPIE, 539, 160.
Goldschmidt, D., Bernstein, T., and Rudman, P.S. (1977). Phys. stat. sol. (a), 41, 283.
Kokado, H., Shimizu, I., and Inoue, E. (1976). J. Non-cryst. Sol., 20, 131.
Mizushima, Y., and Yoshikawa, A. (1981). "Amorphous Semiconductor Technologies and Devices", Ed: Hamakawa (Tokyo: Ohmsha), p.277.
Owen, A.E., Firth, A.P., and Ewen, P.J.S. (1985). Phil. Mag. B, 52, 347.
Polasko, K.J., Pease, R.F.W., Marinero, E.E., and Cagan. M.R. (1985). J. Vac. Sci. Technol., B3, 319.
Tai, K.L., Vadimsky, R.G., Kemmerer, C.T., Wagner, J.S., Lamberti, V.E., and Timko, A.G. (1980). J. Vac. Sci. Technol., 17, 1169.

DEFECT MOTION, ELECTROMIGRATION
AND CONDUCTANCE FLUCTUATIONS
IN METAL NANOCONTACTS

D. C. Ralph, K. S. Ralls, and R. A. Buhrman

School of Applied and Engineering Physics
Cornell University
Ithaca, N. Y. 14853-2501

In recent years, the study of small electronic devices has proved a fruitful area of investigation. Many physical effects can be seen in small devices that, in effect, are averaged away in macroscopic samples so as to be unobservable. In this spirit we are fabricating exceedingly small metal constrictions, with size of order $< (20$ nm$)^3$, between two clean three-dimensional electrodes. These nanoconstrictions are small enough that the motion of one atomic-size defect causes a directly measurable change in the conductance through the device, allowing studies of the dynamics of individual defect motion within the metal. Under high bias the devices may be "nanomachined" by electromigration, which can change their size and introduce defects. We find that by introducing defects in a gradual fashion we can tune the transport properties from a clean regime, to a regime containing time-independent conductance fluctuations, and ultimately to a weak localization regime.

The nanobridges are fabricated using electron beam lithography to pattern a 40 - 100 nm hole in PMMA, and then using a reactive ion etch to transfer the pattern to a 50 nm suspended Si_3N_4 membrane. The etching time is chosen so that the hole just breaks through the far side of the Si_3N_4 membrane. The wafers are then rotated while a 200 nm coating of metal is evaporated to fill the orifice and form the constriction in a single processing step. In this paper we will focus on results from copper constrictions, but recently we have fabricated aluminum and palladium constrictions as well. Scanning electron microscope studies show that while the openings on the patterned sides of the orifices are about 40 nm wide, the openings on the far side of the Si_3N_4 are much smaller. We can estimate the

453

constriction width by approximating the devices as lengthless constrictions of radius r between bulk electrodes. For a free electron metal, the resistance of such a point contact as determined by a Boltzmann formalism is approximately (1,2)

$$R = \frac{h}{e^2} \frac{2}{(k_F r)^2} \left(1 + \frac{3\pi}{8} \frac{r}{l_{tot}} \right).$$

Here k_F is the fermi wave vector and l_{tot} is the electron mean free path. In copper films that were deposited in the same evaporations as our constrictions, the elastic mean free path at 4 K is 180 nm, as determined by residual resistivity. Assuming that this bulk value for l_{tot} holds throughout the copper constriction, the above formula gives an estimate of 3 nm for the radius of a typical 30 Ω sample.

The samples are characterized with point contact spectroscopy (PCS), in which the second derivative of the I-V curve for a clean constriction gives the phonon density of states times a weakly energy-dependent coupling parameter (1,2). High quality PCS spectra, such as figure 1, indicate that the as deposited constrictions are ballistic. Each is probably a single crystallite, without grain boundaries to scatter electrons strongly in the constriction region. Further evidence to support this view comes from the high electric fields and current densities that the samples are able to withstand. One hundred ohm constrictions are stable against a bias of 10 mV, which corresponds to an electric field of 10^4 V/cm and a current density of 10^8 amp/cm^2 in the constriction. Polycrystalline copper films begin to electromigrate at field strengths an order of magnitude less than this due, it is believed, to enhanced diffusion along grain boundaries (3).

As discussed previously by Ralls and Buhrman (4), at low temperatures (T<150 K) and moderate DC biases (< 50 mV), the as deposited constrictions exhibit low-frequency resistance noise dominated by two level fluctuators (TLFs) -- random switching of the resistance between two values, as shown in figure 2a. Fluctuations studied range from 0.005 % to 0.2% of the total resistance. The distribution of times spent in either state is exponential. Above 20 K the two characteristic times τ_{up}, τ_{down} have been measured over three to four decades for many TLFs and they are well described by thermally activated behavior $\tau \sim \tau_0 e^{E/kT}$. Attempt times τ_0 range from 10^{-11} to 10^{-15} s, clustering around 10^{-13} s,

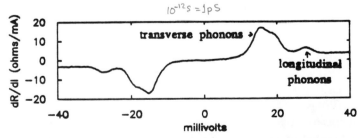

Fig. 1. Point contact spectrum for a 30 Ω Cu constriction at 4.2 K.

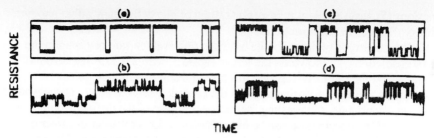

Fig. 2. Resistance vs. time in copper nanoconstriction for T<150 K, showing several types of behavior. Fluctuations studied range from 0.005% to 0.2% of the total resistance. Time scales are somewhat arbitrary, as they depend on the temperature of observation.

a characteristic time for atomic vibration. Measured activation energies range from 30 to 300 meV and are primarily an indication of the measurement temperature, as the temperature determines which fluctuators will be visible in the experimental bandwidth. The change in scattering cross section, $\Delta\sigma$, between states of a fluctuator can be estimated as the constriction area times the ratio of the change in resistance caused by the fluctuator to the magnitude of the constriction resistance. For eight devices with total resistance ranging from 5 to 180 Ω, we find 1 $\text{Å}^2 < \Delta\sigma < 30$ Å^2. The magnitudes of the attempt times and the scattering cross section changes indicate that each resistance fluctuation is due to the reconfiguration of an individual atomic-size defect thermally excited over an energy barrier between two metastable configurations.

Figure 2b shows two TLFs moving in the experimental bandwidth. In this case the fluctuators are independent -- neither the characteristic times nor the amplitude of one fluctuator depend on which state the other occupies. However, we often observe interacting defects. Figure 2c shows amplitude modulation and figure 2d frequency modulation of one fluctuator by another. The observed interactions indicate that one defect can change the detailed potential energy governing the behavior of another. The mechanism of interaction is open to discussion, but it is reasonable to propose, as has been proposed for tunnel

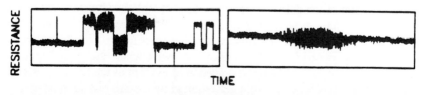

Fig. 3. Resistance noise due to a defect diffusing through the sample volume, superimposed on a normal TLF (left) and with the TLF signal subtracted (right). Total transit time is about 30 s.

junctions (5), that the changes in configuration of one defect may affect another through changes in its strain field.

One other type of noise source has been observed on rare occasion in the nanoconstrictions, as shown on the left in figure 3, with the TLF background subtracted on the right. We ascribe this sort of event to a mobile defect approaching and then diffusing through the constriction. We do not know what the defect is, but we find it interesting that in the presence of the diffusing defect the resistance of the constriction can be either higher or lower than in the absence of the defect. This is a particularly direct indication of the importance of local quantum interference in electron transport.

As either the temperature or the voltage across the constriction is increased, the number of defects that reconfigure on time scales within the experimental bandwidth increases and interactions between defects become more common. At temperatures above about 150 K in copper constrictions, and at slightly lower temperatures for aluminum, there is a gradual qualitative change in the character of the defect dynamics. At low temperatures the defect behavior is stable over long periods, with the same TLFs active over days. At higher temperatures the resistance noise is still composed of discrete fluctuations, but the characteristic times and amplitudes and the number of active fluctuators changes constantly. Fourier transforms of such noise over several minutes typically yield a smooth 1/f power spectrum, even though only a few TLFs are active at any given time. In contrast to standard models of 1/f noise which posit the superposition of noise from many independent defects to explain 1/f behavior (6), in our constrictions the important factor is the interaction between defects, which changes the nature of the TLFs over time and on average brings forth noise with a 1/f power spectrum.

The fluctuator interactions and lack of higher temperature stability indicate that even though our constrictions are crystalline, it is appropriate to picture the systems of defects they contain as "glassy" systems in which no defect is truly independent. Above about 150 K the defect system has "melted" and the overall configuration of defects wanders in a very complicated potential. At low temperatures the system is "frozen" into one configuration in which there are locally only a few low energy barriers that allow "single defects" to form stable two level fluctuators. This view is supported by the fact that when a sample is taken to high temperatures and then recooled, the nature of the TLF noise that is found is in general completely different than that observed before heating.

The melting of the defect system may be accomplished by applying a bias across the constriction as well as by raising the temperature. The voltage required depends on the resistance of the constriction, the temperature, and the material. As the sample bias is raised further, well above the point where the noise signal begins to vary in time, electromigration becomes evident as the sample resistance begins to change significantly. Voltages required for electromigration have ranged from ~ 50 mV for a 5 Ω aluminum sample at room temperature to ~ 500 mV for a 100 Ω copper sample at 4.2 K. During electromigration the resistance dynamics are still dominated by the discrete changes due to interacting TLFs. The qualitative

456

picture that we observe is very different from that of simple diffusion of independent defects. Instead we see a tortuous process of the strongly interacting, "glassy" defect system "annealing" under the influence of the applied field. Since electromigration does not occur until voltages much greater than those sufficient to melt the defect glass, a great many defect modes must be excited before bulk atomic motion may occur.

Electromigration may increase or decrease the resistance of the constriction by more than a factor of 10. In general, when the resistance increases the PCS spectrum quality improves consistent with merely making the sample orifice smaller. When the resistance decreases significantly the PCS signal decreases, indicating the introduction of elastic scatterers to the constriction region, and time-independent conductance fluctuations may appear at low voltage. We currently cannot determine how much of a change in the electron mean free path is caused by the electromigration and therefore cannot determine how much scattering is needed to create the conductance fluctuations. Our samples that exhibit the conductance fluctuations may or may not be fully diffusive, and we note that the scatterers may be distributed non-uniformly in the constriction. However, since in a point contact of radius r the potential drop occurs entirely within a distance of order r about the constriction, all of the changes occurring during electromigration should occur in the region of the constriction. Since our leads are cleaner than the constriction region and branch out 3-dimensionally away from the constriction, the probability that an electron emerging from the constriction into a lead will return to the constriction to cause further interference effects is much less than it would be if the leads had the same dimensionality and were composed of the same material as the constriction. Thus our geometry should allow us to probe interference effects on a scale less than the dephasing length. For comparison, we estimate that the

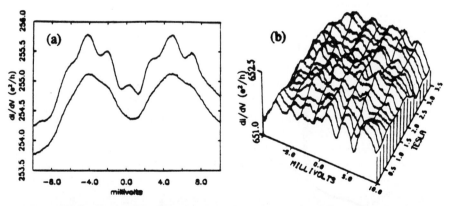

Fig. 4. (a) Time-independent conductance fluctuations in a Cu nanoconstriction at T=4.2 K (lower curve) and T=2 K (upper curve). The sharpest features we observe are characteristically no smaller than 5 kT wide. (b) Conductance fluctuations as a function of bias and magnetic field parallel to the constriction axis in our most magnetic field sensitive copper sample at T=2 K.

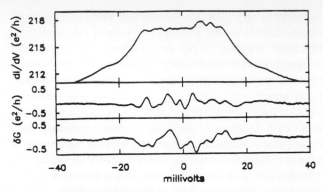

Fig. 5. (Top) Differential conductance for a single defect configuration, measured at 2 K. (Bottom) Deviations from an average of 25 different configurations for two specific defect configurations. Note that the fluctuations die out near 15 mV, where the electron dephasing length decreases rapidly due to phonon emission.

dephasing length in our leads is 1000 nm, using $l_{el} \sim 180$ nm from our bulk measurements and $\tau_\phi \sim 2 \times 10^{-11}$ sec as determined by Bergman in disordered copper films at 5 K (7). We emphasize that we see no conductance fluctuations before defects are introduced at the constriction, so that our effect cannot simply be due to interference in the leads.

Even though the UCF theory for diffusive samples may not apply to our samples, much of what we observe is consistent with the results in UCF theory (8). The peak to peak sizes of the fluctuations we see are 0.2 - 1 e^2/h, close to the UCF rms value $\sim e^2/h$. We see fluctuations both as a function of the applied voltage and the magnetic field (figure 4). By taking samples to voltages that are high enough to reconfigure defects but not high enough to cause electromigration, we can change the defect distribution in the sample, and this too changes the fluctuation pattern. Figure 5 shows the differential conductance as a function of voltage for one defect distribution in a sample in which we altered the defect distribution 25 times. The difference between the average over all 25 conductance traces and the conductance for two specific defect distributions are also shown. Note that at ±15 mV, where significant inelastic phonon scattering begins, causing a sharp decrease in the conductance, the magnitude of the difference between the conductance patterns for different defect distributions also drops to zero. Therefore the presence of a short inelastic scattering length destroys the fluctuations we see.

One aspect in which our observations appear to differ considerably from the ideas in UCF theory is in the voltage correlation length of the fluctuations. The characteristic scale in which our fluctuations may vary is no larger than 1.0 mV. This is approximately the limit allowed by temperature smearing at our current lowest temperature of ~ 2.0 K, so that the true voltage scale may be even smaller. Within UCF theory for diffusive samples, Larkin and Khmel'nitskii (9) predict that the characteristic voltage scale is

$$V_c \approx \frac{\hbar}{e\tau_{transit}} \approx \frac{\hbar D}{eL^2} = \frac{\hbar v_F l_{el}}{3eL^2},$$

where $\tau_{transit}$ is the mean time for an electron to transit the sample, D is the diffusion constant, L is the effective sample dimension, v_F the Fermi velocity, and l_{el} is the elastic mean free path. If we try to apply this estimate to our samples, assuming that they are fully diffusive and taking as an upper bound on our sample dimensions L≤ 15 nm, the largest value close to being consistent with the observed resistance and PCS spectrum, we can use the observed voltage correlation scale to give an upper bound on the elastic scattering length in the constriction. The result is $l_{el} \leq 0.6$ nm. We consider this highly unlikely, indicating that the voltage scale that would be predicted by UCF theory is much too large in our samples. If our samples are not diffusive, the theoretical estimates of the correlation voltage would be even larger and worse (9). Webb et al. (10) have found a similar discrepancy in a diffusive Sb wire, with an observed correlation voltage of about 0.1 V_c. Both our result and the Sb result are currently consistent with a calculation by Hershfield (11) that for diffusive samples at T=0 the characteristic voltage scale is 0.03 V_c rather than V_c. However, Webb et al.'s measurement of the characteristic voltage, as well as our own, is on the scale of the temperature smearing, so it is possible that we are both seeing the resolution limit allowed by our temperatures rather than the true correlation scale. We are now preparing to take our samples to much lower temperatures to see how small the true characteristic voltage scale might be.

If samples which show conductance fluctuations are migrated further they can show strong features with all the characteristics of weak localization (12). Figures 6 a,b show the conductance as a function of voltage for copper and aluminum samples. In these graphs one must understand that the inelastic scattering length will decrease with increased voltage since the density of available states into which an electron may scatter increases. In the weak localization picture the copper and

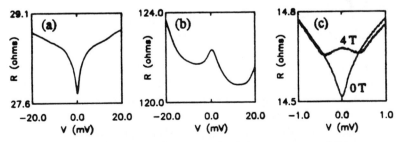

Fig. 6. (a) Cu constriction at 100 mK, showing weak anti-localization signal. (b) Al constriction at 4.2 K showing weak localization. (c) Cu constriction at 70 mK in magnetic fields of 0 and 4 T, showing the suppression of weak localization by a magnetic field.

aluminum plots curve in opposite directions because of the strong spin-orbit scattering in copper which is largely absent in aluminum. Our features are suppressed by the application of a magnetic field (figure 6c). We have just begun to investigate these results, but we find them quite surprising. Estimates of l_{el} based on the sample resistance imply that the samples are small compared to l_ϕ, assuming that l_ϕ is dominated by phonon emission. Therefore it seems that either it may be possible to see significant weak localization in samples smaller than l_ϕ or electromigration can introduce an additional source of electron dephasing into our most abused samples. Our investigation of this effect will continue.

In conclusion, metallic nanoconstrictions have begun to show their worth as a probe into a wealth of physical processes on nanometer length scales. Work is proceeding now in better characterizing the physical size and scattering lengths of our samples and cooling them to lower temperatures. In the future we expect to study the properties of very small samples of a variety of novel materials.

This research was supported by the National Science Foundation both through the Cornell Materials Science Center under Grant No. DMR-85-16616 and through the use of the National Nanofabrication Facility under Grant No. ECS-86-19049. One of us (D.C.R.) acknowledges an NSF fellowship, and one of us (K.S.R.) acknowledges an AT&T Bell Labs GRPW fellowship.

REFERENCES

1. Yanson, I. K., Zh. Eksp. Teor. Fiz. **66**, 1035 (1974) [Sov. Phys. JETP **39**, 506 (1974)].
2. Jansen, A. G. M., Van Gelder, A. P., and Wyder, P., J. Phys C **13**, 6073 (1980).
3. d'Heurle, F. M., and Ho, P. F., in *Thin Films--Interdiffusion and Reactions*, edited by J. M. Poate, K. N. Tu, and J. W. Mayer (1978) p. 243.
4. Ralls, K. S. and Buhrman, R. A., Phys. Rev. Lett. **60**, 2434 (1988).
5. Farmer, K. R., Rogers, C. T., and Buhrman, R. A., Phys Rev. Lett. **58**, 2255 (1987).
6. Weissmann, M. B., Rev. Mod. Phys. **60**, 537 (1988) and references therein.
7. Bergmann, G., Z. Phys. B - Condensed Matter **48**, 5 (1982).
8. Lee, P. A., Stone, A. D., and Fukuyama, H., Phys. Rev. B **35**, 1039 (1987) and references therein.
9. Larkin, A. I. and Khmel'nitskii, D. E., Zh. Eksp. Teor. Fiz. **91**, 1815 (1986) [Sov. Phys. JETP **64**, 1075 (1986)].
10. Webb, R.A., Washburn, S., and Umbach, C.P., Phys. Rev. B **37**, 8455 (1988).
11. Hershfield, S., unpublished.
12. Bergmann, G., Physics Reports **107**, 1 (1984) and references therein.

NANO- AND MICRO-STRUCTURES IN CHEMISTRY, ELECTROCHEMISTRY, AND MATERIALS SCIENCE[1]

C.R. Martin, M.J. Tierney, I.F. Cheng, L.S. Van Dyke,
Z. Cai, J.R. McBride, C.J. Brumlik

Department of Chemistry
Texas A&M University
College Station, Texas 77843

INTRODUCTION

We have recently become interested in the fabrication and characterization of ensembles of nano- and micro-scale structures. In contrast to the majority of the work in this area, our interests do not lie in the microelectronic applications of such structures; rather, we are interested in the possible electrochemical, chemical, optical, and materials applications of nanostructure ensembles. We review these applications and the results of our investigations in this paper.

EXPERIMENTAL

The general procedure used in this laboratory for preparation of ensembles of nanostructures entails electrochemical deposition of metals or plastics into the pores of a microporous filtration membrane (1-5). The key point is that the pores of the host membrane act as templates for the nascent nanostructures. This procedure is shown schematically in Figure 1. Details can be found in references (4,5). We have used this procedure to prepare Pt, Au and electronically conductive plastic nanocylinders with diameters as small as 10 nm (see e.g. Figure 2).

As indicated in Figure 1, membranes with discrete, straight-through micro- or nanopores form the basis of our fabrication methods. These membranes are prepared by tracking and etching plastics (6,8) or by anodizing Al (7). The track-etched plastic (8) and anodized Al (Anopore (7)) membranes are commercially available.

1This work was supported by the Office of Naval Research, the Air Force Office of Scientific Research, and the NASA Johnson Space Center.

Ultramicroelectrodes are electrodes which have at least one characteristic dimension which is smaller than ca. 10 μm. The most obvious example is a disk with a diameter which is 10 μm or smaller. Because ultramicroelectrodes offer electrochemists a myriad of opportunities that are not possible at electrodes of conventional dimensions, ultramicroelectrodes are of considerable current research interest (4,9-11). We have used the procedure outlined in Figure 1 to prepare ensembles of Pt and Au ultramicrodisk electrodes with diameters as small as 50 nm.

One of the most exciting potential applications of ultramicroelectrode ensembles is in the area of chemical sensors. Theory predicts, and we have experimentally demonstrated, that such ensembles can electrochemically sense significantly lower concentrations of electroactive analyte molecules than can electrodes of macroscopic dimensions. That is, put in the jargon of analytical chemistry, ultramicroelectrode ensemble-based sensors show lower detection limits than analogous sensors based on conventional electrodes.

The lower detection limits displayed by ultramicroelectrode ensembles are illustrated in Figure 3. Figure 3A shows a calibration curve associated with the electrochemical detection of an Fe-containing ion at a conventional carbon paste electrode. The detection limit at this conventional electrode is ca. 300 nM. Figure 3B shows the analogous calibration curve at an ultramicroelectrode ensemble. The detection limit at the ultramicroelectrode ensemble is ca. 30 nM. Thus, the ultramicroelectrode ensemble shows a one order of magnitude enhancement in detection limit relative to the conventionally-sized electrode (9,10).

Metal films thicker than several hundred angstroms are usually opaque to visible and infrared radiation. However, manipulating the microstructure of metal films can produce transparency. Effective medium theory predicts that electrically disconnected metal particles which are small relative to the wavelength of the incident light will be transparent (12). This effect occurs when the optical electric field produces a screening charge at the metal-dielectric boundaries. This charge excludes the electric field from the particle and "squeezes" the light into the non-absorbing dielectric between the particles.

The optimal microstructure for transparency to unpolarized light is an ensemble of nanocylinders with their axes oriented parallel to the incident light rays (12). The strategy outlined in Figure 1 was used to prepare an ensemble of transparent metal nanocylinders (5). Gold nanocylinders were electroplated into a silver-backed Anopore (Figure 3B) membrane. The silver was then dissolved, leaving electrically isolated, 200 nm-diameter gold cylinders imbedded in an Anopore membrane. The gold that fills the pores makes up 65% of the gold/Anopore composite surface.

The fourier transform infrared spectrum of an Au/Anopore composite membrane is shown in Figure 4. This spectrum was corrected for the Anopore absorption. At wavelengths greater than ca. 6 μm the composite membrane transmits ca. 75 % of the incident photons. If the Au microstructures were opaque, only ca. 20 % of the light would pass through the membrane. Thus, in agreement with the predictions of effective medium theory, these nanostructures are optically transparent.

A number of plastics are now known which conduct electricity via an electronic mechanism similar to conduction in metals (13). In most cases, however, the conductivities observed are significantly lower than metallic conductivities. We have recently discovered that when electronically conductive plastics are synthesized within the pores of nanoporous host membranes, dramatically enhanced conductivities are observed (14).

Figure 5 shows an example of the conductivity enhancements observed. Figure 5 is a plot of conductivity along polypyrrole or poly(3-methylthiophene) fibers vs. the diameter of the fiber. The large diameter fibers show conductivities roughly equivalent to the conductivities of conventional polypyrrole or poly(3-methylthiophene) (i.e. 10 to 100 S cm^{-1}). However, fibers with diameters on the order of 10's of nm's show dramatically higher conductivities. In the case of poly(3-methylthiophene) a ca. two-order of magnitude enhancement in conductivity is observed.

REFERENCES

1. Spohr, R. U.S. Patent 4,338,164, 1982.
2. Williams, W.D. and Giordano, N. (1984). Rev. Sci. Instrum. 55(3), 410.
3. Possin, G.E. (1970). Rev. Sci. Instrum. 41, 772.
4. Penner, R.M. and Martin, Charles R. (1987). Anal. Chem 59, 2625.
5. Tierney, M.J.; Martin, C.R. J. Phys. Chem. In press.
6. Fleisher, R.L., Price, P.B., Walker, R.M. (1975). "Nuclear Tracks in Solids." University of California, Berkeley.
7. Anopore^{TM} (Alltech Associates, Inc., Deerfield, IL)
8. Poretics Corp,, Livermore, CA.
9. Cheng, I.F. and Martin, C.R. (1988). Anal. Chem 60, 2163.
10. Cheng, I.F., Whiteley, L.D., Martin C.R. Anal. Chem. In press.
11. Martin, C.R. (1987). In "Ultramicroelectrodes" (Fleischmann, M; Pons, S.; Rollison, D.R.; Schmidt, P.P.), pp. 263-272. Datatech Systems, North Carolina.
12. Aspnes, D.E., Heller, A., Porter, J.D., (1986). J. Appl. Phys. 60, 3028.
13. Skotheim, T.A., ed. (1986). "Handbook of Conducting Polymers." Marcel Dekker, New York.
14. Cai, Z. and Martin C.R., J. Am. Chem. Soc. Submitted.

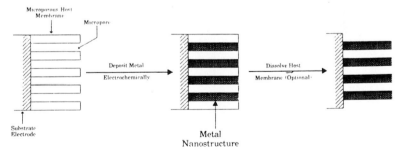

Fig. 1. General approach for preparing ensembles of micro- and nanostructures.

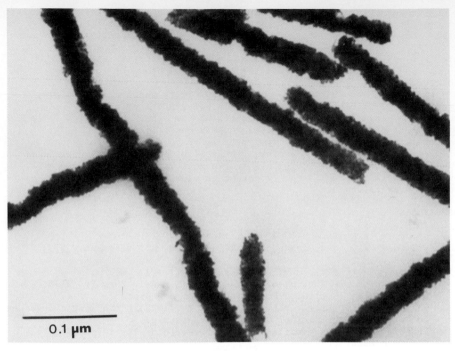

Fig. 2. TEM of 300 angstrom diameter platinum wires.

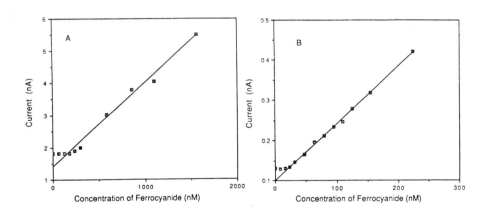

Fig. 3. Calibration curves showing the limit of detection for
$Fe(Cn)_6^{4-}$ at (a) a macro-sized electrode and (b) an
ultramicroelectrode array.

464

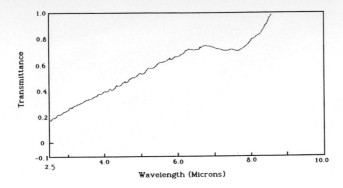

Fig. 4. A fourier transform infrared spectrum of a gold/Anopore composite (spectrum of Anopore subtracted).

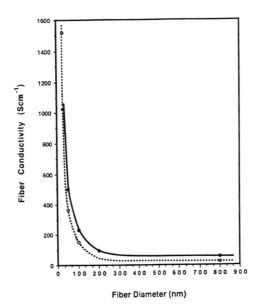

Fig. 5. Plot of conductivity of conducting polymer fibers vs. fiber diameter. Poly(3-methylthiophene) (dashed line). Polypyrrole (solid line).

ELECTRONIC TRANSPORT
IN MESOSCOPIC AuFe SPIN-GLASSES[1]

C. Van Haesendonck
H. Vloeberghs
Y. Bruynseraede

Laboratorium voor Vaste Stof-Fysika en Magnetisme
Katholieke Universiteit Leuven
B-3030 Leuven, Belgium

R. Jonckheere

IMEC v.z.w.
B-3030 Leuven, Belgium

1. INTRODUCTION

Despite numerous theoretical as well as experimental efforts, the exact nature of the spin-glass phase is still unclear. It is not clear whether a phase transition occurs at the temperature T_f where the low-field magnetic susceptibility shows an anomalous cusp (1). Although it seems impossible to get a microscopic picture of the exact ground state, the renormalization group approach as well as Monte Carlo simulations have clearly indicated that the presence of a phase transition for short-range Ising spin-glasses strongly depends upon the effective dimensionality of the magnetic system (2). The lower critical dimension d_l below which $T_f \rightarrow 0$ is likely to be larger than 2 but smaller than 3, in agreement with recent experiments (3).

[1] *Research supported by the Belgian National Fund for Scientific Research (N.F.W.O.)*

When $2 < d_l < 3$, we expect to observe a dramatic change in the low temperature susceptibility of the spin-glass when its thickness becomes smaller than the relevant correlation length ξ_{SG} for the three-dimensional phase transition. In metallic spin-glasses, the basic mechanism causing the freezing of the magnetic moments below T_f is the isotropic long-ranged RKKY interaction which should be described in terms of an Heisenberg spin model. Recent numerical simulations for a d = 3 Heisenberg model with long-range RKKY interactions indicate a totally different behaviour when compared to the d = 3 short-range Ising model (4). It is therefore not at all clear whether the result $2 < d_l < 3$ is also valid for archetypical spin-glasses such as CuMn, AgMn or AuFe. Measurements of the non-linear susceptibility for $T > T_f$ indicate the presence of a phase transition for d = 3 metallic spin-glasses, i.e. $d_l < 3$ (5).

The lower bond $d_l > 2$ has been confirmed experimentally by Kenning et al. for CuMn-Si multilayers (6,7). The multilayer structure allows to increase artificially the magnetization signal of an individual CuMn layer. Kenning et al. observe that T_f drops continuously towards zero between a CuMn thickness t = 500 nm and t = 5 nm.

When a spin-glass is cooled quickly, it will move through various states of quasi-equilibrium. The continuous rearrangement of the spins will induce a 1/f noise in the spin-glass state which is superimposed upon the average magnetization signal (8). The noise can also be detected via resistance measurements, provided the coupling between the spins and the conduction electrons is strong enough. In disordered spin-glasses with a short elastic mean free path, this coupling has been calculated quantitatively by Feng et al. for mesoscopic samples with a sample size comparable to ξ_{SG} (9). The interference between diffusively scattered electrons will give rise to Universal Conductance Fluctuations (UCF) which are very sensitive to the spin fluctuations. The model of Feng et al. predicts a noise corresponding to conductance fluctuations $\delta G \sim e^2/h$ in the mesoscopic limit. In larger samples, the noise will decrease inversely proportional to the square-root of the sample volume.

2. EXPERIMENTAL METHOD

In this paper, we will use electrical transport measurements rather than magnetization measurements to study the spin-glass order in reduced dimensionalities. Usually, one assumes that the electrical resistance of a metallic spin-glass does not show any anomaly near T_f. In the presence of a strong

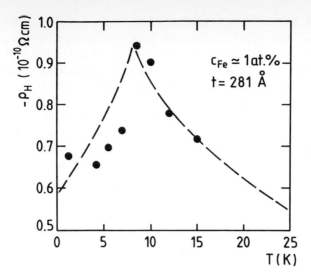

Fig. 1. Temperature dependence of the anomalous Hall resistivity ρ_H near the spin-glass freezing temperature T_f. The dashed curve is only a guide to the eye.

spin-orbit scattering however, an anomalous Hall effect will appear due to the asymmetric scattering of the conduction electrons by the local magnetic moments (10). For bulk AuFe alloys which are characterized by a very strong spin-orbit effect, it has been shown experimentally that the temperature dependence of the Hall resistivity at low magnetic fields ($B \sim 0.01\,T$) is nearly identical to the cusp appearing in the magnetic susceptibility (11). As illustrated in Fig.1 for a film with thickness t = 28.1 nm, the anomalous Hall effect can also be used to determine the spin-glass order in thin AuFe films. The homogeneous, continuous AuFe films (Fe concentration $c_{Fe} \simeq 1.0\pm0.1\,at\%$ as determined from neutron activation measurements) with a thickness ranging between 5 nm and 50 nm are obtained by flash evaporation of small pieces of the bulk alloy ($c_{Fe} = 1.00\pm0.05\,at\%$). In this way, the dimensional crossover can be studied without having to use an artificially modulated structure where the interface may strongly influence the spin-glass freezing process (due to disorder, inter-diffusion, etc...).

The mesoscopic samples have been prepared using a Cambridge Instruments EBMF10.5 e-beam system operating at 20 kV. Using a bilayer resist, lift-off profiles were prepared allowing to obtain AuFe samples with a linewidth comparable to 0.1 μm and a length of the order of 1 μm. The five-terminal pattern allows to obtain high-resolution magnetoresistance and 1/f noise measurements (ac method with a PAR 124A lock-in amplifier).

3. RESULTS AND DISCUSSION

For our bulk AuFe alloy $T_f = 9.0$ K, as confirmed by magnetization measurements in a commercial SQUID magnetometer (BTI). We may therefore conclude that the observed reduction of T_f in films with a thickness $t < 50\,nm$ (see Fig.1) is caused by the thin-film structure. In Fig.2, we have plotted the suppression of T_f for different thicknesses. Our results confirm the experimental observation of Kenning et al. that the spin-glass order is no longer observed within the available temperature range for $t < 50\,nm$ (6).

In the mesoscopic samples, we have tried to determine the 1/f noise caused by the rearrangement of the frozen spins. Although our ac measuring circuit has a resolution better than 1 nV, no 1/f noise is observed when the current is kept low enough to avoid Ohmic heating. This indicates that the coupling between the conduction electrons and the spin system is very weak, in disagreement with the predictions of Feng et al. (9). Conductance fluctuations $\delta G \sim e^2/h$ can easily be detected by our low-noise ac measuring system.

The absence of UCF in our AuFe samples is also confirmed by the magnetoresistance data shown in Fig.3. For a AuFe sample having a width w = 0.15 μm and a length l = 2.0 μm, we observe a smooth, monotonic background. The conductance variation $\delta G \simeq \delta R/R^2 \propto B^2$ is consistent with a freezing-out of the magnetic scattering when the spins are aligned by the magnetic field. On the other hand, aperiodic fluctuations $\delta G \sim e^2/h$ are clearly observed for a pure Au sample which has a width w = 0.15 μm and length l = 3.0 μm.

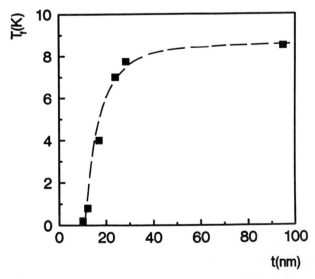

Fig. 2. Variation of the spin-glass temperature T_f as a function of the AuFe film thickness t. The dashed line is only a guide to the eye.

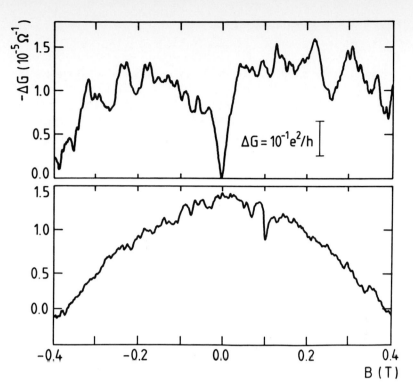

Fig. 3. Magnetoconductance fluctuations ΔG in a pure Au film with $t = 30.6\,nm$ (upper curve) and a AuFe film with $t = 35.1\,nm$ (lower curve).

REFERENCES

1. K. Binder and A.P. Young, Rev. Mod. Phys. **58**, 801 (1986).
2. R.N. Bhatt and A.P. Young, Phys. Rev. Lett. **54**, 924 (1985)
3. C. Dekker et al., Phys. Rev. Lett. **61**, 1780 (1988).
4. J.D. Reger and A.P. Young, Phys. Rev. B **37**, 5493 (1988).
5. L.P. Lévy and A.T. Ogielski, Phys. Rev. Lett. **57**, 3288 (1986).
6. G.G. Kenning et al., Phys. Rev. Lett. **59**, 2596 (1987).
7. D.S. Fisher and D.A. Huse, Phys. Rev. B **36**, 8937 (1987).
8. W. Reim et al., Phys. Rev. Lett. **57**, 905 (1986).
9. S. Feng et al., Phys. Rev. B **36**, 5624 (1987).
10. A. Fert and O. Jaoul, Phys. Rev. Lett. **28**, 303 (1972).
11. R.D. Barnard and I. Ul-Hacq, J. Phys. F: Met. Phys. **18**, 1253 (1988).

NATURE OF THE SUPERCONDUCTING TRANSITION IN VERY THIN WIRES [1]

N. Giordano

Department of Physics, Purdue University,
West Lafayette, Indiana 47907, USA

The characteristic length scale for a superconductor is the coherence length, ξ. In an elemental superconductor, ξ is typically of order a few thousand Å at low temperatures, and diverges as one approaches the critical temperature. Hence, with modern microfabrication techniques it is fairly straightforward to make structures which are one dimensional with regards to superconductivity. In this paper we report a study of the superconducting transition in one dimensional In strips, i.e., wires, with diameters as small as 400 Å.

Our current understanding of the superconducting transition in one dimension is due to the work of Little (1), Langer and Ambegaokar (2), and McCumber and Halperin (3). They showed that there can be dissipation, i.e., a nonzero voltage in the presence of a small current, below T_c as a result of phase slippage; here we refer to the phase of the Ginzburg-Landau (GL) order parameter. Phase slip is a process in which the phase difference between the ends of the system, $\Delta\phi$, changes by 2π. Since, via the Josephson relation, $V \sim d\,(\Delta\phi)\,/\,dt$, each phase slip is accompanied by a voltage pulse. Langer and Ambegaokar (LA), and McCumber and Halperin (MH), calculated the rate of thermally activated phase slippage, and used it to obtain the resistance below T_c. Experiments on relatively large systems (by today's standards; typical diameters were $\sim$5000 Å) were conducted some time ago (4), and shown to be in good agreement with the LA-MH

[1] Supported by National Science Foundation grant DMR-8614862.

theory. In our experiments with much smaller systems we have found evidence that quantum tunneling of the phase degree of freedom may also be important. This behavior is similar in many ways to the phenomena of macroscopic quantum tunneling, which has recently been discussed with regards to tunnel junctions and SQUIDS (5,6).

The thermal activation theory of LA and MH is based on Ginzburg-Landau theory, according to which the system can be described by a complex order parameter, $\psi = f \exp(i\,\phi)$. The problem is analogous to the case of a particle moving in a washboard potential. The minima of the potential correspond to states for which $\Delta\phi = \pm 2n\,\pi$ (where n is an integer), and the slope of the washboard is proportional to the current. Phase slip occurs when the system moves from one (local) minimum of the washboard to an adjacent one. The rate for thermal activation over the energy barriers which separate the minima is given by (2,3)

$$\Gamma_{TA} = \Omega \exp(-\Delta F_0 / k_B\, T)\,, \tag{1}$$

where the energy barrier, in the limit $I \to 0$, is $\Delta F_0 = \sqrt{2}\sigma H_c^2\xi/3\pi$. Here H_c is the critical field, σ is the cross-sectional area, and Ω is an attempt frequency (3). The voltage and hence also the resistance are both proportional to Γ_{TA}. Both H_c and ξ are temperature dependent, leading to a barrier height which vanishes as $\Delta F_0 \sim (\Delta T_c)^{3/2}$, as $T \to T_c$. This enters the exponent in Eq. 1, and dominates the temperature dependence of Γ_{TA}.

Mooij and co-workers (7) have proposed that phase slip can also occur via quantum tunneling *through* the energy barriers in the washboard potential. At present there is no quantitative theory of this process. However, we can work from analogy with the well developed theory of macroscopic quantum tunneling to guess the form which such a theory might take. In general we expect the rate for phase slip via quantum tunneling to have the form (5) $\Gamma_{QT} = A \exp(-B)$. We have discussed likely forms for A and B elsewhere (8). The behavior of Γ_{QT} is dominated by the exponential factor, so we will only consider B here. Calculations (5) show that for an underdamped particle $B \approx 7.2\Delta U/\hbar\omega$, where ΔU is the barrier height, and ω is the frequency for small oscillations about the minima. For our case it is clear that $\Delta U = \Delta F_0$, but estimating ω poses a fundamental problem. The time dependent GL equation, which is the basis of the thermal activation theory, is purely diffusive, i.e., it has no mass term, and hence there is no parameter analogous to ω. We will therefore *assume* that $\omega \sim \tau^{-1}$, where τ is the GL relaxation time, since this is the only time scale in the problem. This leads to

$$\Gamma_{QT} \sim \exp(-\beta \Delta F_0 \tau/\hbar)\,, \tag{2}$$

where β is a numerical factor of order unity. It is interesting to note that essentially the same result is found in the overdamped limit (which other measurements (8) suggest may be the regime relevant here). The transition rates Eqs. 1 and 2 are very similar in form, the essential difference being the presence of τ in the exponent in Eq. 2. Since $\tau \sim (\Delta T_c)^{-1}$, the temperature dependence of Γ_{QT} is weaker than that of Γ_{TA}. This leads to a crossover from thermally activated phase slip near T_c, to phase slip via quantum tunneling at low temperatures.

The samples were very narrow In strips which were fabricated using a step-edge lithographic method (9), which involves directional ion-milling of a metal film which has been deposited onto a stepped substrate. The wires had diameters in the range 400-1000 Å, with normal state resistivities varying from 4 $\mu\Omega$cm for the largest wires, to 12 $\mu\Omega$cm for the smallest ones. Sample cross-sections were approximately right triangular, and the diameters we quote correspond to $\sigma^{1/2}$. Further details of the fabrication and sample properties are given elsewhere (8).

Figure 1 shows results for the resistance as a function of temperature for two samples. The solid lines are the thermal activation theory, while the dashed line is the sum of the resistances due to thermal activation and quantum tunneling. In obtaining these theoretical curves, it was necessary to adjust the barrier height, ΔF_0, downwards by a factor of $\approx 3-10$, from the theoretical prediction. This is reasonable given the uncertainties in the parameters which enter ΔF_0. The coefficient β in Eq. 2 was of order unity, as expected (8). It is seen from Fig. 1 that the thermal activation theory is in good agreement with the results for the large sample, except very near T_c, where this theory is expected to break down because of approximations made is estimating the thermal rate (2,3). This theory is also in agreement with the results for the small sample for $T_c - T \lesssim 0.2$ K, but farther from T_c the data deviate abruptly from the thermal activation form. This deviation suggests that in this region there is another mechanism for phase slip, and as can be seen from Fig. 1, the quantum tunneling rate, Eq. 2, agrees well with the data in this range. The different temperature dependences seen above and below $T_c - T \approx 0.2$ K are thus due to the different exponential factors in Eqs. 1 and 2, as discussed above.

Fig. 2 shows the results of a different type of measurement. Here the sample voltage was monitored while the current was increased continuously starting from zero. All of the temperatures employed here were well below T_c, so that the resistance (Fig. 1) was extremely small, and thus the voltage was very small until the current was near the critical current, I_c. Figure 2 shows the current sweep rate, dI/dt, as a function of I_S, the value of I at which the sample voltage first exceeded a fixed trigger level (which was 1 mV in this case). We note that repeated measurements of I_S yielded a spread of values which was less than the size of the data points in Fig. 2. These results are closely related to the transition rate measurements which have been widely used in studies of macroscopic quantum tunneling (6,10). The results in Fig. 2 can be compared, at least qualitatively, with the rates Γ_{TA} and Γ_{QT} obtained from the appropriate generalizations of Eqs. 1 and

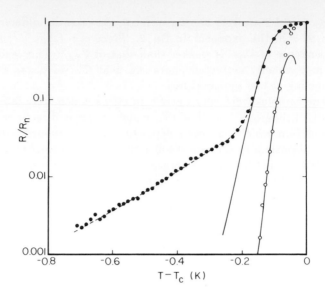

Fig. 1. Resistance, normalized by the normal state value, as a function of temperature for a 410 Å wire (●) and a 745 Å wire (○).

2 (Eqs. 1 and 2 are not applicable in this case since they were obtained in the limit $I \rightarrow 0$). A detailed comparison is given elsewhere (8); here we simply note the main conclusions of that analysis. The thermal rate is a *very* strong function of I, and for a fixed T would be an essentially vertical line in Fig. 2. It is in reasonable agreement with the data at high temperatures, but at low temperatures Γ_{TA} varies much more rapidly than is observed in Fig. 2. The quantum tunneling rate displays a much weaker dependence on I, and is in good qualitative agreement with the low temperature results. Hence, there appears to be a crossover from thermal activation to quantum tunneling. Most importantly, the analysis indicates that quantum tunneling is important in *both* the large and small samples.

Our experiments thus show that thermal activation is not the only mechanism through which the phase can relax in a one dimensional superconductor. The results are consistent with a model of phase slip via quantum tunneling. Unfortunately, there is currently no quantitative theory of quantum tunneling in this system. Work is needed to determine the correct equation of motion, so that this analogy, if indeed it is appropriate, can be made precise. In any case, the study of an "old" problem with very small structures appears to have uncovered some new and interesting physics.

I thank P. Muzikar for critical and enlightening discussions, and E. Sweetland for invaluable assistance with the fabrication of the samples.

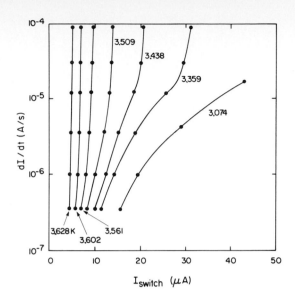

Fig. 2. Speed at which the current is swept, dI / dt , as a function of the current, I_S , at which the sample switches into the finite voltage state, at several temperatures, for a 650 Å wire.

REFERENCES

1. Little, W. A. (1967) Phys. Rev. **156**, 396.
2. Langer, J. S., and Ambegaokar, V., (1967) Phys. Rev. **164**, 498.
3. McCumber, D. E., and Halperin, B. I., (1970) Phys. Rev. B **1**, 1054.
4. Skocpol, W. J., and Tinkham, M. (1975) Rept. Prog. Phys. **38**, 1049.
5. See, for example, Caldeira, A. O., and Leggett, A. J. (1981) Phys. Rev. Lett. **46**, 211; Caldeira, A. O., and Leggett, A. J. (1983) Ann. Phys. (N.Y.) **149**, 374.
6. Washburn, S., Webb, R. A., Voss, R. F., and Faris, S. M. (1985) Phys. Rev. Lett. **54**, 2712. Schwartz, D. B., Sen, B., Archie, C. N., and Lukens, J. E., (1985) Phys. Rev. Lett. **55**, 1547. Devoret, M. H., Martinis, J. M., and Clarke, J. (1985) Phys. Rev. Lett. **55**, 1908.
7. Van Run, A. J., Romijn, J., and Mooij, J. E. (1987) Jap. J. Appl. Phys. **26**, 1765.
8. Giordano, N. (1988) Phys. Rev. Lett. **61**, 2137; and to be published.
9. Prober, D. E., Feuer, M. D., and Giordano, N. (1980) Appl. Phys. Lett. **37**, 94.
10. Fulton, T. A., and Dunkelberger, L. N. (1974) Phys. Rev. B **9**, 4760.

POPULATION INVERSION IN SUPERCONDUCTING QUANTUM WELLS UNDER BALLISTIC CONDITIONS *

Lino Reggiani

Dipartimento di Fisica e Centro Interuniversitario di Struttura della Materia dell'Universita' di Modena, Via Campi 213/A, 41100 Modena, Italy

Vladimir Kozlov

Institute of Applied Physics, Academy of Sciences of the USSR, Gorky, USSR

I. INTRODUCTION

Physical systems exhibiting population inversion of energy levels are of great interest because of their potential possibility in producing self-organizing phenomena [1] such as oscillatory current-voltage characteristics, laser action, and so on. Different solid state systems have been found to fall into this category, and among these insulators and semiconductors can be considered as typical materials to be used. In this work we present a novel physical system, based on superconductor quantum-wells, which is expected to exhibit population inversion. A quantitative estimate of the effect is given within a simple physical model.

II. PHYSICAL MODEL

The proposed system is schematically shown in Fig. 1. It can be considered as a natural extension of the structure used by Giaever in his pioneer experiments [2]

---

* Supported by Progetto Finalizato Materiali e Dispositivi per l'Elettronica a Stato Solido (MADESS) del Consiglio Nazionale delle Ricerche (CNR) and by the Centro di Calcolo of the Modena University (CICAIA).

and consists of three superconductors layers separated by two thin insulators layers. The whole system is equivalent to a three terminal device where we use the label L, I and R for the Left, Intermediate and Right layer, respectively. Because of the geometry we shall apply a one-dimensional treatment and we restrict our approach to the Bardeen-Cooper-Schrieffer (BCS) theory of superconductivity [3].

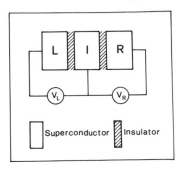

Fig. 1 - Schematic of the superconducting three-terminal device proposed for obtaining population inversion. The symbol L, I, R label the left, intermediate and right layers, respectively; V_L and V_R are the applied potentials.

For our purposes the following conditions should be satisfied. (i) The thickness of the insulator layers should be large enough to avoid tunnelling of Cooper-pairs through the insulator, but small enough to fulfill the two following requirements. Firstly, it should enable tunnelling of quasi-particle excitations between different superconducting layers. Secondly, it should keep the transmission coefficient from the L to the I-layer sufficiently small to exclude the formation of a large density of quasi-particles in the I-region which might decrease the value of the superconducting energy gap 2 Δ till suppressing the superconducting state at all in this region. Typical lengths could be within 10 and 100 Angstroms. (ii) The thickness of the superconducting I-layer should be short enough for the motion of the quasi-particles to occurr in the ballistic regime.

Thus, the probability that quasi-particles will leave the I-layer through one of the insulator layers must exceed their probability of thermalization in the I-layer. Under the above conditions, an appropriate biasing of the system should displace the equilibrium allignement of the Fermi energy E_F (see Fig. 2a) as shown in Fig. 2b. These figures represent the "semiconductor like model" of superconductors in the BCS theory. Accordingly, the L-layer will act as an emitter of quasi-particles into the I-layer. The injected quasi-particles will travel the I-region to eventually be collected from the R-region. Because of the high density-of-states available in proximity of the energy edges, the majority of the injected quasi-particels from L to I-region will be in the high energy region. Furthermore, for the same reason, the R-layer will collect preferably the low energy quasi-particles from the I-layer. **Therefore, under sta-**

tionary conditions we expect that the quasi-particle energy distribution in the I-active region will exhibit a population inversion. A quantitative calculation of the above features can be carried out as follows. From the tunnelling Hamiltonian model the stream of quasi particles from L to I-layer, S_{li}, and from I to R-layer, S_{ir}, are given, respectively by:

$$S_{li} = |T_{li}|^2 N_l(E - eV_{li}) f_l(E - eV_{li}) N_i(E)[1 - f_i(E)] \tag{1}$$

$$S_{ir} = |T_{ir}|^2 N_i(E) f_i(E) N_r(E - eV_{ir})[1 - f_r(E - eV_{ir})] \tag{2}$$

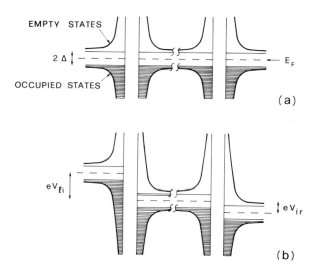

Fig. 2 - Energy diagram of the superconducting three-terminal device representing the behavior of the density-of-states $N(E)$. The energy E is plotted on the vertical scale and the density of states $N(E)$ on the horizontal scale. Case (a) refers to equilibrium conditions; (b) to the appropriate biasing for obtaining a population inversion in the active I-region.

where T_{li} and T_{ir} are the respective tunneling probability, N_l, N_i, N_r the density-of-states of the L, I, R-regions with $f_l(E)$, $f_i(E)$, $f_r(E)$ the respective energy distribution functions, V_{li} and V_{ir} the potential difference between L and I and I and R regions, respectively, and e is the absolute value of the electron charge. For our purposes the S, T and N quantities are taken dimensionless. The distribution function $f_i(E)$ is obtained from the stationary condition $S_{li} = S_{ir}$ as:

$$f_i(E) = \left\{ 1 + \frac{|T_{ir}|^2}{|T_{li}|^2} \frac{N_r(E - eV_{ir})[1 - f_r(E - eV_{ir})]}{N_l(E - eV_{li})f_l(E - eV_{li})} \right\}^{-1} \tag{3}$$

We consider the case $\left| \frac{T_{li}}{T_{ir}} \right|^2 \ll 1$ for which, within our condition (i), $f_i(E) \ll 1$. Thus, because of the small density of quasi-particles in I-region, superconductivity is not suppressed in this region and Eq.(3) can be simply evaluated for the case of zero temperature $T = 0$ within the BCS theory. Indeed, this implies the following expressions for the quantities entering Eq. (3):

$$f_l(E - eV_{li}) = 1 \quad ; \quad f_r(E - eV_{ir}) = 0 \tag{4}$$

$$N_l(E) = N_r(E) = N_N \frac{|E|}{\sqrt{E^2 - \Delta^2}} \quad ; \quad |E| > \Delta \tag{5}$$

where N_N refers to the "normal" density-of-states. Small density of quasi-particles together with their population inversion does not change Δ significantly, according to BCS relation which connects Δ to $f_i(E)$. Thus, one can use the same value of Δ for all superconducting layers. By substituting Eqs. (4,5) into Eq.(3) it is thus obtained:

$$f_i(\epsilon) \simeq \frac{|T_{li}|^2}{|T_{ir}|^2} \frac{|\epsilon - U_l|}{|\epsilon - U_r|} \frac{\sqrt{(\epsilon - U_r)^2 - 1}}{\sqrt{(\epsilon - U_l)^2 - 1}} \tag{6}$$

where we have introduced the dimensionless variables: $\epsilon = E/\Delta$, $U_l = eV_l/\Delta$, $U_r = eV_r/\Delta$ with the constraints: $2 < U_l$, $0 \geq U_r \geq -2$, $1 < \epsilon < (U_l - 1)$.

III. RESULTS AND DISCUSSION

The results of the calculations obtained from Eq. (6) at different values of the applied voltages are reported in Fig. 3. In all cases under consideration the distribution function is found to be inverted in the energy range here considered. Its peculiar peaking at the highest energies reflects the singularity of the quasi-particle density-of-states in the L-region. Furthermore, for the case $U_r = 0$ (see Figs. 3a and 3b) the distribution function goes to zero for $\epsilon \to 1$. This behavior is due to the coincidence of the singularity of the density-of-states in the R-region with the bottom of the energy in the I-region. In other words, all quasi-particles in the lowest energy region of the I-layer are swept into the R-layer because of this singularity. By switching on the U_r potential (see Fig. 3c) the distribution function differs from zero at $\epsilon \to 1$ because of the mismatch between the bottom energy in the I-region and the singularity in the R-region.

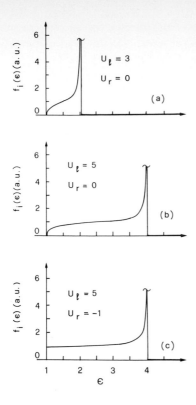

Fig. 3 - Energy distribution function of quasi-particles in the I-region for the given applied voltages. Case (a) refers to $U_l = 3$, $U_r = 0$; (b) to $U_l = 5$, $U_r = 0$; (c) $U_l = 5$, $U_r = -1$.

That is, now not all quasi-particles coming from the I-layer are immediately collected by the R-layer. In conclusion, we have proposed a new physical system which, by taking advantage of the non analytic density-of-states behavior of a superconductor, should exhibit energy population inversion of quasi-particles in a confined space region operating under ballistic regime. The working conditions to achieve such a result should be within the experimetal feasibility available from superconducting technology. Therefore, we hope to have provided the necessary stimulus for experimentalists to look for a practical realization and investigation of this system.

REFERENCES

1. Haken H. (1983). Synergetics, an Introduction, 3rd ed. Springer Sez. Syn. 1 Springer, Berlin-Heidelberg.
2. Giaever I. (1960). Phys. Rev. Lett. **5**, 147; ibidem **5**, 464.
3. Bardeen J., Cooper L.N. and Schieffer J. (1957). Phys. Rev. **106**, 162; ibidem **108**, 1175.

Ali Badakhshan
Christopher Durbin
Adriana Giordana
R. Glosser

Center for Applied Optics and Physics Programs
The University of Texas at Dallas
Richardson, Texas 75083

Steve A. Lambert

VARO Incorporated
Garland, Texas 75042

Jian Liu

Texas Instruments Incorporated
Dallas, Texas 75265

I. INTRODUCTION

The resurgence of interest in photoreflectance (PR) in the last few years has been driven by the nondestructive, contactless nature of the technique which allows the study of semiconductors by a form of electro-reflectance (ER) for those situations in which direct contact is impossible, undesired or simply inconvenient.[1-3]

In this paper, we extend photoreflectance measurements to higher energy critical points. This results in significantly shorter optical penetration depths so that in multilayer structures a top layer can be studied without interference from underlying layers.

Nanostructures represent a particular example for which PR is particularly suited for exploration of electronic structure. For instance, the GaAs/AlGaAs multiple quantum well system has been well studied using this technique. (This is reviewed in ref. 3).

In obtaining a PR signal, there are several considerations to be juggled. The energy of the chopped laser light must be greater than the band gap of the semiconductor of interest. On the other hand, the laser radiation typically generates photoluminescence (PL) which on top of the scattered laser radiation yields an undesired signal that is synchronously

485

detected along with the desired PR signal. Normally the unwanted signal can be minimized by placing a low pass optical filter in front of the detector so that the scattered laser radiation is removed along with as much as possible of the PL without cutting off the PR signal. The remnant PL signal which is independent of the monochromator setting can usually be eliminated by the lock-in DC offset. Even so, if the sample has a high level of doping, the PL may be so large that a practical level of sensitivity for PR detection becomes difficult to achieve.

If one wishes to study critical point structures at energies higher than the photon energy of the exciting laser, then somewhat more care is required to separate the desired PR from PL and scattered laser light. Early examples of this effort are by Cerdeira and Cardona,[4] Nahory and Shay,[5] Nilsson,[6] and Amirtharaj and coworkers.[7] We have explored in some detail, techniques for achieving high energy PR which may be practically applied. Our concern here is obtaining information in the vicinity of a higher energy critical point for those situations where information at the band gap itself is not accessible or easily interpretable. There are a number of technologically important situations where this could occur. For example, in the case of Si and other indirect gap materials with much higher direct gaps, while the surface potential can be modulated with a long wavelength laser line (e.g. 632.8 nm) to obtain PR, we need a practical technique for studying the direct gap which may lie in the ultraviolet (3.4 eV in Si with a higher critical point at 4.2 eV). This has recently been applied to the study of silicon films on sapphire (SOS) by Giordana and coworkers.[8]

Another example occurs in the characterization of semiconductors where it is feasible to use PR in order to unfold parameters which may vary as a function of distance into the semiconductor. (e.g. measure changes in alloy composition with depth). We may exploit the fact that the higher lying critical points have significantly shorter penetration depths compared to that at the band gap. For example, in GaAs at 3 eV (E_1 structure associated with transitions λ_3-λ_1) the penetration depth is approximately 15 nm compared to over 2000 nm at the band gap (E_0 structure associated with the transitions Γ_{15}- Γ_1). Recent work by Amirtharaj and coworkers[7] and Knudsen and coworkers[9] have made effective use of this capability in studying ion implanted semiconductors.

An important application occurs in thin films which may be part of some multilayer device structure. We describe here an epitaxially grown device which includes a thin layer of a III-V alloy whose composition needs to be determined. Because the film is thin, the band gap PR structure may be too weak to measure. Often in these multilayer devices, the PR structure is a tangled superposition of contributions from various layers. We demonstrate how PR measurements at a higher energy critical point structure with a shorter optical penetration depth can be used to provide the desired compositional information.

II. PROCEDURE

In order to demonstrate the capability of PR at the higher energy critical points as a probe of thin ($\sim$10-100 nm) layers, we compare the PR measurements of two multilayer III-V compound devices at the band gap and at the higher energy. The difference between the two devices is that one contains an $In_{0.05}Ga_{0.95}As$ top layer while the other has no In containing layers. We also compare the PR response of the top Si layer in a SIMOX (Separation by IMplanted Oxygen) structure for two different preparation conditions.

III. EXPERIMENTAL

1. Photoreflectance at Higher Energies

The basic description of photoreflectance is well presented in the literature.[1-3] Here we report techniques which facilitate PR exploration of materials at higher lying critical point energies. The laser excitation for the III-V materials was 30 mW of the HeNe 632.8 nm line while 400 mW of the Ar ion 514.5 nm line was used for the Si studies. The method for removal of undesired light, which has so far proved most fruitful, is the use of Corning glass UV transmitting-Visible absorbing filters. For example, the GaAs PR spectra shown in figure 1 were obtained by placing

Corning glass filters CS 7-59 and CS 5-59 over a UV enhanced photodiode. The first filter effectively removed any scattered laser light at the HeNe line 632.8 nm. The second filter, while not as effective at removing the laser lines, acts to remove the photoluminescence while still transmitting the desired signal. The technique is simple and effective, however, one still needs to use the lock-in DC offset to deal with the remnant scattered light or photoluminescence. Another consideration is the removal of light by the filters: both filters are about 65% transmitting at 3 eV so that only about 40% of the desired light is transmitted by the filters. Furthermore, one needs to choose different filters or filter sets for each energy range.

2. Sample Preparation

2.1 InGaAs/GaAs. The samples are MOCVD grown vertical HBT (heterojunction bipolar transistor) structures. Sample #140 with a top n^+-InGaAs emitter layer is described in Table I. We will compare this with sample #143 which is similar except that the In is absent from all layers and so its top layer is now 70 nm of n^+-GaAs. The MOCVD growth technique is similar to that described elsewhere.[10] The indium source was trimethylindium.

2.2 Si-SIMOX. The basic process has been discussed in the recent literature.[11] Our starting material was [100] p-Si of resistivity 15-20 Ω-cm. One sample was subjected to three oxygen implantations, each with a dose between 0.2 and 0.4×10^{18} cm^{-2} and annealed for several hours after each implantation. A second sample was prepared with the top layer left as polycrystalline Si.

IV. RESULTS AND DISCUSSION

1. InGaAs

Figure 1a shows the PR spectrum of sample #143 (top layer GaAs) in the vicinity of E_0 and E_1. The vertical scale of the E_1 structure is 100 times that of the E_0 structure. We use this as a reference against which we

TABLE I. Vertical, emitter--up HBT Structure for #140. The sample
#143 is similar except that the InGaAs layers are replaced by GaAs
layers of the same doping.

Layer	Doping $(cm)^{-3}$	Thickness nm
n^+-InGaAs	5×10^{18}	20
n^+-GaAs	5×10^{18}	50
n-grading	1×10^{18}	30
$n-Al_{0.3}Ga_{0.7}As$	3×10^{17}	100
n-GaAs	1×10^{18}	10
p^+-InGaAs	1×10^{19}	20
p^+-GaAs	1×10^{19}	40
p^+-InGaAs	1×10^{19}	20
p^+-GaAs	1×10^{19}	40
n-GaAs	3×10^{16}	1000
n^+-GaAs	5×10^{18}	500
AlGaAs buffer	Undoped	500
SI SUBSTRATE	Undoped	0.5 mm

compare sample #140 (top layer InGaAs) depicted in figure 1b. For the E_0
structure in the vicinity of 1.4 eV we see that for both samples there is
evidence of at least two contributions: a rather narrow set of oscillations just below 1.42 eV and a set of somewhat wider Franz-Keldysh oscillations extending beyond 1.5 eV. The point here is that for sample #140,
there is no obvious way to recognize the contribution of the InGaAs layer.
The fact that a similar kind of behavior occurs in the spectrum for sample
#143 which contains no InGaAs layers suggests that the two sets of oscilla-
tions are from separate GaAs layers within the structure. Whatever InGaAs
contribution there may be in #140 cannot be distinguished in the structure
around 1.4 eV.

Now we compare the E_1 structure of these samples in the vicinity of
3 eV. (It seems that at these higher doping levels, the E_1 and $E_1 + \Delta_1$ PR
contributions broaden and overlap giving the appearance of a single structure. At this point we simply make note of the complication). The optical
penetration for GaAs in this energy is about 17 nm so that our PR measurements should just correspond to the topmost layer. Using the Aspnes three

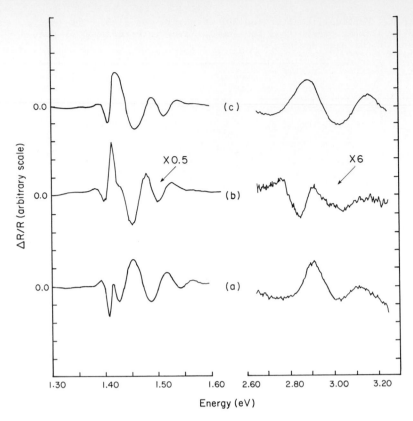

Fig. 1. PR spectra of HBT samples near 1.4 eV (E_0) and 3 eV (E_1). The vertical scale of E_1 is magnified by 100 times compared to E_0. (a) Spectrum of sample #143. (b) Spectrum of sample #140. The magnifications are with respect to the corresponding energy regions in figure 1a. (c) Spectrum of #140 after etch removal of the top of the InGaAs layer.

point method[12] (recognizing the limitations in view of the above-mentioned overlap), we calculate the energy of E_1 to be 2.94 eV for sample #143. This is somewhat higher than the ER value obtained by Williams and Rehn[13] (2.904 ± 0.002) which probably reflects the overlap with $E_1 + \Delta_1$. On the other hand, this energy for #140 is about 2.84 eV, a red shift of approximately 0.1 eV. It is clear in the latter sample that the major contribution is that of the InGaAs layer. Based on the targeted growth parameters, the In concentration is estimated to be 5%. We are currently working to determine this more accurately, taking account of lattice strain. In addition to the InGaAs contribution to the spectrum of figure 1b, it appears that there is also a response from the underlying GaAs layer as judged by the peak around 3.15 eV. This is not surprising since the top layer and α^{-1} are comparable. Since the contributions are evidently superposed, we have simply estimated the energy of the InGaAs structure. Nevertheless, the resolution of this contribution is greatly enhanced compared to what is possible around 1.4 eV.

In order to verify that we are seeing primarily the InGaAs layer, we removed the InGaAs layer with an H_2SO_4, H_2O_2, H_2O etch (1:8:160). The etch rate is 3.8 nm/sec (for pure GaAs which we assumed was approximately the same for the low In concentration InGaAs). An etch time of 5-6 seconds was used to remove the layer. The result is shown in figure 1c. The major structure now appears at 2.93 eV as is to be expected with the uncovering of the GaAs layer and in fact is comparable to the E_1 structure exhibited by #143.

2. Si (SIMOX)

Figure 2 shows the PR spectrum from a triply annealed sample. Modulation was provided by the 514.5 nm line of an argon ion laser. Sufficient optical filtering was obtained with the Corning CS7-54 glass filter. The structure at 3.4 eV (E_0 and E_1) with $\alpha^{-1} \cong 15$ nm corresponds primarily to transitions along the λ and Δ directions and the 4.2 eV (E_2) structure with $\alpha^{-1} \cong 5$ nm is believed to be associated with transitions in the vicinity of ($2\pi/a$) (3/4,1/4,1/4).[14] The essential feature is that optical penetration depths are smaller than the top layer Si thickness (whose value may be in the hundreds of nanometers range) so that there is a clear cut measurement of just this layer without any contribution from the underlying bulk Si. We also attempted a PR measurement on SIMOX with a top layer of polycrystalline silicon which yielded no response also indicating no contribution from the underlying bulk Si. We are currently examining other samples prepared under other implantation and annealing conditions in order to correlate PR results with sample behavior.

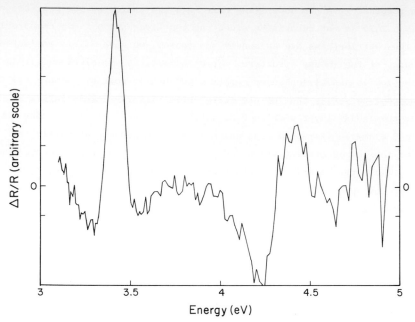

Figure 2. PR spectrum of a triple oxygen implanted SIMOX sample.

V. CONCLUSIONS

The extension of photoreflectance measurements to higher energy critical points further enhances the power of this technique by exploiting the shorter penetration depth at these energies. This is of particular value in those instances where thin layers must be studied in the presence of underlayers of similar materials such as described for the case of a vertical HBT structure or a SIMOX sample. The technique can be extended to thin buried layers using selective etching techniques.

VI. ACKNOWLEDGEMENTS

We appreciate helpful discussions with Prof. A. J. Cunningham and the technical assistance of M. K. Rector and Doug Vass. This work was supported in part by The Organized Research Fund of The University of Texas at Dallas.

REFERENCES

1. Wang, E.Y., Albers, Jr. and Bleil, C.E. (1967). In "II-VI Semiconducting Compounds" (D.G. Thomas, ed.) p. 136. Benjamin, New York.

2. Bottka, N., Gaskill, D.K., Sillmon, R.S., Henry, R. and Glosser, R. (1988). J. of Electronic Materials 17, 161.

3. Pollak, F.H. and Glembocki, O.J. (1988). Proc. SPIE 946, 2.

4. Cerdeira, F. and Cardona, M. (1969). Solid State Commun. 7, 879.

5. Nahory, R.E. and Shay, J.L. (1968). Phys. Rev. Lett. 21, 1969.

6. Nilsson, N.G. (1969). Solid State Commun. 7, 479.

7. Amirtharaj, P.M., Bowman, Jr., R.C. and Alt, R.L. (1988). Proc. SPIE 946, 57.

8. Giordana, A., Glosser, R., Pellegrino, J.G., Qadri, S. and Richmond, E.D. (1989). Materials Lett. (in press).

9. Knudsen, J.F., Bowman, Jr., R.C., Smith, D.D. and Moss, S.C. (1988). Proc. Mat. Res. Soc. Symp. A (Boston) 128 (in press).

10. Bayraktaroglu, B. and Lambert, S.A. (1989). IEEE Electron Device Lett. 10, 120.

11. Kajiyama, K., Izumi, K., and Nakashima, S. (1985). In "Silicon-on-Insulator: Its Technology and Applications" (S. Furukawa, ed.) p. 283. KTK Scientific Publishers, Tokyo.

12. Aspnes, D.E. (1973). Surface Science 37, 418.

13. Williams, E.W. and Rehn, V. (1968). Phys. Rev. 172, 798.

14. Cohen, M.L. and Chelikowsky, J.R. (1988). "Electronic Structure and Optical Properties of Semiconductors," p. 84, Springer, Berlin.

SPACE-CHARGE EFFECTS IN COMPOSITIONAL AND EFFECTIVE-MASS SUPERLATTICES

M. Cahay[1], M. A. Osman[1], H. L. Grubin[1]

Scientific Research Associates, Inc.
Glastonbury, CT 06033

M. McLennan[2]

School of Electrical Engineering
Purdue University
West Lafayette, Indiana 47907

I. INTRODUCTION

In the Effective-Mass Superlattice (EMSL) proposed by Sasaki [1,2], the effective-mass of electrons (or holes) is changed periodically and the conduction (for electrons) or valence band (for holes) is supposedly aligned. This eliminates the potential discontinuities between the respective superlattice layers which are present in doping and compositional superlattices. Previous work [3] neglecting space-charge effects has shown the threshold voltage for negative differential resistance (NDR) in EMSL to be much lower ($\approx 10mV$) and the current density

---

[1]Supported by ONR.
[2]Supported as a fellow by the Semiconductor Research Corporation.

$(10^4\text{-}10^5$ A/cm^2) much higher than in resonant tunneling diodes (RTD). Hereafter, we report preliminary self-consistent calculations of space-charge effects in EMSL and compositional superlattices under zero bias condition.

II. THEORY

We use the technique described in ref. [4] to calculate quantum-mechanically the electron charge density profile in various types of superlattices (see fig. 1). Under the assumption of ballistic transport, the Schroedinger equation throughout the entire structure is solved using the scattering-matrix approach [5] for each electron impinging from the contacts, with the usual boundary conditions for plane-wave solutions, requiring everywhere continuity of the wave function and its first derivative divided by the electron effective mass. Once the Schroedinger equation is solved for electrons impinging from the contacts, the electron density is then calculated for the two streams of electrons incident from the left and right contacts. The calculation for electron density and electrostatic potential are then performed iteratively, until the electrostatic potential converges to a final solution.

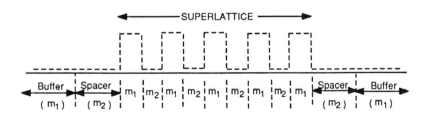

Fig. 1. Spatial variation of the conduction band edges of the device studied in the text. The dashed (full) curve describes a compositional (EMSL) superlattice respectively. Only the buffer regions are assumed to be doped (10^{18}cm^{-3}).

III.   NUMERICAL EXAMPLES

The technique outlined above was applied to calculate self-consistently the electron density and conduction band energy profile of two types of superlattices whose configuration are shown in figure 1. The first superlattice consists of five regions of $In_{0.72}Ga_{0.28}As_{0.86}P_{0.14}$ with width 41Å and effective mass $m_1=0.039m_0$ separated by InP regions of width 29.34Å and effective mass $m_2=0.073m_0$. The current-voltage characteristic of this EMSL was calculated in ref. [3] neglecting space-charge effects. As shown in ref. [1], this special choice of layer thicknesses leads to equal energy gaps between the EMSL energy subbands.

This EMSL is sandwiched between two undoped InP spacer layers (50Å) contacting two highly doped ($10^{18}cm^{-3}$) InP buffer regions (500Å). The second superlattice is a compositional superlattice obtained by replacing the $In_{0.72}Ga_{0.28}As_{0.86}P_{0.14}/InP$ regions in the EMSL described above by $Al_{0.3}Ga_{0.7}As/GaAs$ respectively ($m_1=0.0919m_0$ and $m_2=0.067m_0$). The conduction band discontinuity between $Al_{0.3}Ga_{0.7}As/GaAs$ is taken to be 0.209 eV (see fig. 1).

Figure 2a) shows the self-consistent charge density profile in the two different superlattices under zero bias condition. Also shown for comparison is the charge density profile obtained without iteration, i.e., using the conduction band energy profile such as shown in figure 1. As a result of the doping gradients, internal contact potentials are important (while different) in both types of superlattices as can be seen in figure 2b). This leads to an upward shift of the conduction band energy profile in both superlattices resulting in a substantial decrease of the charge density, when calculated self-consistently, from their non self-consistent values. As shown in fig. 2a), for identical doping concentration in the buffer regions, the charge density is about one order of magnitude higher in the EMSL than in the compositional superlattice.

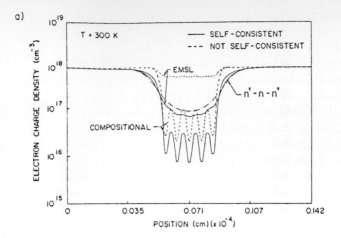

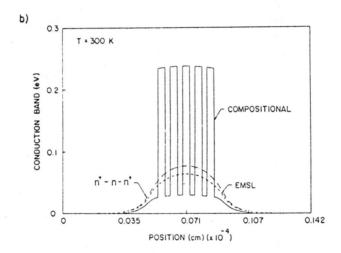

Fig. 2. Self-consistent and non self-consistent
(a) charge density profiles, (b) conduction band
energy profile in an EMSL and compositional
superlattice (see text and fig. 1) with identical
doping concentration in the buffer regions. The
dashed-dotted curves show the self-consistent
results for a structure in which the compositional
superlattice has been replaced by undoped GaAs
(n^+-n-n^+).

IV. CONCLUSIONS

Our preliminary (zero bias) self-consistent calculations have shown that, in superlattices of similar dimensions, space-charge effects are more important in EMSL than in compositional super-lattices. The resulting increased overall capacitance of EMSL should therefore have a detrimental effect on their potentiality for high-speed device applications [3]. Indeed, inclusion of space-charge effects in RTD has been shown to reduce (up to 50%) their calculated peak current and peak-to-valley ratio and to shift to a higher voltage their NDR [4] (to an extent increasing with the importance of space charge). A careful investigation of the superiority of EMSL over RTD for achieving low-power and ultrafast bistable switches must therefore await a detailed self-consistent calculation of their current-voltage characteristic.

REFERENCES

1. A. Sasaki, Phys. Rev. B 30, 7016 (1984).
2. A. Sasaki, Surf. Sci. 174, 624 (1986).
3. A. Aishima and Y. Fukushima, J. Appl. Phys. 61 (1), 249 (1988); Electronic Letters 24, 65 (1988).
4. M. Cahay, M. McLennan, S. Datta and M. S. Lundstrom, Appl. Phys. Lett. 51 (10), 612 (1987).
5. M. Cahay, M. McLennan and S. Datta, Phys. Rev. B37, 10125 (1988).

1/f NOISE IN TWO-DIMENSIONAL MESOSCOPIC SYSTEMS FROM A GENERALIZED QUANTUM LANGEVIN EQUATION APPROACH*

G. Y. Hu
R. F. O'Connell

Department of Physics and Astronomy
Louisiana State University
Baton Rouge, Louisiana

In the $T \rightarrow 0$ limit for a two-dimensional weak localized electronic system, we derive an explicit form for the excess noise, which we show is inversely proportional to the total volume of the sample, the frequency, and the temperature. A model for mesoscopic electronic systems is proposed in which the system is described by many Brownian particles. The spectrum of the auto-correlation function of the velocity fluctuations is calculated by using the non-linear generalized Langevin equation, which has a random force which is a function of the coordinates of the Brownian particle.

I. INTRODUCTION

Recent advances in the study of quantum interference effects in mesoscopic systems [1] have stimulated much interest in 1/f noise [1-3], the origin of which is a long-standing unanswered problem. Two theoretical models which relate the quantum interference effect arising from defect motion to 1/f noise, the local interference (LI) model and the universal conductance fluctuation (UCF) model [2-3], have been proposed. On the other hand, the dynamics of the problem have not been fully explored and no systematic study has yet been performed. In this paper we study the fluctuation spectrum, from a dynamical point of view and basically from a first principles approach.

We approach the problem by adopting the UCF model, and by using the generalized quantum Langevin equation (GLE) method [4-6], which we recently developed to study a variety of transport phenomenon. In Sect. 2, we derive a formula for the excess noise based on the non-linear Langevin equation which considers the fluctuations of the coordinates of Brownian particles due to their environment. In Sect. 3, we treat the mesoscopic electronic system as a system of Brownian particles. We show how to obtain an explicit expression for the excess noise by adopting the UCF model without any assumption of defect motion and by using our previous results obtained in the study of weak localized systems by the GLE formulation [5]. Our results are summarized in Sect. 4.

---

*Supported by Office of Naval Research, Grant number N00014-86-K-0002

II. NON-LINEAR GENERALIZED LANGEVIN EQUATION AND EXCESS NOISE

The generalized Langevin equation method is now widely used in describing many-body systems in many different areas of physics [4-7]. The non-linear GLE is such that the memory function and random force are functions of the coordinates of the Brownian particle. Physically this dependence is due either to a strong external field or to the fluctuations of the coordinates of the Brownian particle due to its environment. In this paper we concentrate on the latter. The relevance of this model for mesoscopic systems will be discussed in Sect. 3.

We study the kinetics of a N Brownian particle system (all particles have the same mass m) by starting from the non-linear generalized Langevin equation (NGLE) of a Brownian particle in an external field f(t)

$$m\ddot{x}(t) + \int_{-\infty}^{t} \mu(x(t)-x(t'); t-t')\ \dot{x}(t')\ dt' = F(x,t) + f(t)\ , \qquad (1)$$

where the site index i (i = 1,2...,N) of the Brownian particle is suppressed as they all obey the same equation. Apart from the inclusion of the external field f(t), (1) is a generalization of the linear GLE in the sense that the memory function and random force are treated as having a dependence on x (no such dependence in the linear GLE), the coordinate of the Brownian particle.

To study the fluctuation phenomenon for a system in a steady state in an external field f(t), it is convenient to introduce the velocity fluctuation v(t) of the Brownian particle, defined by the relation

$$\dot{x}(t) = v_d + v(t), \quad <v(t)> = 0\ , \quad v^2 \equiv <v(t)^2>\ , \qquad (2)$$

where v_d is the drift velocity of the steady state, and the x dependence of v(t) is suppressed.

Now we study (1) by using (2) in the weak field limit ($v_d \to 0$) and in the presence of the lowest order fluctuation contribution v^2. In other words, we expand the integrand in (1) to order v_d, v(t), and $v(t)^2$. Thus terms like v_d^2, $v(t)^3$ or higher orders will be neglected. Next, we decompose (1) into two equations, the steady state transport equation [4] and the velocity fluctuation equation. Those terms proportional to v_d and $v(t)^2$, or both, will not vanish after taking the ensemble average, and they form the steady state transport equation in the weak field limit. The velocity fluctuation equation, to order of $v(t)^2$ and in the weak field limit, is

$$m\dot{v}(t) + \int_{-\infty}^{t} \mu(t-t')\ v(t')\ dt' = F(x,t)\ , \qquad (3)$$

where F(x,t) is understood to be expanded to order $v(t)^2$, and $\mu(t)$ is the $\mu(x,t)$ of (2) at x=0, i.e. it is independent of the coordinates. In obtaining (3) we have used the parity property for the isotropic system for the non-linear memory function $\mu(x,t)$, which implies that it does not contain terms odd in x. Also, we remark that the $v(t)^2$ term in the expansion of $\mu(x,t)$ does not appear in (3) as it contributes a term of higher order than $v(t)^2$.

From (3), it is now straightforward to obtain, to order v^2, the spectrum of the auto correlation function of the velocity fluctuation:

$$S_v(v^2,\omega) = \frac{1}{|m\omega + i\mu(\omega)|^2} \, S_F(v^2,\omega) \, , \tag{4}$$

where the Fourier transform of the auto correlation function of the non-linear random force is

$$S_F(v^2,\omega) = \int_{\infty}^{\infty} dt \, <F(x,t) \, F(o,o)>e^{i\omega t} \tag{5}$$

and the v^2 in the bracket emphasizes the fact that (4) is correct to the order of v^2.

Equation (4) is a key result of the present paper. In the linear GLE treatment, one neglects the x dependence of $F(x,t)$ and then (5) reduces to the linear result $S_F(\omega)$ and so (4) then simply gives the Nyquist noise ($2k_BT\tau/m$, where τ is the relaxation time and T is the temperature). Now, the difference between the non-linear formula (4) and its linear correspondence is, after including the familiar factor of N^{-1} for a N particle system,

$$\Delta S_v(\omega) \equiv S_v(v^2,\omega) - S_v(\omega) = \frac{1}{N} \frac{S_F(v^2,\omega)-S_F(\omega)}{|m\omega + i\mu(\omega)|^2} \, , \tag{6}$$

which is a measure of the excess (relative to the Nyquist $S_v(\omega)$) noise. It is due to the non-linear effect caused by the movement of the Brownian particle and is valid to order v^2. Obviously, in the linear GLE treatment $\Delta S_v(\omega) = 0$.

Since the linear functions $\mu(\omega)$ and $S_F(\omega)$ in (6) are known from the calculation of the linear GLE, it is desirable if we can evaluate the only unknown function $S_F(v^2,\omega)$ in (6) by means of these known functions. Actually, the difference between the non-linear random force $F(\vec{x},t)$ and its linear correspondence $F(t) \equiv F(x=0,t)$ can be seen more clearly if we expand $F(\vec{x},t)$ in a Fourier series:

$$F(\vec{x},t) = \sum_{\vec{q}} F(\vec{q},t)e^{i\vec{q}\cdot\vec{x}} \, , \quad F(t) = \sum_{\vec{q}} F(\vec{q},t) \, . \tag{7}$$

We find that there is a general relationship between the space Fourier transform of (5) and its linear counterpart. In the low frequency limit, we obtain

$$S_F(v^2,\omega) = S_F(\omega) + \sum_{\vec{q},\alpha} \frac{q_\alpha^2 \, v_\alpha^2}{\omega^2} \, [S_F(\vec{q},\omega) - S_F(\vec{q},o)] \, . \tag{8}$$

The above equation is obtained by first deriving the second order time differential equation for the space Fourier transform of (5) and then solving the equation by Fourier transform techniques. Physically, this is similar to applying the non-Markovian law [7] to the diffusion process of the Brownian particles.

III. APPLICATION: 1/f NOISE IN MESOSCOPIC SYSTEMS

We can now evaluate the excess noise $\Delta S_v(\omega)$ of (6) supplemented by (8), if only the Fourier transform of both the linear random force autocorrelation function, $S_F(\vec{q},\omega)$, and the linear memory function, $\mu(\omega) = \sum_{\vec{q}} \mu(\vec{q},\omega)$, in the linear GLE is known.

Recently, we have studied the GLE of an electron system from a center of mass formulation. In our approach, we visualize the center of mass of the electrons as a Brownian "particle", while the phonons and relative electrons act as a heat bath, which is coupled to the center of mass through the electron-impurity and electron-phonon interactions. The only assumption used in obtaining the NGLE is that the number of electrons N_e composing this Brownian particle should be much larger than one. The application of our NGLE method to study the 1/f noise in the $T \to 0$ limit is motivated by a recent work [2] of Feng, Stone, and Lee (FSL), in which they presented compelling arguments that the weak localization mechanism is a source of 1/f noise in the quantum regime. Following FSL's work, we divide the electronic system into equivalent small boxes (each box being a "Brownian particle") and the length of each box is bounded by the dephasing length L_ϕ. Inside each box, there are strong interference effects between electrons. We use our quantum mechanical approach to derive a NGLE by the center of mass description. Between electrons in different boxes, the interference effects are expected to be very weak. Obviously, the total number N of Brownian particles representing the whole system in this description depends on $L_\phi (\sim T^{-\frac{1}{2}}$ for electron electron interactions). We emphasize that the mass m and the coordinates x refer to a "box" and that N is the total number of boxes with (N_e/N) electrons in each box.

Previously, we have derived the GLE of an electron-impurity system at T=0 in the weak-localization regime. For the d=2 case, the memory function we obtained in the low frequency region ($\omega\tau \ll 1$) is[5]

$$\mu(\omega) = \frac{1}{\tau} (1 + \frac{1}{k_F\ell} \ell n \ \omega\tau)^{-1} \tag{9}$$

where τ is the relaxation time in the absence of the back scattering effect and k_F is the Fermi momentum. Also, the form of $\mu(q,\omega)$ defined by $\mu(\omega) = \sum_{\vec{q}} \mu(q,\omega)$ can be found in Ref. 5. It follows that we can obtain the $S_F(\vec{q}\omega)$ by using the fluctuation-dissipation theorem [6] in the T=o limit

$$S_F(\vec{q}\omega) = \hbar\omega \ \text{Re} \ \mu(\vec{q}\omega) . \tag{10}$$

Substituting (9) and (10) into (8) and then using (6), and using the proportional property between the velocity and the current, we obtain the excess noise of the current fluctuation spectrum for a two-dimensional electron-impurity system

$$\mathcal{S}_I(\omega) \equiv S_I(\omega)/I^2 = \frac{L_\phi^2}{\Omega\omega} \ \frac{\delta G^2}{G^2} \ \frac{3k_F\ell}{\hbar} \ \frac{1}{1 + \frac{1}{k_F\ell} \ell n(\omega\tau)} \tag{11}$$

where Ω is the total volume of the sample ($\Omega = N L_\phi^2$), G is the conductance ($G=\sigma=ne^2\tau/m$ for the d=2 case) of the box, the 1/N is a factor that comes from statistics for N boxes, the other terms are the one box contributions, and we have used the relation $v^2/v_d^2 = \delta G^2/G^2$.

Qualitatively, (11) is similar to the general Hooge formula in that $\mathcal{S}_I(\omega)$ is inversely proportional to ω and Ω, and $\mathcal{S}_I(\omega)$ increases when the density or k_F of the electrons decreases (this can be easily observed if we use the result for the universal conductance fluctuation [1,2] viz.

$\delta G^2/G^2 \sim 1/(k_F \ell)^2$ in (11)). Eq. (11) is a T=0 result. An extension of (11) to the finite temperature case is among our plans for the future. Nevertheless, it is clear that the factor N^{-1} in (11) will be kept for the finite temperature version, which indicates a decrease of $S_I(\omega)$ with increase of T (since $L^2 \sim T^{-1}$ i.e. the number of boxes increases with increasing temperature). This predicted decrease of $S_I(\omega)$ with increase of T is opposite to what one finds in the classical high-temperature case (where, around room temperature $S_I(\omega)$ increases with increase of T). Yet it is in conformity with the theoretical results of Feng, Lee and Stone [2], and confirmed by experiments [8,9]. We note that in Ref. 2 the basic idea is that the time variation of the positions of the impurities is the source of the 1/f noise. In our approach, the time variation of the random force exerted on the electrons due to the environment is the source of the 1/f noise. In addition, we have considered a dynamical model based on first principles.

IV. SUMMARY

We have analyzed the phenomenon of excess noise, based on the study of a non-linear Langevin equation which considers the fluctuations of the coordinates of Brownian particles due to the environment. Our basic results, given in (6) and (8), can be used to evaluate excess noise quite generally in the GLE formulation. For mesoscopic systems, we adopt the UCF model and divide the electronic system into equivalent small boxes (each box being a "Brownian particle"). The length of each box is bounded by the dephasing length, and the center of mass of the electrons in each box is designated as a Brownian particle. The memory function (9) of the Brownian particle, together with the fluctuation-dissipation theory (10), are then used to obtain for the 2d mesoscopic system the current fluctuation spectrum (11), which is inversely proportional to the total number of electrons, the frequency and the temperature.

REFERENCES

1. B. L. Al'tshuler and P. A. Lee, Physics Today, Dec., 1988, p 36.
2. S. Feng, P. A. Lee, and A. D. Stone, Phys. Rev. Lett. 56, 1960 (1986); 56, 2772 (E) (1986).
3. M. B. Weissman, Rev. Mod. Phys. 60, 537 (1988).
4. G. Y. Hu and R. F. O'Connell, Phys. Rev. B36, 5798 (1987); ibid., Physica 149A, 1 (1988).
5. G. Y. Hu and R. F. O'Connell, Physica 153A, 114 (1988).
6. G. W. Ford, J. T. Lewis, and R. F. O'Connell, Phys. Rev. A37, 4419 (1988); ibid. Ann. Phys. (N.Y.) 185, 270 (1988).
7. R. Kubo, M. Toda, and N. Hashitsume, Statistical Physics II (Springer Verlag, Berlin, 1985).
8. G. A. Garfunkel, G. B. Alers, M. B. Weissman, J. M. Mochel, and D. J. Van Harlingen, Phys. Rev. Lett. 60 2773 (1988).
9. N. O. Birge, B. Golding, and W. H. Haemmerle, Phys. Rev. Lett. 62 195 (1989).

NPF Symposium Participants

NANOSTRUCTURE PHYSICS AND FABRICATION
INTERNATIONAL SYMPOSIUM AT
TEXAS A&M UNIVERSITY
MARCH 13-15, 1989

509

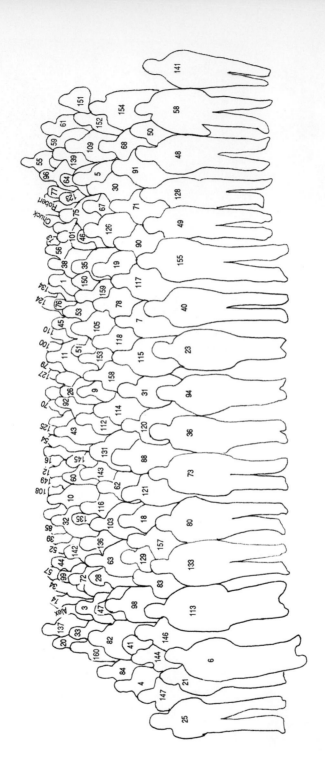

1   Glenn Agnolet
    Texas A&M University
    Department of Physics
    College Station, TX 77843

2   Roland Allen
    Texas A&M University
    Department of Physis
    College Station, TX 77843

3   David Keith Allison
    Texas A&M University
    Department of Nuclear Engineering
    College Station, TX 77843

4   Craig Andrews
    Texas A&M University
    Physics Department
    College Station, TX 77843

5   Raymond Ashoori
    Cornell University
    Dept. of Physics
    Ithaca, NY 14853

6   Gary M. Atkinson
    Hughes Research Lab
    3011 Maliba Canyon Rd.
    Malibu, CA 90403

7   Ali Badakhshan
    University of Texas at Dallas
    P.O. Box 830688, M.S. F023
    Richardson, TX 75083

8   Karl Balke
    Texas A&M University
    Department of Physics
    College Station, TX 77843

9   Supriyo Bandyopadhyay
    University of Notre Dame
    Electrical and Computer Eng. Dept.
    Notre Dame, IN 46556

10 John R. Barker
    University of Glasgow
    Dept. of Electronics & Elec. Eng.
    Glasgow G12 8QQ,
    United Kingdom

11 Robert T. Bate
    Texas Instruments, Inc.
    MS-154
    P.O. Box 655936
    Dallas, TX 75265

12 Steven P. Beaumont
    University of Glasgow
    Dept. of Electronics & Elec. Eng
    Glasgow G12 8QQ, Scotland
    United Kingdom

13 Steven Berger
    AT&T Bell Las
    Rm. 7B-202
    600 Mountain Ave
    Murray Hill, NJ 07974

14 Gary Bernstein
    University of Notre Dame
    Electrical and Computer Eng. Dept.
    Notre Dame, IN 46556

15 Anil Bhatnagar
    Texas A&M University
    Department of Physics
    College Station, TX

16 Suyog Bhobe
    University of Notre Dame
    Notre Dame, IN 46556

17 Matthew Blain
    Texas A&M University
    Department of Chemistry
    College Station, TX 77843

18 Paul Boudreaux
    Physical Sciences Laboratory
    4928 College Ave.
    College Park, MD 20740

19 Alec N. Broers
    University of Cambridge
    Engineering Dept.
    Cambridge CB2 1PZ,
    United Kingdom

20 Charles Brumlik
    Texas A&M University
    Department of Chemistry
    College Station, TX 77843

21 Markus Buettiker
    IBM
    T.J. Watson Research Ctr.
    P.O. Box 218
    Yorktown Heights, NY 10598

22 Robert Burns
    Texas A&M University
    Department of Physics
    College Station, TX 77843

23 Marc Cahay
    Scientific Research Associates, Inc.
    50 Nye Road
    Glastenbury, CT 06033

24 Zhihua Cai
    Texas A&M University
    Department of Chemistry
    College Station, TX 77843

Symposium Scientific
Participants

25   Albert M. Chang
     AT&T Bell Labs
     Electronics Research Lab
     Rm. 4c428, Crawfords Corner Rd.
     Holmdel, NJ 07733

26   Leroy L. Chang
     IBM
     T. J. Watson Research Center
     P.O. Box 218
     Yorktown Heights, NY 10598

27   Frank Cheng
     Texas A&M University
     Department of Chemistry
     College Station, TX 77843

28   Yeongki Cheong
     Texas A&M University
     Dept. of Nuclear Engineering
     College Station, TX 77843

29   James Chepin
     Texas A&M University
     Department of Physics
     College Station, TX 77843

30   Stephen Chou
     Stanford University
     108 McCullough
     Stanford, CA

31   Bright Chu
     Texas A&M University
     Dept. of Electrical Engineering
     College Station, TX 77843

32   Larry R. Cooper
     Office of Naval Research
     Code 1114
     800 North Quincy St.,
     Arlington, VA 22217

33   Harold Craighead
     Cornell University
     National Nanofabrication Facility
     M105 Knight Lab
     Ithaca, NY 14853-5403

34   Gentry Crook
     The University of Texas at Austin
     Microelectronics Research Center
     Austin, TX 78712

35   George Damm
     University of Houston
     Department of Electrical Eng.
     Houston, TX

36   Supriyo Datta
     Purdue University
     School of Electrical Engineering
     West Lafayette, IN 47907

37   Digant P. Dave
     Texas A&M University
     Dept. of Electrical Engineering
     College Station, TX 77843

38   John Davies
     University of Glasgow
     Electronics and Electrical Eng.Dept
     Glasgow, G12 8QQ,
     United Kingdom

39   Alex de Lozanne
     University of Texas
     Department of Physics
     Austin, TX 78712

40   Karim Diff
     Temple University
     Department of Physics
     Philadelphia, PA 19122

41   Elizabeth Dobisz
     Naval Research Lab
     Code 6831
     Washington, DC 20375

42   Christopher Durbin
     University of Texas at Dallas
     P.O.Box 830688, MS F023
     Richardson, TX 75083

43   Laurence Eaves
     University of Nottingham
     Department of Physics
     Nottingham NG7 2RD,
     United Kingdom

44   Edward E. Ehrichs
     University of Texas at Austin
     Department of Physics
     Austin, TX 78712

45   David Ferry
     Arizona State University
     Center for Solid State Sci.
     University Drive
     Tempe, AZ 85287

46   Alex Fong
     University of Houston
     Department of Electrical Eng.
     Houston, TX

47   William R. Frensley
     Texas Instruments, Inc.
     Central Research MS 154
     Dallas, TX 75265

48   James Garris
     Texas A&M University
     Electrical Engineering Dept.
     College Station, TX 77843

49 Amiya R. Ghatak-Roy
Texas A&M University
Department of Electrical Eng.
College Station, TX 77843

50 Ruby Ghosh
Cornell University
Dept. of Applied Physics
Ithaca, NY 14853

51 Adriana Giordana
University of Texas at Dallas
P.O. Box 830688 MS F023
Richardson, TX 75083

52 Nicholas J. Giordano
Purdue University
Physics Department
West Lafayette, IN 47907

53 Robert Glosser
University of Texas at Dallas
Department of Physics
P.O. Box 830688, MS F023
Richardson, TX 75083

54 Wolfgang Hansen
IBM
T.J. Watson Research Center
Rm. 19-122, P.O. Box 218
Yorktown Heights, NY 10598

55 Chris Harmans
Delft University of Technology
Department of Applied Physics
Lorentzweg 1 POB5046
Delft,
The Netherlands

56 Ron R. Hart
Texas A&M University
Department of Nuclear Engineering
College Station, TX 77843

57 Eivind Hiis Hauge
Ohio State University
174 West 18th Avenue
Columbus, OH 43210

58 Mordehai Heiblum
IBM
T. J. Watson Research Center
P.O. Box 218
Yorktown Heights, NY 10598

59 Detlef Heitmann
Max-Planck Institut
Heisenbergstr 1
7000 Stuttgart 80,
West Germany

60 Karl Hess
University of Illinois
Coordinated Science Laboratory
1101 W. Springfield Ave.
Urbana, IL 61801

61 Richard Higgins
Georgia Institute of Technology
Microelectronics Res. Center
800 Atlantic Ave.
Atlanta, GA 30332

62 Toshiro Hiramoto
University of Tokyo
Institute of Industrial Science
7-22-1 Roppongi, Minato-ku
Tokyo 106,
Japan

63 Daniel H. Howard
Georgia Institute of Technology
Microelectronics Research Center
800 Atlantic Drive,N.W.
Atlanta, GA 30332

64 Ben Hu
Ohio State University
Department of Physics
174 W. 18th Avenue
Columbus, OH 43210

65 Chia-Ren Hu
Texas A&M University
Department of Physics
College Station, TX 77843

66 Gen-you Hu
Louisiana State University
Dept. of Physics and Astrophysics
Baton Rouge, LA 70803

67 Toshiaki Ikoma
University of Tokay
Institute of Industrial Science
7-22-1 Roppongi, Minato-ku
Tokyo 106,
Japan

68 Yoseph Imry
Weizmann Institute
Physics Department
Rehovot,
Israel

69 Czeslaw Jedrzejek
Jagellonian University
Institute of Physics
30-059 Krakow 16
Reymonta 4,
Poland

70 Robert Kamocsai
University of Notre Dame
Department of Elec. Engineering
Notre Dame, IN 46556

71 Brian Kinard
Texas A&M University
Electrical Engineering Department
College Station, TX 77843

72 George Kirczenow
Simon Fraser University
Department of Physics
Burnaby, BC V5A 1S6,
Canada

513

73 Wiley P. Kirk
Texas A&M University
Department of Physics
College Station, TX

74 Pawel Kobiela
Texas A&M University
Department of Physics
College Station, TX 77843

75 Jorg P. Kotthaus
Universitat of Hamburg
Institut fur Angewandte Physik
JungiusstraBe 11
D-2000 Hamburg 36,
West Germany

76 Alfred Kriman
Arizona State University
College of Engineering
University Drive
Tempe, AZ 85287

77 Roger K. Lake
Purdue University
Electrical Engineering
BOX 259
W. Lafayette, IN 47907

78 Rolf. Landauer
IBM
T. J. Watson Research Center
P.O. Box 218
Yorktown Heights, NY 10598

79 Yong Lee
Purdue University
Dept. of Electrical Engineering
W. Lafayette, IN 47906

80 Anthony J. Leggett
University of Illinois
Department of Physics
1110 W. Green St.
Urbana, IL 61801

81 Junting Lei
Texas A&M University
Department of Chemistry
College Station, TX 77843

82 Craig S. Lent
University of Notre Dame
Dept. of Electrical Engineering
Notre Dame, IN 46556

83 Anthony F. J. Levi
AT&T Bell Laboratories
Rm 1E-450
Murray Hill, NJ 07974

84 Feng Li
Texas A&M University
Department of Physics
College Station, TX 77843

85 Pavel Lipavsky
Ohio State University
174 W. 18th Ave.
Columbus, OH 43210

86 Xiang Yang Liu
Texas A&M University
Department of Physics
College Station, TX 77843

87 Marshall Luban
Iowa State University
Department of Physics
Ames, Iowa 50011

88 Wayne W. Lui
AT&T Bell Laboratories
Rm. 2C-220
Allentown, PA 18103

89 Charles Martin
Texas A&M University
Department of Chemistry
College Station, TX 77843

90 Kevin Martin
Georgia Institute of Tech
School of Electrical Engineering
Atlanta, GA 30332

91 David Mast
University of Cincinnati
Department of Physics
M.L. 11
Cincinnati, OH 45221

92 Christine M. Maziar
The University of Texas at Austin
Electrical and Computer Engineering
Austin, TX 78712

93 John McBride
Texas A&M University
Department of Chemistry
College Station, TX 77843

94 Michael J. McLennan
Purdue University
School of Elec. Engineering
W. Lafayette, IN 47907

95 Imran Mehdi
University of Michigan
Solid State Electronics Lab
EECS Rm. 2234
Ann Arbor, MI 48109

96 Dain C. Miller
Purdue University
Dept. of Electrical Engineering
W. Lafayette, IN 47907

97 Douglas R. Miller
The University of Texas at Austin
Microelectronics Research Center
ENS 413
Austin, TX 78712

98 George Misium
Texas Instruments, Inc.
P.O. Box, MS 944
Dallas, TX

99 Lars Montelius
IBM
T.J. Watson Research Center
P.O. Box 218
Yorktown Heights, NY 10598

100 George Neofotistos
Temple University
Department of Physics
Philadelphia, PA

101 A. E. Owen
Arizona State University
Center for Solid State Elec.
Tempe, AZ 85287

102 Fredrik Owman
University of Lund
Department of Solid State
P.O. Box 118
S-221 00 Lund,
Sweden

103 Eric Palm
Texas A&M University
Department of Physics
College Station, TX 77843

104 Donald L. Parker
Texas A&M University
Electrical Engineering
College Station, TX 77843

105 Keith Passmore
University of Texas at Dallas
P.O. Box 830688, MS F023
Richardson, TX 75083

106 Rajesh Pathak
Texas A&M University
Dept. of Electrical Engineering
College Station, TX 77843

107 Amarkumar Pathi
Texas A&M University
Dept. of Mechanical Engineering
College Station, TX 77843

108 Wolfgang Porod
University of Notre Dame
Dept. Electrical and Computer Eng.
Notre Dame, IN 46556

109 Daniel Ralph
Cornell University
Department of Physics
Ithaca, NY 14853

110 John N. Randall
Texas Instruments
MS134
P.O. Box 655936
Dallas, TX 85265

111 Asok Kumar Ray
University of Texas at Arlington
Department of Physics
Arlington, TX 76019

112 Umapathi Reddy
University of Michigan
Dept. of Elec. Eng. & Comp. Sci
Ann Arbor, MI 48109

113 Mark A. Reed
Texas Instruments, Inc.
Central Research MS 154
P.O. Box 655936
Dallas, TX 75265

114 Lino Reggiani
Universita Deli Studi di Modena
Dipartimento di Fisica
4110 Modena,
Italy

115 Joseph H. Ross
Texas A&M University
Department of Physics
College Station, TX 77843

116 Michael L. Roukes
Bell Communications Research
Rm NVC 3X-285
331 Newman Springs Rd.
Red Bank, NJ 07701

117 Joe Ryan
Arizona State University
Center for Solid State Elec. Res.
University Drive
Tempe, AZ 85287

118 Andrew Sachrajda
National Research Council
Physics Division
Montreal Rd.
Ottawa, Ontario,
Canada K1A OR6

119 Kayvan Sadra
University of Texas at Austin
Microelectronics Research Ctr.
Austin, TX 78712

120 Wayne Saslow
Texas A&M University
Department of Physics
College Station, TX 77843

121 Axel Scherer
Bell Communications Res.
NVC 3X-373
331 Newman Sprigs Rd.
Red Bank, NJ 07701

122 Howard Schmidt
Schmidt Instruments
2476 Bolsover Suit 234
Houston, TX 77005

123 Sudippo Sen
University of Houston
Department of Electrical Eng.
Houston, TX 77004

124 Robert H. Silsbee
Cornell University
Department of Physics
Ithaca, NY 14853

125 Srinivas Sivaprakassam
University of Notre Dame
Notre Dame, IN 46556

126 Henry I. Smith
MIT
Room 13-3061
77 Massachusetts Ave
Cambridge, MA 02139

127 Eric Snow
Naval Research Laboratory
Code 6875
Washington, DC 20375

128 Fernando Sols
University of Illinois
Department of Physics
1110 W. Green St.
Urbana, IL 61801

129 Takashi Soma
Texas A&M University
Department of Physics
College Station, TX 77843

130 Joseph S. Spector
AT&T Bell Labs
Rm. 1D-470
600 Mountain Ave
Murray Hill, NH 07974

131 Gregory F. Spencer
Texas A&M University
Department of Physics
College Station, TX 77843

132 Capp Spindt
SRI International
Menlo Park, CA 94025

133 A. Douglas Stone
Yale University
Applied Physics
P.O. Box 2157
New Haven, CT 06520

134 Jon Andreas Stovneng
Ohio State University
174 W. 18th Avenue
Columbus, OH 43210

135 Tomasz Szafranski
Texas A&M University
Department of Physics
College Station, TX 77843

136 Wieslaw Szott
Texas A&M University
Department of Physics
College Station, TX 77843

137 Michael Tierney
Texas A&M University
Department of Chemistry
College Station, TX 77843

138 Chris Tigges
Sandia National Laboratories
Org. 1150
P.O. Box 5800
Albuquerque, NM 87185

139 Gregory L. Timp
AT&T Bell Labs
4G-318
Crawford Corner Road
Holmdel, NJ 07733

140 Philippe Tissot
Texas A&M University
Department of Nuclear Eng.
College Station, TX 77843

141 Ricardo N. Tsumura
Texas A&M University
Department of Physics
College Station, TX 77843

142 Sergio E. Ulloa
Ohio University
Dept. of Physics and Astronomy
Athens, OH 45701

143 Corwin P. Umbach
IBM
T.J. Watson Research Center
P.O. Box 218
Yorktown Heights, NY 10598

144 Dick van der Marel
Delft University of Technology
2628 CJ Delft,
The Netherlands

145 Chris van Haesendonck
University of Leuven
Department of Physics
Celestijnenlaan 200D
B-3030 Leuven,
Belgium

146 Henk van Houten
Philips Research Lab
Physics & Material Science Div.
Briarcliff Manor, NY 10510

147 B.J. van Wees
Delft University of Technology
Department of Applied Physics
P.O. Box 5046
2600 GA Delft,
The Netherlands

148 Annette Vandervort
Texas A&M University
Department of Electrical Eng.
College Station, TX 77843

149 Mike Walsh
University of Houston
Dept. of Electrical Engineering
Houston, TX 77004

150 Chia-Lai Wang
Texas A&M University
Department of Physics
College Station, TX 77843

151 Xiao-Fang Wang
Columbia University
P.O. Box 40, Pupin
New York, NY 10027

152 Ziqiang Wang
Columbia University
Department of Physics
P.O. Box 88
New York, NY 10027

153 Morag Watt
University of Glasgow
Electronics and electrical Eng.
Glasgow G12 8QQ,
United Kingdom

154 Richard A. Webb
IBM
T.J. Watson Research Ctr.
P.O. Box 218
Yorktown Heights, NY 10598

155 Mark H. Weichold
Texas A&M University
Electrical Engineering Department
College Station, TX 77843

156 Michael Weimer
Texas A&M University
Department of Physics
College Station, TX 77843

157 Joseph Weiner
AT&T Bell Laboratories
600 Mountain Ave
Murray Hill, NJ 07974

158 Joel Wendt
Sandia National Laboratories
Division 1141
P.O. Box 5800
Albuquerque, NM 87185

159 John C. Wolfe
University of Houston
Dept. of Electrical Engineering
Houston, TX 77004

160 Wing K. Yue
Texas A&M University
Dept. of Electrical Engineering
College Station, TX 77843

Author Index

Index